WASTE MANAGEMENT

Waste Management

Editors

A.L. Juhasz
Centre for Environmental Risk Assessment and Remediation
University of South Australia
Mawson Lakes, SA
Australia

G. Magesan
Forest Research Institute
Rotoura
New Zealand

R. Naidu
Centre for Environmental Risk Assessment and Remediation
University of South Australia
Mawson Lakes, SA
Australia

CRC Press
Taylor & Francis Group
Boca Raton London New York

CRC Press is an imprint of the
Taylor & Francis Group, an **informa** business
A SCIENCE PUBLISHERS BOOK

First published 2004 by Science Publishers Inc.

Published 2019 by CRC Press
Taylor & Francis Group
6000 Broken Sound Parkway NW, Suite 300
Boca Raton, FL 33487-2742

First issued in paperback 2019

ISBN 13: 978-0-367-44660-4 (pbk)
ISBN 13: 978-1-57808-323-7 (hbk)

Visit the Taylor & Francis Web site at
http://www.taylorandfrancis.com

and the CRC Press Web site at
http://www.crcpress.com

LOC data available on request.

This book is dedicated to the memory of Dr Nicholas Charles McClure. Nick's contributions to this field of research were broad, unique and given with enthusiasm and humour. His characteristic generosity, support and inclusive nature will be sadly missed.

Nicholas Charles McClure
24-8-1961 to 20-9-2002

Preface

The disposal of urban and industrial wastes is a challenging and expensive problem confronting environmental regulators, local councils and industries in many countries throughout the world. The production of these wastes is, and will continue to be an ongoing problem as long as human civilization persists. The current rate of world population growth is paralleled by a global increase in waste production, placing stresses on both the demand for agricultural produce and the finite world resources. Since the late 1970s, the management of urban and industrial wastes has received much attention on recognizing the implications of the presence of potentially toxic substances (PTS) to environmental and human health. Maintenance of environmental quality is imperative since sustainable development, health and well-being of life is completely dependent on ecosystem health.

Urban and industrial wastes can have both beneficial and detrimental effects on the soil environment. These conflicting effects have led to significant division amongst the scientific and general public on the potential land application of wastes. Problems associated with land-based disposal arise because in most metropolitan areas, there is no separation between domestic and industrial waste and wastewater. Although industrial contamination is a major problem with sewage sludge, domestic sewage alone is by no means completely free of a range of potential contaminants, including for example, heavy metals such as copper and zinc, various pathogenic organisms and a wide range of organic chemicals.

While numerous techniques have been developed toward remediation and management of industrial and urban contaminants, some wastes are now being recognized as valuable resources for application to agricultural land. Waste products such as sewage sludge are rich in plant nutrients and also have the potential to improve soil structure. Land application of wet sludge is, therefore, a method of disposal that is both economical and beneficial because organic matter, nutrients and water in the sludge are recycled back to the land. However, both metals (Cu, Cd, Ni, Cr, Zn and Pb) and organics (dioxins, PCBs and endocrine disruptors), which are potentially toxic substances present in excess concentrations, could persist in soils for long periods. The accumulation of PTS in soil may have detrimental implications on both the ecosystem and human health. Elevated concentrations of contaminants may influence plant nutrition, crop yield and quality and decrease soil health through the inhibition of beneficial microbial populations. Human health may be compromized via soil-human, soil-plant-human, soil-animal-human and soil-plant-animal-human exposure scenarios. An understanding of the fate, ecotoxicity, bioavailability and exposure pathways of PTS, in addition to remediation and waste management strategies, is essential to maintain and improve the ecosystem and human health.

This book addresses the important issue of waste management. Chapters are based on selected papers presented during the 2nd International Conference on Contaminants in the Soil Environment, held on 12–17 December 1999 in New Delhi, India and the Towards Better Management of Wastes and Contaminated Sites in the Australasia-Pacific Region workshop, held on 3–5 May 2000 in Adelaide, Australia. Eighteen chapters are presented in this book, which is divided into two sections: *Managing Wastes—Implications of Waste Disposal to Environmental Quality* and *Waste Management Technologies—Case Studies*. The first section focuses on the impact of waste disposal to land, providing an outline of the current underpinning knowledge of processes associated with contaminant sorption, transport and plant uptake. In addition, problems associated with and future improvements to waste application to land and other remediation and waste management strategies are outlined. The second section provides a number of case studies highlighting waste management technologies that are currently being used in the Australasia-Pacific region.

Contents

Section A

WASTE MANAGEMENT IN THE AUSTRALASIA-PACIFIC REGION

1

Contaminants in the Soil of the Rootzone: Transport, Uptake and Remediation

B.E. Clothier[1], *S.R. Green*[1], *I. Vogeler*[1], *B.H. Robinson*[1], *T.M. Mills*[1], *C. Duwig*[2], *J.K.F. Roygard*[1], *M. Walter*[1], *C.W. van den Dijssel*[1] and *D.R. Scotter*[3]

INTRODUCTION

Soil is the fragile yet productive skin of our planet. Soil also occupies a critical position between the atmosphere and the Earth for it lies astride the main thoroughfare along which water enters our subterranean and surface reservoirs. A clean, healthy and productive soil is a social and economic imperative. Scientific research and development needs to work towards providing the underpinning knowledge to permit sustainable management of our soil resource, and to develop tools and strategies to remediate soil that has already become degraded or contaminated.

A key to improving our understanding of soil contamination processes and problems, and their remediation, is a better underpinning knowledge of the processes of transport, exchange and uptake in the soil of the rootzone. In this introductory chapter, we outline the current state of that underpinning knowledge, the prospects for future improvements, and how these might be achieved.

Our balanced goal must be to have a clean and healthy soil environment that sustains both nature and people's needs and aspirations (Ministry for the Environment, 1997).

FORCES FOR A BALANCED GOAL

It is a clearly acknowledged fact that science is only one part of the process of seeking solutions to problems of soil contamination. Indeed, it

[1]Environment and Risk Management Group, HortResearch, Private Bag 11 030, Palmerston North, New Zealand 5301. (bclothier@hortresearch.co.nz)
[2]Laboratoire d'étude des transferts en hydrologie et environnement, CNRS, Grenoble, France
[3]Institute of Natural Resources, Massey University, Palmerston North, New Zealand

has been agreed upon that some sections of the community believe science to be part of the problem, rather than a means to its solution. This is, however, probably a case of what has been called 'Sevareid's Law', namely that '... the chief source of problems is solutions' (Martin, 1973). Undoubtedly, though, scientific research and technology offer the best hopes for developing sustainable strategies for managing land, for generating technologies for rehabilitating degraded environments, and remediating contaminated sites. Yet, the underpinning provided by science is just one of three forces that can work towards achieving our balanced goal (Figure 1.1). The other two forces come either from within socio-legal frameworks, or though economic drivers. Any imbalance between these spheres of influence will, of course, lead to further pollution of the environment, and a continuing lack of remedial action.

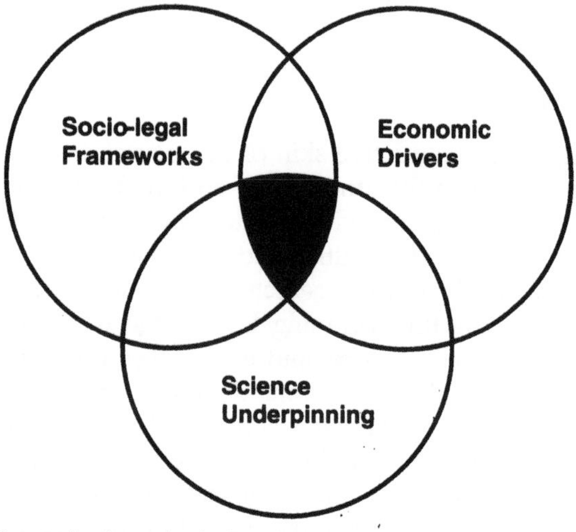

Fig. 1.1: The forces that require balancing to produce the goal of a clean, healthy and productive soil, and remediation of soil that has already been contaminated.

Economic Drivers

A bountiful land is a source of economic development, whether through primary production, or by way of extraction of resources. As nations develop, exploitation of the land's resources can lead to degradation or contamination. Increasingly, however, in both developed and developing countries, enterprises are under increasing pressure, both internally and externally, to ensure that their activities maintain cleanliness, health and productivity. There are increasing economic drivers steering enterprise towards our balanced goal (Figure 1.1), covering both economic and political 'push', as well as market 'pull'. Internationally, the spectre of non-tariff trade barriers based on inadequate environmental performance looms as a means of curtailing some unsustainable practices. Already in Australia, and soon to be enacted elsewhere, it is compulsory for registered companies to report their environmental performance annually. Environmental Management Systems (EMS), such as the ISO 14001 system, are becoming increasingly common in business. EMS demand continuous improvement in environmental performance through a system of 'plan, do, monitor, act and improve'. Not only do companies benefit through EMS via waste minimization, recycling, lowered cost of clean resources, and reduced liabilities, there is also the market advantage to be gained. The number of 'green' customers is increasing, and they are using 'environmental friendliness' as a purchase criterion.

Socio-legal Frameworks

As economies develop, societal values change, and so, eventually, does the governing legislation within a jurisdiction. The last few decades have seen the growth of environmentalism, or the 'green' movement, which has been followed by the passing of laws that serve to protect the environment better, and enactment of regulations that demand remediation of sites where degradation has occurred (Clothier, 1997). The general growth of a green ethos has lead to a corresponding development of a widespread land ethic. In both New Zealand and Australia, this changing spirit has spawned 'Land Care Groups' that seek to develop and employ sustainable systems of primary production. Meanwhile, organic production, an extension of this ethos, is a growing business with an expanding niche market. Likewise, in the urban and industrial sectors, there are moves towards recycling, and the lofty goal of 'zero waste'. Legislators have responded with laws that reflect these societal changes. New Zealand's omnibus environmental legislation, the Resource Management Act, has as its purpose (RMA, Part II, 5(2), 1991):

> '... the use, development and protection of natural and physical resources in a way, or at a rate, which enables people and communities to provide for their social, economic, and cultural well being and for their health and safety while:
>
> - sustaining the potential and physical resources (excluding minerals) to meet the reasonably-foreseeable needs of future generations;
> - safeguarding the life-supporting capacity of air, water, soil, and ecosystems; and
> - avoiding, remedying, or mitigating any adverse effects of activities on the environment.'

Alliterative phrases such as 'polluter pays', and the 'Precautionary Principle' have entered our language. The latter is embodied as Principle 15 in the Rio Declaration on Environment and Development following the United Nations Conference on Environment and Development in 1992:

'In order to protect the environment, the precautionary principle shall be applied widely.

Where there are threats of serious or irreversible damage, lack of scientific certainty shall not be used a reason for postponing cost-effective measures to prevent environmental degradation.'

The Precautionary Principle, thus, sets the challenge. Science and technology must go beyond being used a negative delimiter, as in the Precautionary Principle, to becoming a positive force by providing the knowledge and tools for achieving a clean, healthy and productive environment.

SCIENCE AND SOIL CONTAMINATION

The first Australasia-Pacific Conference on Contaminants and the Soil Environment in the Australasia-Pacific Region (SCRAP) was held in Adelaide, Australia in February 1996. The book of the proceedings contains 25 chapters, of which only four were on remediation, or rehabilitation. This is because the main focus was on establishing an inventory of degraded sites, lists of contaminants, and descriptions of the contamination processes (Naidu et al., 1996).

In the chapter on contaminant transport in the soil environment in the proceedings of the first SCRAP conference (Brusseau and Kookana, 1996), of the eight figures and three tables, just one figure and one table related to measurements made outside the laboratory in the field. This identifies one of the key challenges facing scientists working with contaminated soil—the translation of our detailed understanding of the microcosm to solve problems at the field scale. This chasm is probably the reason that Fowler (1996) began his chapter in the same book, with a quote from the Director of the Victorian Environmental Protection Agency, that:

'... [a] look at the current state of practice in contaminated site management cannot help but arouse some grave doubts as to whether we are any more able to deal with the complexity of evaluating environmental impacts of chemical contaminants theory than we were thirty years ago.'

At the second SCRAP meeting, we saw a movement beyond the detailed knowledge of the microcosm, to the application of technologies at the complex scale of the field. In this chapter, we outline a path along which we feel scientific understanding can be moved in order to achieve our stated goal of a clean, healthy and sustainably-productive soil environment.

Innovative techniques and recent knowledge tend to be first generated in the office with a pencil, or by empiricism in the laboratory. Also, improved understanding or new tools are often developed during studies at the microcosm scale where good experimental control can be maintained. Although some of these developments would initially appear to offer good prospects for developing sustainable management strategies, and remediating contaminated sites, the road to realization is long, tortuous, and littered with cul-de-sacs (Figure 1.2). Jolly (1999) identified five crucial steps along the path from innovation to market:

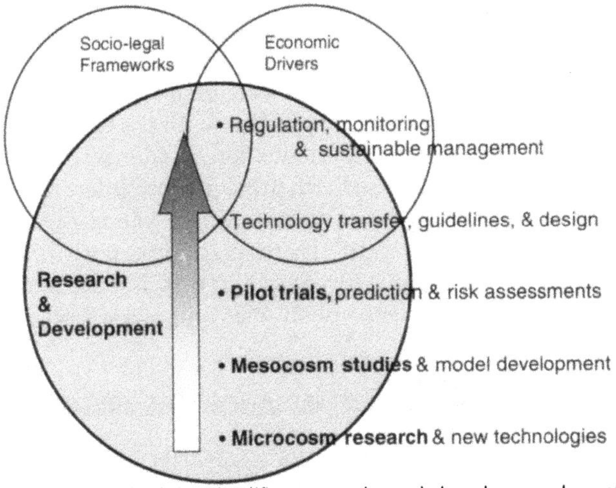

Fig. 1.2: A pathway along which scientific research and development must move so that underpinning knowledge and innovation are employed for sustainable management of soil, or its remediation from contamination.

1. Imagining—the initial insight for a particular development
2. Incubating—nurturing the technology
3. Demonstrating—building prototypes and getting feedback
4. Promoting—persuasion for adoption of the innovation
5. Sustaining—ensuring the product or process has a long life

Traditionally, scientists have been most heavily involved in the first, and maybe the second phase, with 'technology transfer' heralded as the means of progressing innovation through the final three steps. However, it has been mooted that:

> '... the term "technology transfer" is an oxymoron. Real innovations do not move from the laboratory to shopfloor as patents, research reports or even working prototypes. To stand any chance of success, they have to be transferred as concepts embedded in people's heads' (*The Economist*, 1999).

So, it is imperative that the scientist, as the innovator, stay closely involved with all the five steps, as knowledge moves along the path in Figure 1.2 towards achievement of the balanced goal of a clean and healthy soil environment. This process and passage can be achieved when the scientist is an integral part of a multiskilled team.

Innovation and understanding needs to pass on from in vitro studies and microcosm-scale (≈ mm^3) experiments, a process which is being facilitated with new measurement and monitoring technologies. Data logging, with on-line analyses in real-time, from both physico-chemical devices and biosensors, is providing new insights into soil processes (Figure 1.2). This new understanding is increasingly being aggregated into comprehensive models of soil functioning that are being tested at the mesocosm scale (≈ m^3). Models can then be used to examine the success of 'proof-of-concept' pilot trials (≈ 10's m^3), so that predictions and risk assessments can be made in conjunction with the engineers and planners. Moving from the 'imagining/incubation/demonstration' stages through to 'promoting and sustaining' will involve more interaction with elements in the socio-legal frameworks and responses to economic forces.

There is a 'market pull' from elements within the socio-legal frameworks, and economic forces add to this. For the rest of this chapter, we will focus on the 'science push' that will work towards achieving the balanced goal.

TRANSPORT, UPTAKE AND REMEDIATION: THE BRIGHT FUTURE

Three innovative factors will work towards a bright future aimed at maintaining a clean and healthy soil environment, and cleaning up contamination of those sites where there is a social and political will to do so. Improved understanding and prediction of contaminant transport through soil, and of the role played by roots in the uptake of chemicals, is expected to come about through new developments in the areas of:

- new measurement devices and novel monitoring technologies;
- improved modelling techniques and quantitative risk assessments; and
- better understanding of the link between the biological form and biophysical functioning.

Before considering future prospects, it is worthwhile to briefly reflect on the past. A theory describing water flow through soil was developed long ago. Significant descriptions of water flow through a porous medium began in 1856 with Henry Darcy's observations of the saturated flow of water, J_w (m s^{-1}), through a filter bed of sand in Dijon, France (Philip, 1995). L.A. Richards (1931) combined a mass-balance expression with an unsaturated Darcian description of the water flux J_w in order to arrive at the general equation of soil water flow, which Klute (1952) wrote in the following form:

$$\frac{\partial \theta}{\partial t} = \frac{\partial}{\partial z}\left(D(\theta)\frac{\partial \theta}{\partial z}\right) - \frac{\mathrm{d}K}{\mathrm{d}\theta}\cdot\frac{\partial \theta}{\partial z} - U(z,t,\theta). \qquad [1]$$

Here, θ is the volumetric water content of the soil ($m^3\ m^{-3}$), z is the soil depth, and t is time. The two hydraulic properties of the soil are: D, the soil-water diffusivity function ($m^2\ s^{-1}$) which characterizes the capillary processes; and the hydraulic conductivity function $K(\theta)$ ($m\ s^{-1}$), which parameterizes the convective flow through the soil. Included at the end on the right hand side of Equation 1 is a simple sink term U ($m^3\ m^{-3}\ s^{-1}$), which accounts for the temporally changing pattern in the depthwise uptake of water by roots. Locally, the uptake of water by the roots will depend on θ.

The transport of contaminants and uptake of chemicals are processes controlled by the water flow in soil that is described by Equation 1. Following on from the miscible displacement description of solute transport in soil (Nielsen and Biggar, 1961), it is possible to describe the transport of a reactive solute through soil, and passive plant uptake (Van Genuchten and Wierenga, 1976) by:

$$\left(1 + \frac{\rho}{\theta}\frac{\mathrm{d}f}{\mathrm{d}C}\right)\frac{\partial C}{\partial t} = D_s\frac{\partial^2 C}{\partial z^2} - \frac{J_w}{\theta}\frac{\partial C}{\partial z} - U \cdot C \qquad [2]$$

Here, the exchange relationship of the chemical in solution C ($mol\ m^{-3}$) with that sorbed on the soil S_c ($mol\ kg^{-1}$) is the isotherm $S_c = f(C)$. In Equation 2, ρ is the soil's bulk density ($Mg\ m^{-3}$) and D_s is the diffusion-dispersion coefficient of the solute ($m^2\ s^{-1}$). Equation 2 succinctly describes the key processes of contaminant transport and uptake. Sorptive exchange can be seen on the left-hand side to be governed by the exchange isotherm. The first two terms on the right-hand side show that transport of chemical through the soil is by diffusion within the water, and convection with the flow J_w. Finally, the last term is a sink which accounts, in this case, for the passive uptake of solute by the plant's distributed root system. Thus, the 'water drive' of the flow J_w, and the uptake U, are the two key processes in the transport of chemicals through soil.

So advanced was our theoretical understanding of flow and transport processes in soil, that in 1968, E.C. Childs commented that:

> '… further investigations to throw yet more light on the basic principles … tend to be matters of crossing *t*'s and dotting *i*'s … theory has developed more rapidly than its application and until this balance is corrected, further advances must be somewhat academic … serious difficulties remain in the path of practical application of theory … [we are] held back by the inadequate development of methods of assessment of the relevant parameters.'

NEW MEASUREMENT TECHNIQUES AND DEVICES

Hydraulics and Mobility

Water is the vehicle for contaminant transport in soil, as well as being the substrate for the exchange reactions. It is, therefore, important that water flow be understood. Doing so requires that the parameters describing the flow be known. Water enters the soil and is transported through it, in response to two forces: capillarity and gravity. The soil's capillarity is characterized by D, and the flow in response to gravity is parameterized by the conductivity K (Equation 1). Both these functions are strongly dependent of the soil's water content θ, and furthermore, there is great variation between the soils. Finer textured soils have greater capillarity rather than coarse-textured soils, whereas sands and loams have higher conductivities than silts and clays.

Measurement of the soil's hydraulic character is, therefore, necessary in order to predict the transport of contaminants in the soil.

Ever since Darcy's experiments over 150 years ago, it has been possible to measure the saturated hydraulic conductivity, K_s. Here, $K_s = K(\theta = \theta_s)$, where θ_s is the saturated water content of the soil, i.e. the water content when the soil-water pressure head, h (m), is zero. Measurement of the soil's capillarity, which is embodied in the water diffusivity function D, has proved more difficult. In order to proceed, it is useful to note here that, through the definition of D, the integral of the soil's diffusivity is equal to the integral of the conductivity:

$$\phi_s = \int_{\theta_n}^{\theta_s} D(\theta)\,\mathrm{d}\theta = \int_{-h_n}^{0} K(h)\,\mathrm{d}h\,. \qquad [3]$$

Here, ϕ_s is known as the matric flux potential, and the subscript n refers to the antecedent conditions.

Philip (1957a) noted that capillarity, in the absence of gravity, results in water entering the soil in such a way that the cumulative infiltration, I (m), is proportional to the square-root of time, in such a manner that

$$I = St^{1/2}. \qquad [4]$$

Here, S is the sorptivity (m s$^{-1/2}$). Thus, by simply recording the entry of water into the soil, it is an easy matter to measure S (Talsma, 1969). Sorptivity (squared) is related to the integral of the diffusivity function, and White and Sully (1987) showed that:

$$\phi_s = \frac{bS^2}{(\theta_s - \theta_n)}\,. \qquad [5]$$

Theoretically, it is known that b can only vary between $½<b<\pi/4$. So, if we know S, we can infer ϕ_s (Equation 5), which tells us about the integral of the diffusivity D, and the magnitude of the soil's capillarity (Equation 3).

New devices have been developed over the past decade to measure both K and S in the field. Two of the most popular, the Guelph permeameter (Reynolds and Elrick, 1985) and the disc permeameter (Perroux and White, 1988), use multi-dimensional infiltration so that as per the flow geometry, S and K can be measured simultaneously.

The disc permeameter (Figure 1.3) not only operates at saturation, but it can also work in the unsaturated region down to negative pressure heads of about $h_o \approx -200$ mm. Thus, the disc permeameter can be used to explore the matrix-macropore dichotomy in that crucial zone of near-saturated pressure potentials where water flow and contaminant transport is likely to be most rapid, preferential and far-reaching (Clothier and White, 1981; Clothier, 2002). The permeameter is circular, and good hydraulic contact with the soil is maintained using contact sand. Wooding (1968) found that the flow from a circular source of radius r_o was influenced both by capillarity and gravity such that a steady-state flux density v_∞ (m s^{-1}) soon becomes established. This he found to be:

$$v_\infty = K_o + \frac{4\phi_o}{\pi r_o}. \qquad [6]$$

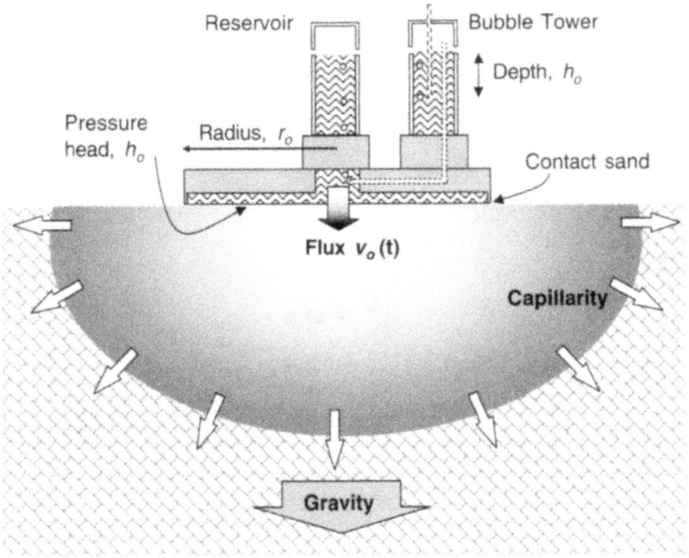

Fig. 1.3: The disc permeameter, or tension infiltrometer, which can be used to measure the unsaturated hydraulic conductivity of soil in situ. As well, it can be employed to obtain a measure of the mobile fraction of the soil's water that is active in chemical transport (see Fig. 1.5).

Here, the subscript '*o*' refers to the unsaturated conditions at the soil surface. Wooding's equation contains two unknowns, the gravity term K_o and the capillary term ϕ_o, and so at least two independent observations are required so as to determine both soil properties.

The method of Ankeny et al. (1991) achieves this by using observations of flow from the same permeameter at two or more pressure heads. Thus, the unsaturated hydraulic conductivity function can be built up in the region close to saturation. The results found using this techniques for a Swedish soil by Messing and Jarvis (1993) are shown in Figure 1.4. The role of the macropores can be seen to result in a rapid drop-off in K as the soil becomes unsaturated, and the macropores drain and cease to play a role in the water flow. Beyond the break point, there is a more gradual decline in K as the smaller pores in the matrix sequentially drain. Furthermore, a temporal change can be seen in the relative roles of the macropores and matrix throughout the year as the soil's structure changes with the seasons.

As alluded to above, the slope of the $K(h)$ relationship is a measure of the soil's macroporosity. We can make this more explicit. This exposition is simplified if we consider that the $K(h)$ relationship is exponential, and of the form:

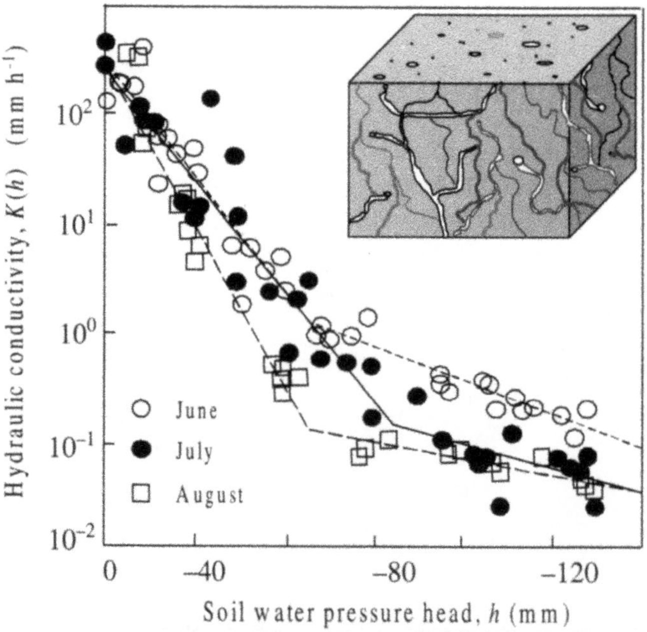

Fig. 1.4: The hydraulic conductivity function of a Swedish soil measured using disc permeameters (from Messing and Jarvis, 1993). These data, measured three times during the year, reveal a different hydraulic separation due to the changing relative influence of the macropores and the matrix.

$$K = K_o \exp[\alpha (h - h_o)]. \qquad [7]$$

Thus, up to some pressure head h_o, a 'mean' capillary length, λ_c (m), can be defined (Philip, 1985) as:

$$\lambda_c = \frac{\int_{-\infty}^{h_o} K \, \mathrm{d}h}{K_o}. \qquad [8]$$

Therefore, on the basis of Equation 7, $\lambda_c = \alpha^{-1}$. By putting this capillary length into Laplace's equation for the rise of water in a capillary tube, Philip (1987) arrived at a mean 'pore' size λ_m (mm):

$$\lambda_m = \frac{\sigma}{\rho g} \frac{1}{\lambda_c} \cong 7.4 \ \alpha. \qquad [9]$$

This relationship assumes that appropriate values are taken for the surface tension σ, the density of water ρ, and the acceleration due to gravity g. White and Sully (1987) called λ_m a '... physically plausible estimate of the flow-weighted mean pore dimensions'. We can apply this analysis to the data of Messing and Jarvis (1993) in Figure 1.4. For the August measurements, the decline in K down to the break point at –42 mm returns a slope of $\alpha = 0.19$ mm^{-1}, which from Equation [9] would indicate a mean macropore size of $\lambda_m = 1.4$ mm. The flatter slope beyond the break point has an $\alpha = 0.015$ mm^{-1}, suggesting that the matrix pores are of a much smaller size, $\lambda_m = 0.1$ mm.

Macropores, through their large size and high degree of connectedness, strongly influence the pattern of flow and the transport of chemicals. So great has this effect been, that the uniform soil description that lies behind Equation 2 has been called into question (Clothier, 2002). Alternative formulations have been sought. One approach has been to consider mechanistic-less transport, and describe the transport of chemicals using transfer functions (Jury, 1982). However, it is possible to develop a mechanistic description in order to try and account for the matrix-macropore dichotomy. Occam's Razor can be used to cut the soil's pore water into a mobile fraction, θ_m, and a complementary portion, θ_{im}, that is not involved in the transport of contaminants: $\theta = \theta_m + \theta_{im}$. Thus, Equation 2 can be rewritten to account only for convective-dispersive solute flow in the mobile domain, and also to allow a diffusive exchange between the two fractions:

$$\theta_m \frac{\partial C_m}{\partial t} + \theta_{im} \frac{\partial C_{im}}{\partial t} = \theta_m D_s \frac{\partial^2 C_m}{\partial z^2} - v_m \theta_m \frac{\partial C_m}{\partial z}. \qquad [10]$$

Here, the subscripts 'm' and 'im' refer to the mobile and immobile domains, respectively (Van Genuchten and Wierenga, 1976). The diffusive-like exchange between the domains can be described by a first-order mass transfer coefficient, $\beta(\mathrm{s}^{-1})$,

$$\theta_{im}\frac{\partial C_{im}}{\partial t} = \beta(C_m - C_{im}). \quad [11]$$

By restricting the transport to just some part of the soil's pores, for example the macropores, the formulations of Equations 10 and 11 mimic the oft-observed preferential flow. However, obtaining direct measures of θ_m was a problem, until Clothier et al. (1992) proposed a method that relied on using permeameters whose reservoirs were filled with a tracer solution. Jaynes et al. (1995) then proposed a multiple-tracer method that allowed for a resolution of the mass transfer coefficient β, as well.

If infiltration into a soil is from a permeameter whose reservoir contains a tracer at some concentration C_p, then after a period of time, during which the hydraulic properties are measured, if β is not too large, the concentration of solute in the mobile domain C_m will be C_p (Clothier et al., 1992). So, if samples are taken in the soil under the permeameter, and if the concentration of tracer measured in the sample is C^*, then:

$$\theta C^* = \theta_m C_p + \theta_{im} C_{im}. \quad [12]$$

On the other hand, if a tracer were used for which none were present initially ($C_{im} = 0$), then:

$$\theta_m = \theta\frac{C^*}{C_p}. \quad [13]$$

So, if $C^* < C_p$, then $\theta_m < \theta$ as some of the initial water must have been immobile and remained in place to account for the dilution from C_p. This new measurement technique allows parameterization of the mobile-immobile water formulation (Equation 10) for better prediction of chemical transport through structured soil.

The results in Figure 1.5 show that different soils can have different mobile water fractions, θ_m. The Ramiha silt loam is an aggregated soil, yet despite these aggregates, the transport of chemical through the soil is through the entire pore space. The slope of the $C(z)$ profile could be used to infer the solute dispersion coefficient D_s (Equation 2). The pattern of solute transport in the Manawatu fine sandy loam is quite different (Figure 1.5), as there it seems that some 30 % of the pore space is not available to the invading tracer. Furthermore, the greater smearing of the profile indicates greater dispersion in this case.

Although it is possible to use tracer-filled disc permeameters to infer the soil's mobile fraction, θ_m, it still remains somewhat tedious to carry out the chemical analyses. In a general sense, Bouma (1989) proposed the usage of key measures of the soil's characteristics to infer other traits that might be more difficult to measure directly. The vehicle in this process of

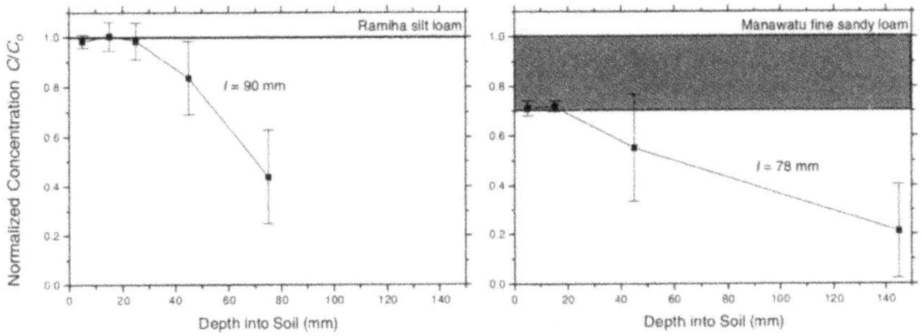

Fig. 1.5: Penetration of bromide tracer into two soils from a disc permeameter after about 80-90 mm of infiltration at the unsaturated head of h_o= -40 mm. The Ramiha silt loam (left) is, therefore, a 100% mobile-water soil, whereas some 30% of the water in the Manawatu fine sandy loam (right) appears immobile (following Clothier et al., 1998).

inference, he termed the 'pedotransfer function', for it allows existing knowledge to be transferred to deduce something that is unknown, or difficult to determine. We have collected additional measures of θ_m in such cases where we also have determined the flow-weighted mean pore size, λ_m. These results are shown in Figure 1.7. For those soils with the larger pores, the fraction of mobile water is smaller, as might be expected due to preferential flow processes moving the chemical more rapidly through just a small fraction of the soil's water-filled pore space. Thus, as new measurement techniques are applied more widely and our database of information grows, it should be possible to extrapolate our knowledge through the use of well-chosen pedotransfer functions such as the one shown in Figure 1.6.

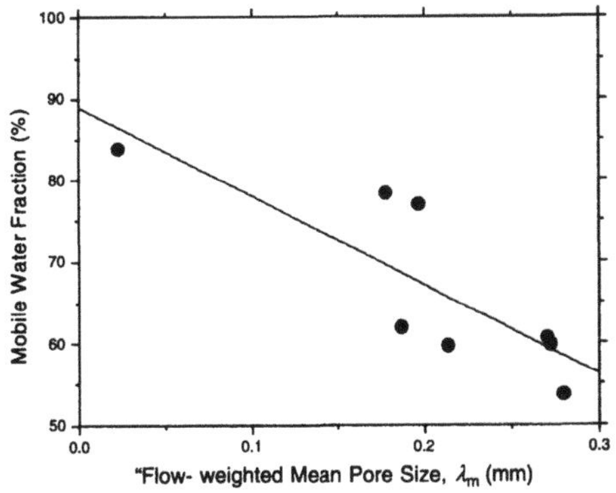

Fig. 1.6: The pedotransfer function relating the flow-weighted mean pore size, λ_m, as found by permeametry (Eq 9) for six New Zealand soils, to the measured mobile water measured under the permeameter using a single tracer (Eq 13).

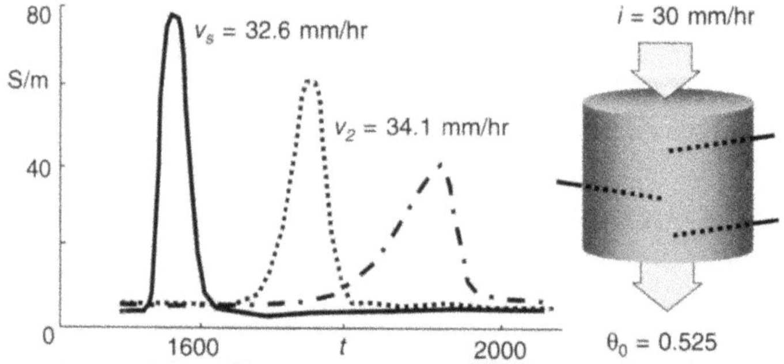

Fig. 1.7: Measurement of the velocity of a nitrate pulse travelling through Maréan soil by using the peak-to-peak changes in electrical conductivity from TDR rods located at depths of 30, 130 and 230 mm (following Vogeler et al., 2000). The time *t* is in seconds, and the unsaturated steady flow though the soil, *i*, is 30 mm/hr.

Soil Water Content and Chemical Transport

The new measurement device that has had the greatest impact in moving our focus from microcosm to the field (Figure 1.2) is TDR—Time Domain Reflectometry. Topp et al. (1980) developed the TDR technique, whereby high frequency electromagnetic pulses are sent along waveguides of length *L* that can be inserted in the soil. Nowadays, three-wire waveguides are most commonly used. The travel time down and back along the waveguides, t_p, depends on the soil's dielectric ε,

$$\varepsilon = \left[\frac{ct_p}{2L}\right]^2 \qquad [14]$$

where c is the speed of light (m s^{-1}). Since ε is most strongly influenced by the volumetric water content of the soil, θ, TDR provides a robust and useful means of determining θ, both in the laboratory and the field. Sometimes, especially for soils high in organic matter, or volcanic clays, a special calibration for $\theta(\varepsilon)$ is required. In the next step, by observing the ratio of the voltage of the input signal, V_o, to that finally realized after all the echoes have died away, V_f, a measure of the soil's bulk electrical conductivity σ_e (S m^{-1}) can be obtained (Dalton et al., 1984). From this bulk electrical conductivity, it is possible to obtain a measure of the soil solution conductivity, σ_s, so that some inference can be made about the concentration of electrolytes (Nadler et al., 1991). Deconvoluting the σ_e signal to infer electrolyte concentration would be very useful, for it would allow the use of TDR to monitor the C of tracers moving the soil. However, at present, it is difficult to use TDR with confidence in this inverse way to measure directly C. TDR can detect, through σ_e, that rate of passage of chemicals through soil. Kachanoski et al. (1992) used TDR in the field to measure the travel times of inert tracers through the rootzone.

Here, we present an example of how TDR can be used to measure the travel times of a reactive tracer so that the exchange characteristics of the chemical and root uptake can be inferred (Vogeler et al., 2000; Vogeler et al., 2001a).

Many soils in Asia and the Pacific are variably charged in such a manner that the anions undergo an exchange reaction with the soil (Duwig et al., 1999). If a chemical in solution at concentration C, results in an amount S_c being sorbed, then the exchange relationship is termed as the isotherm (see Equation 2). For simplicity, we consider here the linear case of:

$$S_c = K_D C \qquad [15]$$

where K_D is called the distribution coefficient ($m^3\ kg^{-1}$). Vogeler et al. (2000) carried out a rainfall simulator experiment on a column of a variably-charged Ferrasol from Maré, New Caledonia (Figure 1.7). The steady rainfall rate i was 30 mm/hr, and the TDR measured the soil's water content at 0.525. Thus, the velocity of the water moving through the soil would be v = 57 mm/hr. Three-wire TDR rods were inserted horizontally at depths of 30, 130 and 230 mm, and the flow out base at 300 mm was maintained unsaturated at $h_o = -20$ mm. The soil has a bulk density of 0.8 Mg m^{-3}. A pulse of nitrate, equivalent to 100 kg-N/ha. was added to the soil surface during the steady rain, and its passage, as evidenced by the changing σ_e, can be seen in Figure 1.7. From the peak-to-peak changes in σ_e, a solute velocity v_s of 33 mm per hour can be calculated. Thus, the nitrate pulse is moving slower than the water flow, in such a way that the retardation R of v/v_s is 1.7. Across all experiments, we found on an average that R was 1.9. It is possible to relate the retardation R to the exchange reaction, since

$$R = 1 + \frac{\rho K_D}{\theta}. \qquad [16]$$

Thus, since we know from our TDR-measured v_s that R is 1.9, we can infer that the strength of the anion-exchange reaction that caused this retardation must be characterized by a K_D of 0.6 $m^3\ kg^{-1}$. Thus we have, in this microcosm experiment, been able to determine the exchange reaction in the soil by observing solute transport with TDR.

Since we directly measured the flux concentration of nitrate quitting the base as drainage from the core, we can compare the measured leaching of chemical with that predicted using this value of K_D deduced by TDR (Figure 1.8). This is possible by modelling the flow and exchange processes via a numerical solution of the coupled Equations 1 and 2. There is good agreement in Figure 1.8 between the measured flux concentration

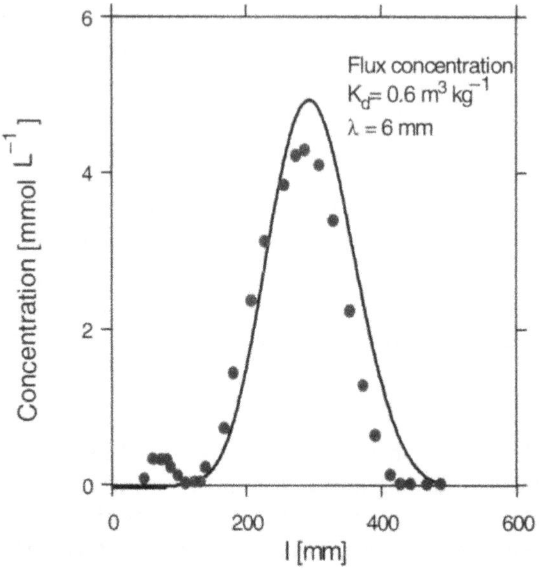

Fig. 1.8: The measured and modelled flux of nitrate out of the base of a core of Maréan soil. The parameterization of the linear isotherm (K_D) came from the TDR observations of Fig. 1.6 (following Vogeler et al. 2000).

of nitrate in the drainage, and that predicted by solving Equations 1 and 2 using parameters found independently by TDR. This modelling exercise demonstrates the value of our better measurements of those parameters that control the flow of water and exchange of chemicals. Modelling then allows the transition to be made from study at the microcosm level, to prediction and risk assessment at the mesocosm and field scale.

MODELLING

Science has forever been an exercise in modelling. Scientists have always developed conceptual and abstract constructs of the linked processes they study. Such abstractions are termed models. An inability to analytically solve the coupled partial differential equations that underpin the models of flow and transport in the environment previously meant that such models were of little general utility. Nowadays, fast personal computers, in conjunction with inexpensive software codes, have resulted in easy access to solutions of flow and transport models for a wide variety of scenarios. Generally, these models are an encapsulation of the best theoretical understanding that we have currently. But the application of such models demands that the parameters describing the processes be aptly measured, or at least confidently estimated. It is, as quoted by the famous mathematician J.W. Tukey,

'... far better an approximate answer to the right question which is often vague, than an exact answer to the wrong question which can always be made precise'.

Models permit questioning, and they can provide answers to 'what if' questions. Models can also be used to allow extrapolation to the field of the meagre data obtained in microcosm or mesocosm experiments. This extension can provide a risk assessment of the effects of current practices, or to quantify the impacts that could result from a change in land management. As well, models can be used to explore avenues by which current practices can be improved in order to ameliorate deleterious consequences.

In order to develop guidelines for the sustainable application of agrichemicals, we have developed a mechanistic model for water-borne pesticide transport through soil. Our model (Green et al., 2002), called SPASMO (Soil Plant Atmosphere System Model), is similar in form to LEACHM (Wagenet and Hutson, 1986), for it accounts for mobile and immobile water flow (Hutson and Wagenet, 1995), plus the exchange of pesticide with organic carbon in the soil, and the degradation of the pesticide with time. In Figure 1.9, we show the results from the application of the model to a drip-irrigated vineyard on a stony soil of low organic carbon content in the Hawkes Bay area of New Zealand. Here, the model results can be used to reduce groundwater contamination through selection of pesticide type, and by appropriate timing of the spraying. SPASMO was run in daily time-steps using a 27-year sequence of weather data taken from a site nearby. Three herbicides were considered—simazine,

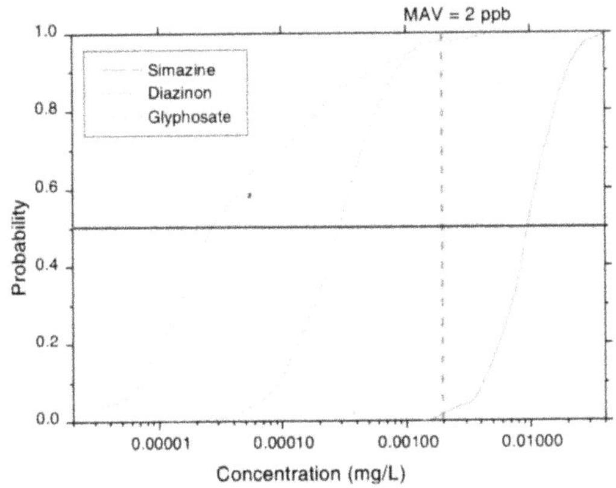

Fig. 1.9: The modelled probability density functions for the concentration of 3 pesticides at a depth of 3 m in a stony soil growing grapes in the Hawkes Bay of New Zealand. The model was run in daily time-steps using 27 years of meteorological data and the pesticides were applied on the same date in spring and autumn of each year.

diazinon and glyphosate—and they were all 'applied' in the model on 1 April and 1 October every year.

The more than 9800 daily values for the simulated concentration in these three pesticides—at a depth of three metres—are presented in Figure 1.9 in the form of a probability density function (pdf). The average values for the three concentrations at this depth, just above groundwater, are markedly different. Simazine, a leacher, is on average simulated to be around 10 ppb, whereas diazinon is at about 0.3 ppb, and glyphosate is 0.03 ppb. Therefore, simazine would not be a good choice for a pesticide here. The Maximum Allowable Value (MAV) is currently 2 ppb, and our pdf for simazine suggests that for 98 % of the time, simazine concentrations would be expected to exceed this. On the basis of their respective average concentrations, a risk assessment of the pdfs would favour glyphosate as the best herbicide here.

However, due the occasional timing of the day application with heavy rain, and the impact of a high mobile water fraction (macropores) in this soil, both chemicals exceed the MAV for the same amount of time, about two %, or 195 days over the 27 years. So, they both present the same level of risk in terms the MAV, despite their quite different average behaviour. This average behaviour is a result of their different chemical characteristics and the level of the soil's organic carbon, whereas the risk posed by both is, in terms of the MAV, determined by the local weather conditions and the soil's hydraulic properties. The model could now be used to explore a shift in the date of spraying to 1 March (cf. 1 April) that might serve to reduce the impact of the autumnal rains in leaching the herbicides. As well, the model could be used to examine the role that irrigation might play in exacerbating the leaching of the various pesticides. So, without even conducting an experiment, it would seem that such modelling could provide a decision-support tool to improve pesticide practices that act towards limiting environmental contamination. Lichtenberg and Zimmerman (1999) explored the farmers' willingness to pay for groundwater protection, and they found that farmers were more willing to pay for leaching prevention than non-farm groundwater consumers. A modelling tool such as SPASMO can be used to provide the groundwater consumers with information so that the environmental impacts of their decisions are better understood. Models are effective tools that can be used in 'participatory learning and action' approaches to environmental issues.

But, can we have faith in these simulated results, and should we validate the model?

In the 1999 Langbein Lecture to the American Geophysical Union, John Bredehoeft noted that comprehensive transport models are an abstraction of reality and, by their very nature, require assumptions about the real system. Many of the flow and transport parameters cannot be independently measured. In an effort to engender faith in the results generated in a model, its results are often compared with measurements. The unknown parameters are then derived from best model-fit to the observations. Although this process is sometimes called 'validation' or 'verification', this is not really the case, for such optimization does not necessarily provide a unique set of parameter values. Rather than glorify this process by terming it 'validation' or 'verification', Bredehoeft and Konikow (1993) damned it with faint praise by simply calling it 'history matching'. Oreskes et al. (1994) pessimistically concluded that '... verification and validation of numerical models of natural systems is impossible. So it would seem futile to waste both time and cost to attempt to justify the veracity of comprehensive models via direct comparison with full-scale measurements'. Rather, we should work towards enhancing the modelling procedure through realization of better observations of the processes, and testing of model components. Obtaining improved measures of the key parameters using new devices and techniques will complement this.

Who are the model developers, and who are the model users? Kutilek and Nielsen (1994) added a postscript to their text on soil hydrology, and noted that following a survey of the modelling literature, they were reminded of the 'rule of 90%'. They concluded that:

> '... ninety per cent of the models in papers are not developed sufficiently for others to use them. Of the remaining 10%, 90% of the models are used only by their creators ... [this] is perhaps sad or cynical [but] it offends neither innovative modellers nor productive scientists.'

Democratic access to software code potentially provides a tool for everyone to predict the transport of contaminants in the environment. However, there is an obligation for the user of models to adhere to a code of good modelling practice. More will be achieved through wise use of appropriate models than could be realized by futile attempts to validate or verify comprehensive schemes. The basic objective of the Code of Good Modelling Practice (GMP) that was used for leaching models in the process of European Union registration (EU-directive 91/414, 1997) is 'transparency in all steps' (see Vanclooster et al., 2000). A grave concern highlighted by Vanclooster et al. (2000) was user subjectivity. They found output variability due to user-dependent interpretation of the model's input set, and defining parameters. Adherence to a code of GMP should eliminate much of this subjectivity so that greater faith can be placed on the modelled results.

The basic tenets in this code of GMP are listed in Table 1.1 (Vanclooster et al., 2000)

Table 1.1 Tenets in the code of good modelling practice

Objective: Transparency in all steps
• User is responsible for understanding the model and its appropriate usage
• User is responsible for parameter estimation and the input data for each scenario
• User maintains up-to-date version and documentation
• User reports sufficient information for independent reproduction of simulation results

Vanclooster, M. et al., 2000. *Ag. Water Manage.*

What value can be placed on the results obtained from appropriate models applied under the code of GMP? In a review of the 1997 Chapman Conference on solute transport, Corwin et al. (1998) concluded somewhat pessimistically that:

'... although policy makers and environmental modellers are converging, the policy-making process is necessarily about politics. Models in this realm are most likely to be applied as political weapons, not as unbiased tools. The notion that science and technology will mitigate environmental problems on a "truth wins" basis is probably illusionary.'

Moreover, Bredehoeft (1999) noted that:

'... often these models have become part of a court battle [and] are attacked on the basis of their abstraction ... [Indeed] knowledgeable lawyers comment "most, if not all, models can be destroyed in court"! [However] models are most useful in helping to better understand systems and how they work, and exploring policy to administer [these] systems.'

To use models in a policy context, Corwin et al. (1998) suggested that we require:

'... ultimately, the integration of scientific, economic and political considerations to make pollution assessments with advanced information technologies to become a decision maker's [tool], rather than a purely scientific [one]'.

However, in their review on the use of models for simulating pesticide movement and fate in groundwater, Barbash and Pesek (1996) noted that many transport models currently neglect key biochemical processes. Prime amongst these is the direct and synergistic role played by the plant in contaminant transport and the remediation of degraded environments.

BIOPHYSICS

Globally, an average of 720 mm of rain falls on the earth's surface each year. About 410 mm of this is evaporated back to the atmosphere, either directly from the soil or, more often, by evaporation from plants. This transpired water must first have been captured from the soil by the plants' roots (Figure 1.10). Soil physicists, partly because of their mathematical

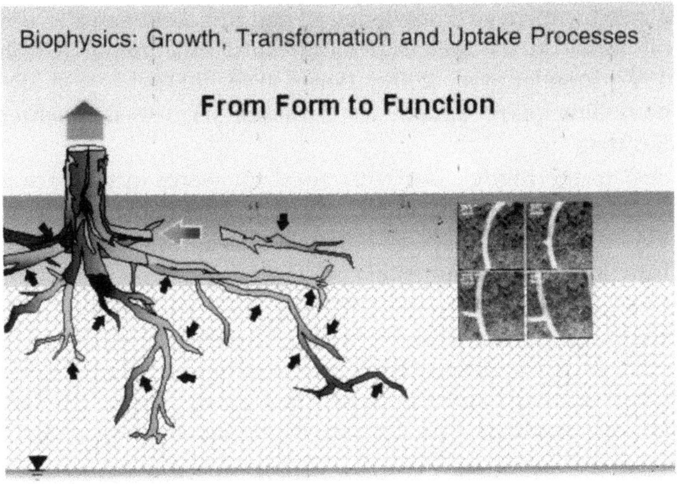

Fig. 1.10: Biophysical processes are the key to a clean, healthy and productive soil, and they are likewise critical for the bioremediation of contaminated sites. The scientific challenge is to translate our better observations of the biological form through to a prediction of biophysical functioning.

bent, have obtained descriptions of, and solutions for, water and chemical movement through biology-free, unsaturated soil. Yet, in Equation 1, we can see that plant uptake *U* can be an important sink for water, and this term remains in Equation 2 for plant uptake of chemicals. Such compounds may either be nutrients which could become pollutants, or contaminants that are already resident in the soil. Harper et al. (1991) concluded that '... plant root systems present the research worker with many of the greatest unresolved problems in the plant sciences'.

Root water uptake, *U* (Equation 1), was for a very long time ignored by physicists because it could not be easily described analytically in a mathematical form that was amenable to solution. Gardner (1960) and Philip (1957b) developed solutions for *U* that considered the root as an infinitesimally small line-sink of semi-infinite length. Twenty-five years later, Gardner (1985) warned that '... while this was probably a useful start, I think it has led us to a dead end!'

Now, numerical solutions of the partial differential equations describing flow and transport (Equations 1 and 2) are easily available, thanks to fast personal computers. Nonetheless, we have not seen commensurate advances in our understanding of how we can mechanistically link our improving observations of the root system form to the uptake functioning of plants. This uptake functioning is often facilitated by, or enhanced through, mycorrhizal activity and microbial actions. If we are to better understand the transport, uptake and remediation of soil contaminants, then we need improved knowledge of the link between the observable biological forms in the soil, and biophysical functioning (Clothier and Green, 1997; Vogeler et al., 2001a; Green et al., 2002).

For modelling plant uptake, *U* (Equation 1), van Genuchten (1995) concluded that:

> '... most of the models currently used for water uptake are essentially empirical ... [and] much research is still needed to derive process-based descriptions of root growth and root water uptake ... as a function of environmental conditions'.

Impetus for this improvement in modelling will, in part, come from '... illuminating comprehensive [model] failures that will stimulate us to change the way we think about the workings of the [plant] and its interactions with its environment' (Passioura, 1996). The means of sustaining this momentum will come from new techniques and modern devices.

Better measurements of both root-system form and uptake functioning are continuously being realized. Minirhizotrons are providing us with better pictures of the changing depthwise pattern of root-length density (Böhm, 1974). On the right of Figure 1.10 is a sequence of minirhizotron observations of root growth at a depth of about 300 mm. The individual frames are 12 × 18 mm in size, and clockwise from top right, the images were recorded 1, 2, and 7 days apart. The challenge is to interpret these observations to know what fraction of the roots 'seen', are indeed active movers of water and chemical. By way of miniaturization of the sensors, Green and Clothier (1995) were able to use the heat-pulse technique in the roots of trees to measure directly the sap flowing inside the root that must have come from uptake. So, these sap-flow measurements permit direct observation of *U* (Equation 1). We are now trying to develop techniques to directly measure the chemical concentration of the sap-stream so that we might also be able to measure *UC* (Equation 2). Also, TDR is providing us with clearer vision of the pattern of root uptake of both water and chemical (Green and Clothier, 1999; Vogeler et al., 2001b).

Technology is rapidly leading us to a deeper understanding of how roots act as the big movers of water and chemical in soil. Plants need to take up nutrients from the soil in order to grow. However, some plants can readily remove chemicals that we consider to be contaminants (Brooks, 1998).

PHYTOREMEDIATION

Phytoremediation is the use of plants, including trees, grasses and aquatic plants, to remove, destroy or sequester hazardous substances from the environment. It is an emerging technology (Glass, 1999), that is moving into the realm of the usable (Watanabe, 1997). Glass (1999) estimates that the phytoremediation business in the United States is already at US$30–50 million, and worldwide it could be US$40–60 million. With there being over 400 000 contaminated sites in Western Europe alone, this business can be expected to grow. Nyer and Gatliff (1998) were so moved:

> '... to predict that phytoremediation will be the next 'hot' technology for the environmental remediation field ... plant species can be selected to extract and assimilate or extract and chemically decompose target contaminants ... the proper application of these properties can lead to a very powerful remediation technique. [However] there are very specific design requirements for the proper application of this technology.'

In this chapter, we shall review the potential for phytoremediation, and then we examine what science needs to do in order that we move along the path described in Figure 1.2 so that phytoremediation is a viable technology.

Trees and plants can be used to clean up both contaminated soil and water (Table 1.2). For soil, plant processes foster a range of mechanisms to remove contamination. Plants, through phytoaccumulation, or just phytotolerance, can extract contaminants directly from the soil—U_C in Equation 2 (Figure 1.11). Noisome chemicals might be immobilized by plant exudates so that contamination is phytostabilized. In other cases, plants might engender microbial decay of contaminants so that remediation is by phytostimulation. Contaminants might be degraded of metabolized within, or by the plant, so that soil clean up is by phytotransformation, or it could be that the chemical taken from the soil by the plant is transpired so that the load on the soil is reduced by phytovolatilization. For hyperaccumulating plants, there might even be the possibility to phytomine valuable chemicals (Robinson et al., 1999). For cleaning up contaminated waters, plants can be used as either filters or barriers, and they can also be used to cap degraded sites. All of these phytoremediation processes require a good understanding of plant-soil-water interactions.

Table 1.2 Phytoremediation processes for the clean up of contaminated water and soil

Soil	*Water*
Phytoextraction	Rhizofiltration
Phytostabilization	Hydraulic barriers
Phytostimulation	Vegetative caps
Phytovolatilization	Constructed wetlands
Phytotransformation	
Phytomining	

A technology reliant upon an understanding of plant-soil-water interactions

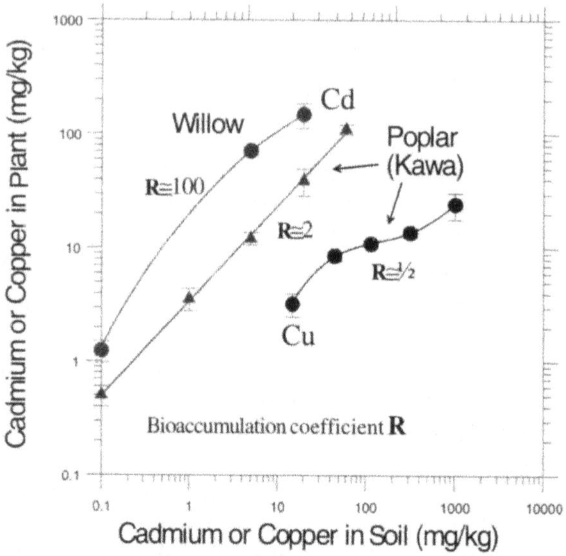

Fig. 1.11: Phytoaccumulation of copper or cadmium by either poplars or willows. The measurements relate to one season's growth for young trees growing in 20 L buckets of contaminated soil.

The advantage of phytoremediation is that, relative to industrial clean-up, the use of plants is low cost, and permanent (Table 1.3). With gentle remediation, in situ treatment, it is possible to enhance uptake through the use of mobilizing agents such as EDTA. Ex situ treatment of piles of contaminated soil is also possible. With phytoremediation, a wide variety of plants are available to treat a range of contaminants. However, this process is limited to moderate levels of contamination, for hyperaccumulators tend not to have high enough biomass production to reduce contaminant loadings. Phytoremediation is slow and cannot be that successfully applied to clean up contamination in shallow soils.

Table 1.3 The strengths and weaknesses of phytoremediation as a technology to clean up environmental contamination

Strengths	*Weaknesses*
Low cost	Limited to moderate contamination
Permanent	Slow remediation
In situ	Limited on shallow soil
Management enhancement	Limited with shallow groundwater
Variety of contaminants	Biological limitations
Variety of plants/fungi	Implementation difficulties
Public 'green' appeal	Incomplete scientific understanding

Poplars and willows are effective phytoremediants (Burken and Schnoor, 1998), both because of their propensity to extract contaminants from the soil, and also because their high rate of water use reduces the hydraulic loading on groundwater. We have carried out some pot trials to determine, in microcosm, the ability of hybrid poplars and willows to extract heavy metal contaminants (Figure 1.11). For the hybrid poplar 'Kawa', the bioaccumulation factor R (C_{plant}/C_{soil}) is just 1/2 across a range of copper concentrations, where $R \approx 2$ for cadmium. For the hybrid willow 'Tangoio', R was found to be about 100 for cadmium (Robinson et al., 1999). Thus, both of these varieties of poplar and willow offer real potential as phytoremediants of cadmium, whereas for copper, the clean-up task will be harder. In such cases, some form of mobilizing treatment will assist phytoremediation. However, the use of chelates and acids to release the soil's grip on the contaminant is a two-edged sword (Thayalakumaran, 2002). Mobilization can be evidenced through pot trials to enhance the uptake by phytoextraction. However, mesocosm experiments reveal that mobilization treatments can make the contaminant more mobile so that it can move more freely through the soil and then pollute groundwater. The linked roles of water transport and plant uptake are critical for the success of phytoremediation.

Phytoremediation can also be used to clean up sites contaminated by polycyclic aromatic hydrocarbons (PAHs) (Table 1.4). Here, the clean up is predominantly by phytostimulation due to microbial activity. Aerating soil contaminated by phenanthrene and letting the indigenous microbiota work on remediating the soil resulted in 93% clean up. However, phytostimulating this by poplar increased the six-month clean up to 98 %.

Table 1.4 The clean up of phenanthrene ($C_{14}H_{12}$) in aerated soil by indigenous microbiota and poplar

Initial concentration (mg/kg)	*Aerated soil and indigenous microbiota*	*Same soil and poplar*
308 (± 111)	19 (± 9)	7 (± 3)
Clean up after 6 months	93%	98%

Space-planted trees, by virtue of their deep-rooting systems and their large leaf areas, consume much greater volumes of soil water than do shallow-rooted full-cover plants. Remediation of contaminated soil with poplars and willows is enhanced by the trees' high rates of water use. Not only will these trees enhance the soil through phytoextraction, but the environment is protected for there is less hydraulic drive pushing contaminants deeper towards groundwater. In a mesocosm experiment in 200 L lysimeters, we compared leaching losses under pasture and poplar by using bromide tracer (Figure 1.12). Irrigation was used to ensure

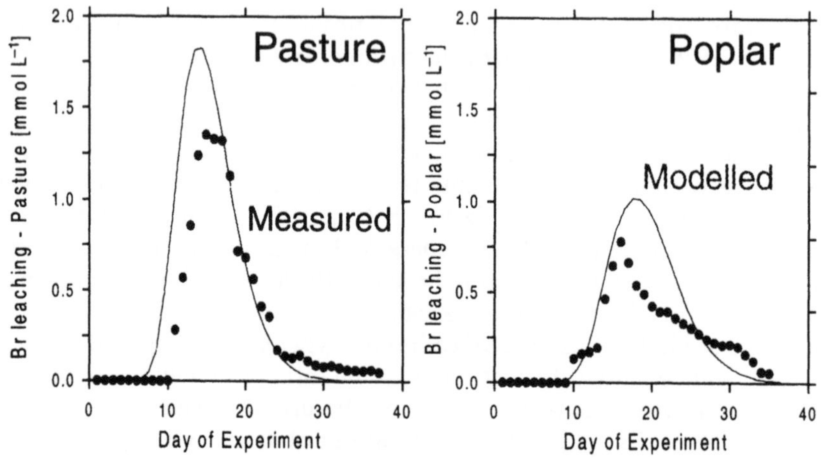

Fig. 1.12: The measured passage of a pulse of bromide tracer through the 1 m deep rootzone of either pasture (left) or a poplar tree. The results of this mesocosm leaching experiment were also simulated using a mechanistic model of chemical transport and uptake (solid line).

leaching at the base of the lysimeter at a depth of one metre. The pasture transpired at an average rate of around 0.8 L d^{-1}, whereas the tree drew water from the soil at 5.9 L d^{-1}. A pulse of bromide tracer was added to the soil, and the breakthrough of chemical measured in the leachate. Under the pasture, the bromide peak came through after 15 days, and the peak concentration of bromide was 1.3 mmol L^{-1}. The greater evaporative pumping of the poplar trees meant a day's delay in the peak, the concentration of which was just 0.8 mmol L^{-1}. The measured results in this mesocosm experiment are well predicted by our SPASMO model of contaminant transport. Phytoremediation is a clean-up technology that offers great potential. There will be a bright future for phytoremediation as we develop new monitoring technologies and improve our modelling techniques, especially those related to biophysical functioning.

FUNGAL BIOREMEDIATION

Bacteria and fungi have an ability to naturally attenuate soil contaminants, and the future for such bioremediation is also bright. Nyer and Duffin (1998) define bioremediation as the biochemical reactions of natural attenuation. Natural attenuation has been defined (USEPA, 1990) as the:

> '... biodegradation, dispersion, dilution, sorption, volatilization and/ or chemical and biochemical stabilization of contaminants to effectively reduce contaminant toxicity, mobility, or volume to levels that are protective of human health and the ecosystem .

White-rot fungi (*Trametes versicolor*) can be found just about everywhere there is dead wood. These fungi exude the enzyme ligninase to degrade the lignin in the wood in order to support their growth (Walter et al., 1998; Potter and Walter, 1998). The ligninase exuded by the fungus is a powerful chemical that can break down soil contaminants such as pentachlorophenol (PCP). PCP contamination is the legacy of old timber mill practices, and in New Zealand alone, there are over 800 PCP-contaminated sites.

Thus, by mixing contaminated soil with a substrate containing white-rot fungi, it is possible to remediate the contamination. Results from a microcosm experiment in the laboratory (Figure 1.13) confirm that white-rot fungi can be effective at breaking down PCP. Here, after just 40 days, some 30 % of the labelled PCP has been broken down. The challenge now is to take this technology from the laboratory bench, through proof-of-concept pilot studies, so that it can end up as a tool for environmental remediation. This transfer will require improved understanding of the conditions that favour ligninase production by the fungi, and knowledge of the ambient conditions that favour vibrant fungal growth.

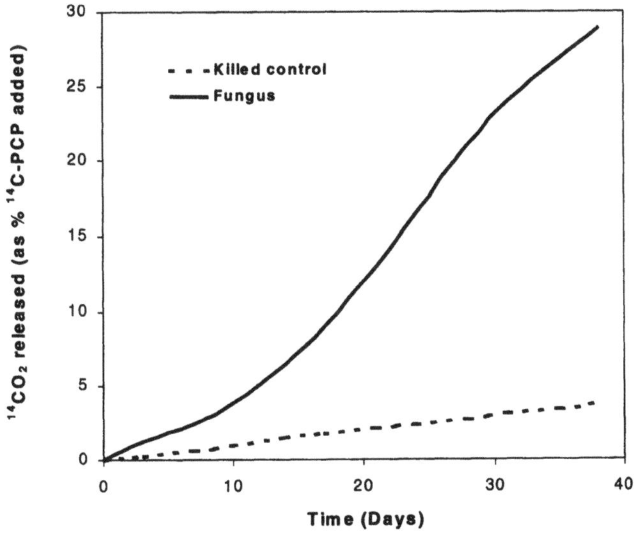

Fig. 1.13: The breakdown of pentachlorophenol (PCP) by an isolate of white-rot fungi (*Trametes versicolor*), in relation to that of a sterile control.

In a provocative article that highlights this imperative, Nyer (1998) noted that:

'... although bacteria [or fungi] are responsible for the biochemical reactions, their natural degradation rate is controlled by chemical and physical factors. An in situ bioremediation design requires

> identification of the rate-limiting factors and the delivery of those factors. [Indeed] I have never seen the bacteria [or fungi] be the cause of an unsuccessful in-situ project ... A hydrogeologist is the expert who will deliver these factors to the bacteria [or fungi] .

What, in essence, Nyer is suggesting, is that as science moves along the path from microcosm to the field (Figure. 1.2), there needs to be a transfer of responsibility and 'control'. The microbiologist and mycologist are the people to deliver the product and microcosm recipe, whereas the engineer and hydrogeologist then become responsible for ensuring that the recipe is followed in the field.

These are the scientific challenges. However, we need to acknowledge the fact that there are also political challenges to ensure the application of effective remediation technologies.

POLITICAL CHALLENGES

There are over 300 000 contaminated sites in the United States, and their annual clean up bill is around US$9 bn. Yet, ironically, clean-up regulations (Figure 1.1) perversely encourage polluters to dither and delay, and serve to limit the use of remediation technologies. So, whereas there is a remediation push from underpinning science, the socio-legal frameworks and economic forces conspire to push against the application of these technologies (Figure 1.2). Suresh Rao, formerly of the University of Florida, chaired a National Research Council study on remediation technologies, and he concluded that '... our collective experience is that aggressive [remediation] technologies meet with resistance' (see Gibb, 1999). Kent Udell from the University of California-Berkeley explains that:

> '... if the technology really works, then regulators may force [polluters] to clean up their sites within five years. So, obviously, they don't want anything to do with it. [So] given the choice between spending $25 million on a risky but fast clean-up that will deplete a company's cash reserves and hurt its stock price, or paying lawyers $1 million a year to delay, any sensible CEO will choose the latter.'

CONCLUSION

Science needs to push its environmental-protection ideas and its remediation tools beyond the microcosm, through the mesocosm, to the field. In moving along this path, research and development will encounter economic exigencies and powerful socio-legal frameworks. These counter-forces may either be attractive or repulsive. The developing force exerted by science and technology must become irresistible, so that we can achieve a clean, healthy and sustainably productive environment.

REFERENCES

Ankeny, M.D., Ahmed, M., Kaspar, T.C. and Horton, R., 1991, Simple field methods for determining unsaturated hydraulic conductivity. *Soil Science Society of America Journal* 55: 467–470.

Barbash, J.E. and Pesek, E.A., 1996, *Pesticides in Groundwater: Distribution, Trends and Governing Factors*. Ann Arbor Press, Chelsea, Michigan, 588 pp.

Böhm, W., 1974, Minirhizotrons for root observations under field conditions. *Z. Acker. Pflanzenbau*. 104: 282–287.

Bouma, J., 1989, Using soil survey data for quantitative land evaluation. *Advances in Soil Science* 9: 177–213.

Bredehoeft, J.D., 1999, Use of ground-water models. The Langbein Lecture, Spring Meeting, San Francisco, American Geophysical Union (Abstract H21D-01). http://earth.agu.org:80/meetings/waissm99.html

Bredehoeft, J.D. and Konikow, L.F., 1993, Groundwater models: Validate or invalidate. *Groundwater* 31: 178–179.

Brooks, R.R., 1998, *Plants that Hyperaccumulate Heavy Metals: Their Role in Archaeology, Microbiology, Mineral Exploration, Phytomining and Phytoremediation*. CAB International, Wallingford, England.

Brusseau, M.L. and R.S. Kookana, 1996, Transport and fate of organic contaminants in the subsurface. In: Contaminants and the Soil Environment in the Australasia-Pacific Region. Proceedings of the First Australasia-Pacific Conference on Contaminants and Soil Environment in the Australasia-Pacific Region, Adelaide, Australia 18-23 February, 1996, Kluwer Academic, Dordrecht, pp. 95-124.

Burken, J.G. and Schnoor, J.L., 1998, Predictive relationships for uptake of organic contaminants by hybrid poplar trees. *Environmental Science and Technology* 32: 3379–3385.

Childs, E.C., 1968, Soil moisture theory. *Advances in Hydroscience* 4: 73–117.

Clothier, B.E., 1997, Can soil management be regulated to protect water quality? *Land Contamination and Reclamation* 5: 337–342.

Clothier, B.E., 2002, Rapid and far-reaching transport through structured soils. *Hydrological Processes* 16: 1321–1323.

Clothier, B.E. and White, I., 1981, Measurement of the sorptivity and soil water diffusivity in the field. *Soil Science Society of America Journal* 45: 241–245.

Clothier, B.E., Kirkham, M.B. and MacLean, J.E., 1992, In situ measurement of the effective transport volume for solute moving through soil. *Soil Science Society of America Journal* 56: 733–736.

Clothier, B.E. and Green, S.R., 1997, Roots: The big movers of water and chemical in soil. *Soil Science* 162: 534–543.

Clothier, B.E., Vogeler, I., Green, S.R. and Scotter, D.R., 1998, Transport in unsaturated soil: aggregates, macropores and exchange. In: Selim, H.M. and Ma, L. (eds), *Physical Nonequilibrium in Soils: Modeling and Application*. Sleeping Bear Press, Michigan, US: 273–295.

Corwin, D.L., Loague, K., and Ellsworth, T., 1998, Assessing nonpoint source pollution in the vadose zone. *Eos May* 5: 210–220.

Dalton, F.N., Herklerath W.N., Rawlins D.S. and Rhoades J.D., 1984, Time-domain reflectometry: Simultaneous measurement of soil water content and electrical conductivity with a single probe. *Science* 224: 989–990.

Duwig, C., Bequer, T., Clothier, B.E. and Vauclin, M., 1999, A simple dynamic method to estimate anion retention in an unsaturated soil. *C.R. Acad. Sci., Sciences de la terre et des planètes* 328: 759–764.

Fowler, R.J. 1996. Legal responses to the problem of soil contamination—The Australian experience. In: Naidu, R., Kookana, R.S., Oliver, D.P., Rogers, S. and McLaughlin, M.J. (Eds), *Contaminants and the Soil Environment in the Australasia-Pacific Region*, Kluwer Academic Publishers, London: 267–280.

Gardner, W.R., 1960, Dynamic aspects of water availability to plants. *Soil Science* 89: 63–73.

Gardner, W.R., 1985, Dynamic aspects of water availability to plants. *Current Contents* 35: 20.

Gibb, W.W., 1999, Not cleaning up. *Scientific American* 280 (2): 22–23.

Glass, D.J., 1999, US and International Markets for Phytoremediation 1999-2000, D. Glass Associates, Needham, MA. http://www.channel1.com/dglassassoc/info/phy99exc.htm

Green, S.R. and Clothier, B.E., 1995, Root water uptake by kiwifruit vines following partial wetting of the root zone. *Plant Soil* 173: 317–328.

Green, S.R. and Clothier, B.E., 1999, The rootzone dynamics of water uptake by a mature apple tree. *Plant Soil* 206: 61–77.

Green, S.R., B.E. Clothier, H.W. Caspari and Neal S.M., 2002, Rootzone processes, tree-water-use and the equitable allocation of irrigation water to olives. In: Environmental Mechanics: Water, Mass and Energy Transfer in the Biosphere, Geophysical Monograph 129, American Geophysical Union pp. 337-345.

Harper, J.L., Jones, M. and Sackville Hamilton, N.R., 1991, The evolution of roots and the problems of analysing their behaviour. In: Atkinson, D. (Ed.), *Plant Root Growth: An Ecological Perspective*. Spec. Pub. 10. British Ecological Society, Blackwell Scientific Publications: 3–25.

Hutson, J.L. and Wagenet, R.J., 1995, A multi-region model describing water flow and solute transport in heterogeneous soils. *Soil Science Society of America Journal* 59: 743–751.

Jaynes, D.B., Logsdon, S.D. and Horton R., 1995, Field method for measuring mobile/immobile water content and solute transfer rate coefficient. *Soil Science Society of America Journal* 59: 352–356.

Jolly, V. 1999. Cited in 'Leaps of Faith' Survey—Innovation in Industry, *The Economist*, 350 (8107), 20 February 1999: 12–16.

Jury, W.A., 1982, Simulation of solute transport with a transfer function model. *Water Resources Research* 18: 363–368.

Kachanoski, R.G., Pringle, E. and Ward, A., 1992, Field measurement of solute travel times using time domain reflectometry. *Soil Science Society of America Journal* 56:47–52.

Klute, A., 1952, Some theoretical aspects of the flow of water in unsaturated materials. *Soil Science Society of America Journal* 16: 144–148.

Kutilek, M. and Nielsen, D.R., 1994, *Soil Hydrology*, Catena Verlag, Cremlingen, 370 pp.

Lichtenberg, E. and Zimmerman, R., 1999, Farmers' willingness to pay for groundwater protection. *Water Resources Research* 35: 833–841.

Martin, T.L., 1973, *Malice in Blunderland*. McGraw-Hill, New York.

Messing, I. and Jarvis, N.J., 1993, Temporal variation in the hydraulic conductivity of a tilled clay soil as measured by tension infiltrometers. *Journal of Soil Science* 44: 11–24.

Ministry for the Environment, 1997, *The State of New Zealand's Environment*. GP Publications, Wellington, New Zealand.

Nadler, A., Dasberg, S. and Lapid, I., 1991, Time domain reflectrometry measurements of water content and electrical conductivity of layered soil columns. *Soil Science Society of America Journal* 55: 938–943.

Naidu, R., Kookana, R.S., Oliver, D.P., Rogers, S. and McLaughlin, M.J., 1996, *Contaminants and the Soil Environment in the Asia-Pacific Region*. Kluwer Academic Publishers, 717 pp.

Nielsen, D.R. and Biggar, J.W., 1961, Miscible displacement in soils. I. Experimental information. *Soil Science Society of America Proceedings* 25: 1–5.

Nyer, E.K., 1998, Hydrogeologists should manage enhanced biological in-situ remediations. In: Nyer, E.K. *Groundwater and Soil Remediation: Practical Methods and Strategies*. Ann Arbor Press, Chelsea, Michigan: 135-142.

Nyer, E.K. and Duffin, M.E., 1998, The state-of-the-art of bioremediation. In: Nyer, E.K. *Groundwater and Soil Remediation: Practical Methods and Strategies*. Ann Arbor Press, Chelsea, Michigan: 155-170.

Nyer, E.K. and Gatliff, E.G., 1998, Phytoremediation In: Nyer, E.K. *Groundwater and Soil Remediation: Practical Methods and Strategies*. Ann Arbor Press, Chelsea, Michigan: 189–197.

Oreskes, N., Shrader-Frechette K. and Belitz, K., 1994, Verification, validation and confirmation of numerical models in the earth sciences. *Science* 263: 641–646.

Passioura, J.B., 1996, Simulation models: Science, snake oil, education or engineering. *Agronomy Journal* 88: 690–694.

Perroux, K.M. and White, I., 1988, Designs for disc permeameters. *Soil Science Society of America Journal* 52: 1205–1215.

Philip, J.R., 1957a, The theory of infiltration: 4. Sorptivity and algebraic infiltration equations. *Soil Science* 84: 257–267.

Philip, J.R., 1957b, The physical principles of soil water movement during the irrigation cycle. Proc 3rd Annual Conference of the International Commission on Irrigation and Drainage, San Francisco 8: 125–154.

Philip, J.R., 1985, Reply to Comments on "Steady infiltration from spherical cavities". *Soil Science Society of America Journal* 49: 788–789.

Philip, J.R., 1987, The infiltration joining problem, *Water Resources Research* 23: 2239–2245.

Philip, J.R., 1995, Desperately seeking Darcy in Dijon. *Soil Science Society of America Journal* 59: 319–324.

Potter, D. and Walter, M., 1998, Ravenous fungus eats toxic waste. *New Zealand Science Monthly*, 4 March.

Reynolds, W.D. and Elrick, D.E., 1985, In situ measurement of field saturated hydraulic conductivity, sorptivity and the α-parameter using the Guelph permeameter. *Soil Science* 140: 292–301.

Richards, L.A., 1931, Capillary conduction of liquids through porous mediums, *Physics* 1: 318–333.

Robinson, B.H., Brooks, R.R., Gregg, P.E.H. and Kirkman, J.H., 1999, The nickel phytoextraction potential of some ultramafic soils determined by sequential extraction. *Geoderma* 87: 293–304.

Talsma, T., 1969, In situ measurement of sorptivity. *Australian Journal of Soil Research* 7: 269–276.

Thayalakumaran, T., 2002, EDTA-enhanced transport of copper from contaminated soil and its implications. Ph.D. thesis, Massey University, New Zealand, 173 pp.

The Economist, 1999, Innovation in Industry, 350(8107):58-59.

Topp, G.C., Davis, J.L. and Annan, A.P., 1980, Electromagnetic determination of soil water content: Measurements in coaxial transmission lines. *Water Resources Research*. 16: 574–582.

US Environmental Protection Agency, 1990, National Pesticide Survey, Phase I Report, Report PB91-126765, US Department of Commerce, National Technical Information Service, Springfield, Virginia.

Vanclooster, M., Boesten, J.J.T.I.,Trevisan, M., Brown, C.D., Capri E., Eklo, O.M., Gottesbüren, B., Gouy, V. and van der Linden, A.M.A., 2000, A European test of pesticide leaching models: Methodology and major recommendations. *Agricultural Water Management*. 44: 1–19.

Van Genuchten, M. Th. and Wierenga, P.J., 1976, Mass transfer studies in sorbing porous media. I Analytical solutions. *Soil Science Society of America Journal* 40: 473–480.

Van Genuchten, M. Th., 1995, New issues and challenges. Proceedings of the 15th World Congress of Soil Science, Acapulco, Mexico, July 10–16, 1995, ISSS. I: 5–27.

Vogeler, I., Duwig, C., Clothier, B.E. and Green, S.R., 2000, A simple approach to determine reactive solute transport using Time Domain Reflectometry. *Soil Science Society of America Journal* 64: 12–18.

Vogeler, I., S.R. Green, D.R. Scotter and Clothier B.E., 2001a, Measuring and modelling the transport and root uptake of chemicals in the unsaturated zone. *Plant and Soil* 231: 161–174.

Vogeler, I., S.R. Green, A. Nadler and C. Duwig, 2001b, Measuring transient solute transport through the vadose zone using time domain reflectometry. *Australian Journal of Soil Research* 39: 1359–1369.

Wagenet, R.J. and Hutson, J.L., 1986, Predicting the fate of nonvolatile pesticides in the unsaturated zone. *Journal of Environmental Quality* 15: 315–322.

Walter, M., Christeller, J., Boul, L., Chong, R. and Slade, A., 1998, Bioremediation using white-rot fungi: Update and future prospects in New Zealand. Proceedings of the New Zealand Land Treatment Collective 17: 90–98.

Watanabe, M.E., 1997, Phytoremediation on the brink of commercialisation. *Environmental Science and Technology* 31(4): 182–186.

White, I. and Sully, M.J., 1987, Macroscopic and microscopic capillary length and time scales from fieid infiltration. *Water Resources Research* 23: 1514–1522.

Wooding, R.A., 1968, Steady infiltration from a shallow circular pond. *Water Resources Research* 4: 1259–1273.

2

Issues in Waste Disposal and Management

R.G. McLaren, K.C. Cameron and *H.J. Di*

INTRODUCTION

The realization that both controlled and uncontrolled disposal of wastes can result in the contamination of soils, surface waters and groundwaters has led to the development of a wide range of diverse regulatory measures and guidelines aimed at reducing or eliminating waste-related pollution. This practice applies not only to urban industries, but also to agriculture, horticulture and forestry and their associated populations. Even so, waste disposal and its associated environmental contamination is an ongoing issue, with the list of potential pollutants in wastes increasing as new chemicals and processes are developed. In addition, many sites contaminated in the past by waste disposal operations, or by a particular agricultural or industrial activity carried out at the site, continue to pose a serious threat to the environment.

SOURCES OF WASTE

Many different types of waste are produced by the modern society as byproducts of domestic, agricultural and industrial life, each of which has its own set of potential contamination problems. However, irrespective of the source of the pollution, the main issues discussed in this chapter relate to the potential contamination of soil, water or the atmosphere and the subsequent effects on environmental or human health. This chapter provides an overview of the predominant types and characteristics of wastes generated in Australia and New Zealand, highlighting the problems with current waste disposal practices, including landfilling, incineration, discharge to water, and land treatment.

For convenience, waste sources can be categorized into:

- industrial;

Centre for Soil and Environmental Quality, Lincoln University, Canterbury, New Zealand

- domestic and urban; and
- agricultural.

However, in practice, it is recognized that there may well be overlap between these categories.

Industrial Wastes

The definition of industrial waste varies among countries, but it generally includes wastes generated in any process of industry, manufacturing, trade or business, including mining. In the late 1980s, New Zealand generated about 300,000 tonnes of industrial waste annually, equivalent to 15 tonnes per US$1 m GDP (OECD, 1991). Australia generates about 20 million tonnes annually, equivalent to 146 tonnes per US$1 m GDP, which is the OECD average.

The composition of industrial wastes varies, depending on the industrial structure of a country or region. It consists of general rubbish, packaging, food wastes, acids, alkalis, oils, solvents, resins, paints, mine spoils and sludges. A proportion of the industrial waste is classified as hazardous waste because it contains materials that are actually or potentially hazardous to humans and other living organisms. New Zealand produced about 86,000 tonnes of hazardous waste annually in the late 1980s (Statistics NZ, 1993).

The exact streams of industrial wastes in many countries, including Australia and New Zealand, are not well documented. The major industries in New Zealand and Australia (Statistics NZ, 1993; Australian Bureau of Statistics, 1992) are :

- meat and dairy processing;
- food, beverages and tobacco;
- textiles, clothing and footwear;
- wood processing;
- paper and paper products;
- printing and publishing;
- chemicals and petroleum products;
- non-metallic minerals;
- basic metal industries;
- fabricated metal products; and
- machinery and equipment.

Possible contaminants contained in various industrial waste sources were summarized by Barzi et al. (1996).

In addition to solid wastes, many industries also produce large volumes of wastewater that require disposal. Although most factory wastewaters are treated before disposal, many nutrients, metals and organic chemicals remain in significant concentrations in the treated sludges and effluents.

For example, wastewaters from dairy, tannery and pulp and paper factories contain high concentrations of sodium ions. Tannery wastewaters also contain undesirable constituents such as chromium, aluminium, polyphenolics and aldehydes (Carnus and Mason, 1994). Wastes from pulp and paper mills contain metals and a range of toxic organic compounds.

Combustion of coal to generate electricity produces large amounts of residues which require disposal, the main material being fly ash. Currently, Australia produces approximately 7.7 million tonnes of fly ash annually (Beretka and Nelson, 1994). Fly ashes tend to be dominated by particles in the silt to fine sand range. They contain many different trace elements, some of which (such as boron) may be potentially toxic to plants (McLaren and Smith, 1996). Many fly ashes are also characterized by high alkalinity (Aitken et al., 1984) and high concentrations of soluble salts (Carlson and Adriano, 1993).

Mining wastes constitute another category of industrial waste. Mining is not a major industrial sector in New Zealand, but it is in Australia where diamonds, gold, iron ore, lead, manganese ore, nickel, titanium, tungsten, uranium, zinc and zircon are mined (Australian Bureau of Statistics, 1992). Mine tailings are waste materials generated by the grinding and processing of ores (Hossner and Hons, 1992). The physical properties and chemical composition of particular mine tailings depend on the minerals mined and the techniques used. Many tailings, however, have either extremely low or extremely high pH, and suffer from salinity or sodicity problems. Many tailings also contain toxic metals and other compounds which may contaminate the wider environment (Dean and Fogel, 1982; Ward, 1987). However, significant progress has been made in recent years to reclaim mine tailings and wastes and to minimize their impact on the environment (Ritcey, 1989; Taylor, 1996).

Domestic and Urban Wastes

The major urban waste materials that require disposal are undoubtedly sewage and solid waste in the form of domestic and industrial garbage. Sewage sludge and sewage effluent are produced in the treatment of domestic and industrial wastewater and sewage. The aim of sewage treatment is to remove solids, pathogens and other contaminants from wastewater streams and so to produce relatively 'clean' water that can be recycled. The actual processes by which this is achieved vary greatly between treatment works, but usually involve a combination of physical separation and aerobic or anaerobic biological treatment.

Sewage sludge is the solid material left behind when the treated water (sewage effluent) leaves the works. It contains a complex mixture of organic and inorganic compounds of biological and mineral origin along with significant concentrations of nutrients, particularly N and P, and is recognized as having potential as a fertilizer material and soil conditioner. Concentrations of nutrients vary greatly between sludges. Sommers (1977) reported ranges for some American sludges of <0.1 to 17.6% for total N, <0.1 to 14.3% for total P, and 0.02 to 2.6% for total K. Data for Australian and New Zealand sludges also tend to fall within these ranges. For example, Ross et al. (1991) quote ranges of 2.1 to 8.2% for total N, 1.1 to 8.9% for total P and 0.09 to 0.34% for total K in sludges from the Sydney region. Similarly, mean values of 1.46 and 3.98% N, 0.87 and 1.52% P and 1.2 and 0.24% K have been quoted for sludges from Auckland and Christchurch, respectively.

Sewage sludge can also contain a range of potentially toxic contaminants, including heavy metals (Table 2.1), pesticide residues, other organic compounds such as polychlorinated biphenyls (PCBs), and a range of pathogenic organisms. Actual sludge composition varies with time, and between treatment plants, depending on the types of sewage and wastewater received and the nature of the treatment process. Many urban sewage treatment works receive a mixture of domestic and industrial sewage as well as urban runoff from roads and other sealed surfaces. This makes prediction of sewage sludge composition extremely difficult. Composition of sludge from rural areas is likely to be less variable and less contaminated with sundry contaminants rather than urban sludges.

In New Zealand, approximately 52,000 tonnes of dry sludge solids are produced every year, with some 70% of this generated at three of the main population centres of Auckland, Wellington and Christchurch (NZ Department of Health, 1992). In Australia, the total annual production is 250,000 tons dry weight (Agriculture and Resource Management Council of Australia and New Zealand, 1995).

Table 2.1 Mean metal concentrations (mg/kg) in some sewage sludges from Australia and New Zealand (adapted from McLaren and Smith, 1996)

Metal	*Location*		
	Adelaide[(a)]	*Sydney*[(b)]	*Christchurch*[(c)]
Arsenic	4.0	5.5	11.9
Cadmium	3.0	3.9	6.2
Chromium	925	81	1618
Copper	888	1427	572
Lead	190	140	443
Mercury	1.7	5.1	5.2
Nickel	61	25	117
Zinc	834	839	1648

Sources: [(a)]Personal comm.; [(b)]Ross et al. 1991; [(c)]NZ Department of Health, 1992

The quality of treated urban sewage effluent depends on the nature of the sewage and wastewater streams supplied to the treatment works and the type of treatment carried out. Treated sewage effluent differs from normal 'clean' water in many ways (McLaren and Smith, 1996), including higher:

- BOD and COD;
- inorganic salt concentrations;
- N and P concentrations;
- metal concentrations; and
- levels of pathogenic microorganisms.

Agricultural Wastes

The agricultural sector constitutes a major part of the Australian and the New Zealand economies. In 1991, farming in New Zealand occupied about 17.5 million hectares, about 64% of the total land area (Statistics NZ, 1993). In Australia, about 466 m ha., or about two-thirds of the land surface, is used for farming activities (Australian Bureau of Statistics, 1992). The biggest proportion of the farming land is used for animal grazing: about 8 m cattle and 55 m sheep in New Zealand in 1991; and 23 m cattle and 170 m sheep in Australia in 1990. Large numbers of other livestock, for example pigs and poultry, are also raised in both countries. Large quantities of waste are generated from these agricultural production and processing industries; for example, statistics (which are generally regarded as incomplete) put the waste arising from the NZ agricultural sector at 530 000 tons in 1991 (Centre for Advanced Engineering, 1992). It is estimated that the New Zealand meat processing industry contributes a pollution load that is equivalent to that produced by the total human population of the country (3.5 m) (Cooper and Russell, 1982).

Most agricultural wastes contain valuable nutrients that could be recycled in order to improve soil fertility and increase the sustainability of farming systems. For example, in South Australia, there are some 400,000 pigs that produce about 2400 million litres of waste annually (Brechin and McDonald, 1994) and this waste contains enough nitrogen to fertilize 200,000 ha. of wheat or barley. The fertilizer value of dairy shed effluent, pig slurry and poultry manure in New Zealand is estimated to be NZ$36 million per year (Roberts et al., 1992).

WASTE DISPOSAL PRACTICES

It is clear from the above discussion that wastes can contain a wide range of potential contaminants, including:

- heavy metals, such as As, Cd, Cu, Cr, Hg, Ni, Pb and Zn;
- nutrients such as N and P;

- soluble salts;
- pesticides (herbicides, insecticides, fungicides);
- pathogens;
- fuel hydrocarbons;
- other organic chemicals, such as PAHs and PCBs;
- organic matter (BOD); and
- various gases (for example, methane).

The exact nature and concentrations of contaminants will vary greatly between wastes from different sources. Wastes from a single source are also likely to vary with time.

Currently, there are four major waste disposal practices commonly used worldwide: landfilling, incineration, discharge to water and land treatment. All four practices have potential problems associated with them and these are discussed in detail below.

Landfilling

Disposal of solid wastes in landfill sites is a common disposal option worldwide. For example, Australia disposed of about 12.8 m tonnes of solid wastes in 1989, or 776 kg per person per year (van den Broek, 1994). By comparison, 2.3 m tonnes are deposited each year in New Zealand, or 733 kg per person per year. Domestic garbage, mainly paper, cardboard and vegetable matter, was the main component of the waste (49–55%) deposited in landfills (van den Broek, 1994; Willmot, 1994a). The other major contributor was industrial waste. Shortage of landfill sites and community opposition are forcing the authorities to seek integrated waste management options and alternative disposal methods. Such approaches include waste minimization, waste avoidance, recycling and resource recovery before resorting to disposal. Within the landfill, uncontrolled fermentation reaction breaks down the solid wastes. The main environmental problems associated with landfills are the impact of leachate on surface and groundwaters and the production of landfill gas (Willmot, 1994b).

The chemical composition of the leachate is highly variable (Table 2.2) and site-specific due to the diverse chemical composition of the material deposited in the landfill. However, the major determinant of the leachate quality is the stage of fermentation of the material. In general, the leachates have a high organic strength (high biological and chemical oxygen demand) and, in many ways, resemble sewage. Contamination of groundwater by landfill leachate is an extensive and increasing problem. For example, data from Finland showed detrimental changes in the groundwater quality downgradient of the landfills (Assmuth and Strandberg, 1993). Common contaminants in the leachate were heavy

Table 2.2 Typical compositions of leachates (g/m^3) from landfill sites (Smith, 1995)

Species	*UK*	*NZ*
pH	8–8.5	6–8
COD	850–1350	500–5000
BOD	80–250	100–4000
Volatile acids	20	20–200
Ammonium-N	200–600	100–1000
Chloride	3400	200–2000
Sulphate	340	200
Sodium	2185	200–2000
Potassium	888	100–1000
Magnesium	214	50–400
Calcium	88	100–1000
Chromium	0.05	0.05–0.5
Manganese	0.5	5–50
Iron	10	50–400
Nickel	0.04	0.1–1
Copper	0.09	0.05–0.5
Zinc	0.16	1–100
Cadmium	0.02	0.01–6
Lead	0.1	0.05–0.5
Monocyclic phenols	0.01	n.d.
Total cyanide	0.01	n.d.
Organochlorine pesticides	0.01	n.d.
Organophosphorus pesticides	0.05	n.d.
PCBs	0.05	n.d.

Note: n.d. = not determined

metals and chlorinated hydrocarbons such as dichloromethane and 1,2-dichloroethane. In addition, the concentration of Fe, Mn and Cl in the leachate exceeded drinking water standards. Not all of the organic pollutants deposited in the solid waste will be leached. Data from landfills in Florida would suggest that some pollutants are metabolized by microorganisms or otherwise degraded, sorbed onto particulate matter, or diluted to such an extent as to be non-detectable in the leachate (Hallbourg et al., 1992).

One method of leachate control is to use a low permeability topcover so as to minimize water entry into the landfill (Willmot, 1994b; Weeks et al., 1992). Possible choices include clay or manufactured liners. The containment of toxic material, salts and other pollutants is achieved through the use of an underlying liner, which collects and directs the leachate generated as a result of fermentation to drainage points. Clays are frequently chosen because of their low permeability. However, high concentrations of some soluble organic material can alter the permeability of water through the clay, therefore, limiting their usefulness as a liner material.

Furthermore, the liners may fail as a result of waste subsidence. In that case, not only does the topcover not work, but it may make the situation worse by channelling water into the landfill (Schulz et al., 1992). The permeability of liners, possible effect of chemicals on their permeability, and the ability to predict the changes over a 100-year period require further investigation.

Problems associated with landfill gas are its accumulation to explosive quantities, smell, groundwater acidification, and reduced plant growth following final restoration of the site (Willmot, 1994b). The net emission of methane (CH_4) is a function of the biodegradable organic C content, the proportion of the organic C that degrades, and the CH_4 component of the landfill gas. The average CH_4 emission rates from municipal solid wastes in Canberra was 49 g m^{-2} d^{-1}, corresponding to an annual emission of 10 kg CH_4 per ton of municipal solid waste (Denmead, 1995, unpublished data). Methane emissions have been found to be seasonal, and dependent on water content (Jones and Nedwell, 1993).

Incineration

Incineration is a controlled-flame combustion process for the decomposition of wastes. It can effectively reduce the volume of waste by upto 90% and weight by upto 75% (Petts, 1994). The process is particularly suitable for organic and other combustible wastes with a high energy content. Energy generated during incineration can be harvested for generating electricity. The process is obviously not suitable for non-combustible wastes, including metals and highly explosive materials.

A survey in New Zealand showed that 109 incinerators were in use (Abbott, 1991). Most of the facilities were suitable for incineration of hazardous wastes, but not for stable compounds such as PCBs and dioxins. In Australia, about 143,000 tonnes of wastes were incinerated in 1989, accounting for about one per cent of the total waste disposal. In Japan and Switzerland, more than 75% of municipal waste is incinerated (United Nations Environment Programme, 1993) and in the United States, the figure is about 16%.

Although incineration is a versatile method for the destruction of many types of wastes, the process itself also creates environmental problems. The issues of concern relate mainly to the emissions of pollutants to the atmosphere and the disposal of contaminated residue ash. Pollutants emitted to the atmosphere include particulates, heavy metals (for example, Hg, Cd and Pb), acidic gases (for example, hydrogen chloride, sulphur dioxide), and trace organics such as polychlorinated dibenzo-p-dioxins (PCDDs) and polychlorinated dibenzofurans (PCDFs). Compounds that are not combusted will accumulate in the incinerator and are collected as bottom ash. The main components of bottom ash are metals or metal oxides, and the ash is classified as hazardous waste in some countries (for example, Germany and the Netherlands). In Europe, the ash is disposed of in landfills that receive only one type of waste, without biodegradable material that may mobilize the metals.

Discharge to Water

Discharge of sewage waters, sludges and other wastes such as dredged spoils and hazardous wastes into the marine environment (rivers, lakes and sea) is practised in many countries (United Nations Environment Programme, 1993). In New Zealand, about 60% of sewage is discharged to coastal waters after secondary treatment. The treated effluents still retain high concentrations of organic matter, suspended solids, nutrients (particularly N and P) and other contaminants (Hauber, 1995). It is estimated that Australia produces about 100,000 tons of N and 10,000 tons of P in sewage effluent annually, much of which is discharged into coastal waters (Brodie, 1995). In low river-flow conditions, sewage effluent may be the major source of nutrients for many rivers.

Discharging sewage and other nutrient-rich wastes to waterways can result in the depletion of dissolved oxygen, eutrophication, chemical toxicity and salinity. Eutrophication and salinity are considered to be the two major water quality problems in Australia (Sumner and McLaughlan, 1996). Eutrophication is produced by an excess concentration of nutrients in the water, leading to accelerated plant growth and changes in plant species composition. The critical nutrients responsible for eutrophication are N and P, although other nutrients, for example Fe, Mo, Mn and Si, may also contribute to the process (Australian Environmental Council, 1987). Eutrophication can have a dramatic impact on the coastal or inland aquatic ecosystems, including blooms of phytoplankton (algal blooms) and loss of seagrass. Some of these algae are toxic and sometimes cause death of fish and livestock.

In the State of the Marine Environment Report for Australia, coastal eutrophication has been identified as one of the most serious and large-scale threats to Australia's nearshore marine environment (Zann, 1995). Large areas of seagrass in southern Australia have suffered dieback from eutrophication, and nuisance and toxic blooms of phytoplankton have affected many of Australia's bays and estuaries (Zann, 1995). Although the dynamic coastal waters of New Zealand generally disperse the discharges effectively, serious eutrophication problems are becoming more common. For example, in January and February 1993, there were widespread incidences of algal blooms and shellfish poisoning around New Zealand's coastlines, which resulted in the temporary shutdown of the entire coastline from shellfishing, and the temporary cessation of shellfish exports. The direct economic consequences and remediation costs of eutrophication are very high; it is estimated that eutrophication may cost Australia $A 10–50 m per year (Cullen, 1996).

Sewage effluent contains various pathogens and even secondary treated sewage could still pose a significant microbiological health risk (Alexander et al., 1992; Ashbolt, 1995). Pathogen concentrations can be particularly high near major sewage outfalls. For example, near the sewage outfall in the harbour of Wellington, the annual median faecal coliform count in 1992 was 55,000 per 100 mL (Statistics NZ, 1993), although the count decreased rapidly away from the major outfall. Acceptable counts of faecal coliforms are set at 200 for bathing water and 14 for shellfishing gathering. If the concentration exceeds these limits, swimming or shellfishing may not be safe.

Heavy metals and organics present in sewage sludges and effluents also pose threats to the marine environment. About 16,000 tons of oil enter the marine environment annually from sewage systems or drains in Australia. Organic compounds such as PCBs, DDT, polycyclic aromatic hydrocarbons (PAHs) and organochlorine pesticides, and heavy metals such as mercury, have all been found at elevated concentrations in marine organisms, particularly near effluent discharge sites (ANZECC, 1991; Richardson, 1995; Batley, 1995).

World, regional and national organizations have, or are imposing, increasingly strict regulations on the discharge of wastes to the sea. The 1972 London Dumping Convention specifies the ban of sea dumping of certain hazardous wastes unless it is proven that the hazardous substance is in trace amounts and would be made harmless in the sea (United Nations Environment Programme, 1993). In the EU, discharge of untreated sewage to the sea will be phased out in the next few years; except in special circumstances, the sewage is required to be secondary treated, and in areas sensitive to eutrophication, the sewage is required to be tertiary treated, and wherever possible, the sewage should be reused.

Land Treatment

The continuing problems associated with the disposal of wastes to landfills and the removal of the aquatic disposal option have, together with other issues, resulted in an increased interest in the disposal of wastes to the land. Ideally, the objective of land treatment of wastes is to utilize the chemical, physical and biological properties of the soil/plant system to assimilate the waste components without adversely affecting soil quality or causing contaminants to be released into water or the atmosphere (Loehr, 1984), or accumulate in produce for human or animal consumption. There are indeed many wastes, particularly from agricultural sources, that have considerable benefits for soil fertility and crop growth (Cameron et al., 1997). However, others such as sewage sludge, sewage and industrial effluents, municipal composts and various industrial wastes, although containing some beneficial nutrients and organic matter, also contain a range of potential contaminants. These can include heavy metals, high salt concentrations, pathogens, and numerous organic chemicals. Clearly, the management of land treatment of wastes requires a good understanding of the physico-chemical and biological processes that determine the fate of waste contaminants in the soil.

Overall, there is a need to consider land application of wastes as an opportunity to recycle and reuse waste constituents such as nutrients rather than simply as a means of waste disposal.

This clearly means matching application rates to crop requirements. However, there are some major issues to be considered including:

- potential loss of soil fertility (toxicity to plants and soil organisms) as a result of the accumulation of certain contaminants;
- introduction of contaminants into the food chain;
- contamination of surface or groundwater; and
- adverse effects on animal and human health.

It is vital to ensure that sites used for the land treatment of wastes do not in time themselves become contaminated sites, requiring remediation.

CONCLUSIONS

It is clear that the management and disposal of wastes is an extremely complex issue, and that there is unlikely to be a single answer to the problems faced by countries worldwide. Of the current disposal options, discharge to water is clearly no longer an acceptable option and landfilling is under increasing pressure because of lack of suitable sites for new landfills. The problems experienced with some existing landfill sites do not help the public perception of this form of disposal. However, even with increasing attention being paid to waste reduction programmes generally, for the foreseeable future, landfilling will almost certainly remain an important component of overall waste management. Improvements in landfill design such as the construction of liners and caps will require a substantial research effort.

With the exception of Japan, incineration as a waste disposal option is probably underutilized in the Australasia-Pacific Region. However, do we really have a good understanding of the emission of pollutants to the atmosphere, or the problems associated with the disposal of incinerator ashes? Perhaps these are other areas where research efforts should be directed.

The land treatment option clearly has considerable potential for some types of waste, but much research is still required to establish the limits of this method of waste management. For example, we need a better understanding of waste characteristics and factors affecting nutrient and chemical release rates. Waste characterization is essential in order to determine the most suitable, environmentally-safe and agronomically-appropriate land application practices. We also need a better understanding of the physico-chemical and biological processes that determine the fate of wastes in the soil, for example:

- leaching of nutrients in macropores;
- leaching of suspended solids;
- leaching of microorganisms;
- degradation of sludge-borne organics; and
- the long-term bioavailability of metals and trace organics fixed by soil organic matter.

Many of these processes are greatly affected by a wide range of soil and environmental factors. A quantitative perspective of the processes is particularly important in optimizing waste application rate, frequency, time and application location. Improved models of the fate of land-applied wastes need to be developed and tested in order to better describe the processes involved and to assist with the prediction of environmental impacts. This will require an integrated knowledge of the processes operating in the soil-plant-water-atmosphere (and occasionally animal) system. This can only be achieved by interdisciplinary teams, including pedologists, soil physicists, soil chemists, soil biologists, plant scientists, hydrologists, engineers and farm managers.

Indeed, the interdisciplinary nature of waste management overall should be stressed. Only by the joint efforts of teams from many different disciplines will we learn how to deal with this important issue.

REFERENCES

Abbott, N.H.C., 1991, New Zealand Incineration Facilities. Final report, NECAL Report S91/968C, NECL Laboratory, DSIR, Auckland.

Aitken, R.L., Campbell, D.J., and Bell, L.C., 1984, Properties of Australian fly ashes relevant to their agronomic utilisation. *Australian Journal of Soil Research*, 22: 443–453.

Agriculture and Resource Management Council of Australia and New Zealand, 1995, Guidelines for sewerage systems—Biosolids management. ARMCANZ Water Technology Committee, Occasional Paper WTC No. 1/95.

Alexander, L.M., Heaven A., Tennant A., and Benton W.H., 1992, Symptomatology of children in contact with sea water contaminated with sewage. *Journal of Epidemiology and Community Health*, 46: 340–344.

ANZECC, 1991, Persistent chlorinated organic compounds in the marine environment. Public information paper, Australian and New Zealand Environment and Conservation Council, Canberra.

Ashbolt, N., 1995, Human health risk from microorganisms in the Australian marine environment. In: Zann L.P. and Sutton D.C. (Eds), *Pollution, State of the Marine Environment Report for Australia, Technical Annex 2*, Department of the Environment Sport and Territories, Canberra, pp. 31–40.

Assmuth, T.W. and Strandberg T., 1993, Groundwater contamination in Finnish landfills. *Water, Air and Soil Pollution*, 69: 179–199.

Australian Bureau of Statistics, 1992, Australia's Environment. Issues and Facts. Australian Bureau of Statistics: Canberra.

Australian Environmental Council, 1987, Nutrients in Australian waters. Australian Environmental Council, Report No. 19, Department of Arts, Heritage and Environment, AGPS, Canberra.

Barzi, F., Naidu R., and McLaughlin M.J., 1996, Contaminants and the Australian soil environment. In: Naidu R., Kookana R.S., Oliver D.P., Rogers, S., and McLaughlin M.J. (Eds), *Contaminants and the Soil Environment in the Australasia-Pacific Region*, Kluwer Academic Publishers, Dordrecht, pp 451–484.

Batley, G.E., 1995, Heavy metals and tributylin in Australian coastal and estuarine waters. In: Zann L.P., and Sutton, D.C., (Eds), *Pollution, State of the Marine Environment Report for Australia, Technical Annex* 2, Department of the Environment Sport and Territories, Canberra, pp. 63–72.

Beretka, J. and Nelson P., 1994, The current state of utilisation of fly ash in Australia. Proc. 2nd Int. Symp. Ash—A Valuable Resource, Vol. 1. South African Coal Ash Association, Johannesburg, pp. 51–63.

Brechin, J. and McDonald G.K., 1994, Effect of form and rate of pig manure on the growth, nutrient uptake, and yield of barley (cv. Galleon). *Australian Journal of Experimental Agriculture*, 34: 505–510.

Brodie, J., 1995, The problems of nutrients and eutrophication in the Australian marine environment. In: Zann L.P., and Sutton D.C. (Eds), *Pollution, State of the Marine Environment Report for Australia. Technical Annex* 2. Department of the Environment, Sport and Territories, Canberra, pp. 1–29.

Cameron, K.C., Di. H.J., and McLaren. R.G., 1997, Is soil an appropriate dumping ground for our wastes? *Australian Journal of Soil Research*, 35: 995–1035.

Carlson, C.L. and Adriano D.C., 1993, Environmental aspects of coal combustion residues. *Journal of Environmental Quality*, 22: 227–247

Carnus, J.M. and Mason I., 1994, Land treatment of tannery wastes. In: *New Zealand Land Treatment Collective Technical Review* No. 10. pp. 31–39.

Centre for Advanced Engineering, 1992, Our Waste—Our Responsibility. *Towards Sustainable Waste Management in New Zealand*. Centre for Advanced Engineering, University of Canterbury, New Zealand.

Cooper, R.N. and Russell J.M., 1982, The current pollution control status of the New Zealand meat industry. Proceedings of the 4th National Water Conference, Auckland. pp. 339–345.

Cullen, P., 1996, Managing nutrients in aquatic systems: The eutrophication problem. In: P. De Deckker and W.D. Williams (Eds), *Limnology in Australia*, CSIRO, Melbourne, Vic., pp. 539–554.

Dean, S.A. and Fogel M.M., 1982, Acid drainage from abandoned metal mines in the mountains of southern Arizona. Symposium on surface mining hydrology, sedimentology and reclamation, University of Kentucky, Lexington, Kentucky, pp. 269–276.

Hallbourg, R.R., Delfino, J.J. and Miller, W.L., 1992, Organic priority pollutants in groundwater and surface water at three landfills in north central Florida. *Water, Air and Soil Pollution*, 65: 307–322.

Hauber, G., 1995, Wastewater treatment in NZ: Evaluation of 1992/93 performance data–ORGD. *Water and Wastes in New Zealand*, 85: 28–34.

Hossner, L.R. and Hons. F. M., 1992, Reclamation of mine tailings. *Advances in Soil Science*, 17: 311–350.

Jones, H.A. and Nedwell, D.B., 1993, Methane emission and methane oxidation in land-fill cover soil. *Microbial Ecology*, 102: 185 –195.

Loehr, R.C., 1984, *Pollution Control for Agriculture.* (2nd edition.) Academic Press: Orlando, Florida.

McLaren, R.G. and Smith, C.J., 1996, Issues in the disposal of industrial and urban wastes. In: R. Naidu, R.S. Kookana, D.P. Oliver, S. Rogers and M.J. McLaughlin (Eds), *Contaminants and the Soil Environment in the Australasia-Pacific Region*, Kluwer Academic Publishers, Dordrecht., pp. 183–212.

NZ Department of Health, 1992, Public health guidelines for the safe use of sewage effluent and sewage sludge on land. Department of Health, Wellington, New Zealand.

OECD, 1991, Environmental indicators, a preliminary set. Organisation for Economic Co-operation and Development, Paris.

Petts, J., 1994, Incineration as a waste management option. In: Hester R.E. and Harrison R.M. (Eds), *Waste incineration and the Environment*, Royal Society of Chemistry, Cambridge, UK, pp. 1–25.

Richardson, B.J., 1995, The problem of chlorinated compounds in Australia's marine environment. In: L.P. Zann and D.C. Sutton. (Eds), *Pollution, State of the Marine Environment Report for Australia, Technical Annex 2*, Department of the Environment Sport and Territories, Canberra, pp. 53–61.

Ritcey, G.M., 1989, *Tailings Management, Problems and Solutions in the Mining Industry.* Elsevier, Amsterdam.

Roberts, A.H.C., O'Connor M.B. and Longhurst R.D., 1992., Wastes as plant nutrient sources: Issues and options. In: Gregg, P.E.H. and Currie, L.D. (Eds), *The Use of Wastes and Byproducts as Fertilizers and Soil Amendments for Pastures and Crops*, Occasional Report No. 6, Fertilizer and Lime Research Centre, Massey University, Palmerston North, pp. 44–55.

Ross, A.D., Lawrie, R.A., Whatmuff, M.S., Keneally, J.P. and Awad, A.S., 1991, *Guidelines for the use of Sewage Sludge on Agricultural Land*. NSW Agriculture, Sydney, Australia.

Schulz, R.K., Ridky, R.W. and O'Donnell, E., 1992, Control of Water Infiltration into Near Surface LLW Disposal Units. Progress Report on Field Experiments at a Humid Regional Site, Beltsville, Maryland. NUREG/CR 4918. Vol. 6.

Smith, G., 1995, The effect of garden waste division on landfills—part II. *Water and Wastes in New Zealand*, 86: 50–52.

Statistics NZ, 1993, *Measuring up, New Zealanders and the Environment*. Statistics New Zealand, Wellington.

Sommers, L.E., 1977, Chemical composition of sewage sludges and analysis of their potential use as fertilizers. *Journal of Environmental Quality*, 6: 225–232.

Sumner, M.E., and McLaughlan M.J., 1996, Adverse impacts of agriculture on soil, water and food quality. In: Naidu, R., Kookana, R.S., Oliver, D.P., Rogers, S. and McLaughlin, M.J. (Eds), *Contaminants and the Soil Environment in the Australasia-Pacific Region*. Kluwer Academic Publishers, London, pp. 44–55.

Taylor, G.F., 1996, Exploration, mining and mineral processing. In: Naidu, R., Kookana, R.S., Oliver, D.P., Rogers, S. and McLaughlin M.J., (Eds), *Contaminants and the Soil Environment in the Australasia-Pacific Region*, Kluwer Academic Publishers, Dordrecht, pp. 213–266.

United Nations Environment Programme, 1993, *Organic Contaminants in the Environment, Environmental Pathways and Effects*, Jones K.C. (Ed.), Elsevier Applied Science, London, pp. 275–289.

van den Broek, B., 1994, Solid waste management in Australia. In: Eastwood N.A.M., (Ed). *The Interdata Environmental Resource Management Handbook*. I.D.P. Inter Data Pty. Ltd. pp. 27–33.

Ward, S.C., 1987, Reclaiming bauxite disposal areas in south-west Australia. In: T. Farrel (Ed.), *Mining Rehabilitation*, Australian Mining Industry Council, Canberra, pp. 61–70.

Weeks, O.L., Mansell R.S., and McCallister S.W., 1992, Evaluation of soil top-cover systems to minimize infiltration into a sanitary landfill: A case study. *Environmental Geology and Water Sciences*, 20: 139–151.

Willmot, C.D., 1994a. Refuse data—New Zealand. Solid wastes. In: *The Interdata Environmental Resource Management Handbook*. N.A.M. Eastwood (Ed.). I.D.P. Inter Data Pty. Ltd. pp. 96–101.

Willmot, C.D., 1994b, The environmentally sound management of landfill sites. In: N.A.M. Eastwood (Ed.), *The Interdata Environmental Resource Management Handbook*. I.D.P. Inter Data Pty. Ltd. pp. 160–163.

Zann, L.P., 1995, Our sea, our future, major findings of the state of the marine environment report for Australia. Department of the Environment, Sport and Territories, Canberra.

3

Environmental Auditing of Land Contamination—Experience in South Australia and Northern Territory

Adrian Hall

INTRODUCTION

Environmental auditing of contaminated land was introduced in Victoria by the *Environment Protection Act 1970*. In 1995, the South Australian Environment Protection Authority (EPA) effectively introduced the scheme in South Australia, by advising planning authorities such as councils to use environmental auditors (accredited by EPA Victoria) to review site assessment and remediation work, and to prepare Site Audit Reports. The author was appointed as an auditor in 1997, and has since completed some 50 audits in South Australia, and two in the Northern Territory.

New South Wales introduced a similar audit system in 1998, and the National Environment Protection (Assessment of Site Contamination) Measure (NEPM) of December 1999 provides guidelines on the competencies and acceptance of environmental auditors, thus foreshadowing acceptance of the environmental auditing system at a national level.

The value of the auditing system as an independent certification that a site is fit for its intended use, and does not pose unacceptable human health and/or environmental risks has been amply demonstrated. The environmental audit system is designed to provide:

- assurance to stakeholders, including purchasers, developers, planning authorities and financial institutions; and
- expert opinion on which others can rely in making decisions.

Relevant Provisions of the *Environment Protection Act 1970* (Victoria)

The following is an excerpt from Section 57AA—Certificates of Environmental Audit:

Tonkin Consulting, 5 Cooke Terrace, Wayville SA, Australia, 5034

(2) *Before determining whether or not to issue a certificate of environmental audit, the... auditor must prepare an environmental audit report which must include:*
   (a) *an evaluation of the quality of the relevant segment of the environment;*
   (b) *an assessment of whether any clean up is required to that segment of the environment; and*
   (c) *if any clean up is necessary, any recommendations relating to the carrying out of the clean up.*

If the auditor cannot issue a Certificate of Environmental Audit, he should issue a Statement of Environmental Audit.

The wording of the Pro Forma for the Certificate of Environmental Audit is:

I, —— of ——, a person appointed ... as an environmental auditor ...in making a total assessment of the nature and extent of any harm or detriment caused to, or the risk of any possible harm or detriment which may be caused to, any beneficial use made of the site by any industrial processes or activity, waste or substance (including any chemical substance) ... HEREBY CERTIFY that I am of the opinion that the condition of the site is neither detrimental nor potentially detrimental to any beneficial use of the site ...

The wording of the Pro Forma for the Statement of Environmental Audit is:

I, —— of ——, a person appointed ... as an environmental auditor ... in making a total assessment of the nature and extent of any harm or detriment caused to... any beneficial use made of the site ... HEREBY CERTIFY that I am of the opinion that:

1. *The site is suitable for the following beneficial uses subject to the conditions attached thereto:*
2. *The condition of the site is detrimental or potentially detrimental to any (one or more) beneficial uses of the site. Accordingly, I have not issued a Certificate of Environmental Audit for the site in its current condition, reasons for which are presented in the environmental audit report and are summarised as follows: ...*

Auditing Methodology

Notwithstanding the particular legislative requirements of each jurisdiction, the fundamental principles of environmental auditing give rise to the following methodology.

In order to assess whether the primary consultant's investigation and remediation work is acceptable, the Auditor has to determine whether the:

- site history adequately defines the potential contaminants;
- sample density and testing frequency gives a representative picture of site conditions;
- selection of analytes adequately represents the potential contamination;
- selection of acceptance criteria is appropriate; and
- remediation works have been effective in ensuring that the site is fit for its intended use.

In determining the current environmental condition of the site, the Auditor has to give consideration to defining the beneficial uses of the site. This includes issues relating to:

- the health and wellbeing of humans, on or off the site;
- environmental impacts to flora and fauna; and
- effects of soil contamination on surface water and groundwater.

The South Australian Environment Protection Authority (SA EPA) stipulates that the Site Audit Report should generally provide a concluding statement incorporating one of the following:

- the condition of the site is such that the site is suitable for unrestricted use (similar to EPA Victoria's 'any beneficial use');
- the condition of the site is such that it is suitable only for certain stated uses; any conditions pertaining to the use of the site must be specified; or
- the condition of the site presents an unacceptable health and/or environmental risk, and is not suitable for any use unless remediated.

The general environmental auditing process is shown in Fig. 3.1.

In South Australia, the Environment Protection Authority recently provided the following notes for the guidance of auditors:

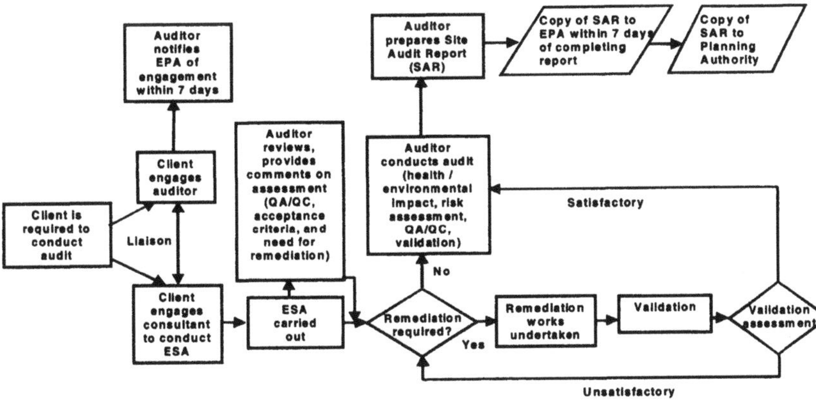

Fig. 3.1: Flowchart of environmental auditing process

- *Audits are required for redevelopment of potentially contaminated land to residential or more sensitive land use;*
- *Audits are generally not needed for proposed industrial or commercial purposes;*
- *Both surface water and groundwater must be considered—the auditor must state his opinion of any impacts arising from contamination or surface water or groundwater;*
- *All surface water and groundwater is potentially potable; and*
- *The auditor must consider any groundwater impacts extending off site.*

Auditing in the Northern Territory

For some sites undergoing remediation in the Northern Territory, the Department of Infrastructure, Planning & Environment (DIPE), which is the equivalent of the EPA, requires that the remediation processes be reviewed by an auditor and that the site be certified to the same standards as those of EPA Victoria. Conduct of audits in the Northern Territory is a non-statutory process; activities of DIPE are governed by legislation such as the Waste Management and Pollution Control Act and the Water Act. The Northern Territory Government is also guided by the provisions of the National Environment Protection (Assessment of Site Contamination) Measure (NEPM).

CASE STUDIES

Case Study 1—33 Vassall Street, Semaphore, South Australia

A former furniture factory site, built on fill materials contaminated with Polycyclic Aromatic Hydrocarbons (PAHs) and heavy metals, was to be redeveloped as four residential allotments. Groundwater was encountered at 5.5 m below surface. Remediation was accomplished by excavation of about 1000 m^3 of fill materials (to 0.5–1.0 m depth), removal and off site disposal as Low Level Contaminated Soil (LLCS). Following a satisfactory validation testing of the base and sides of the excavations, clean imported fill materials were used to build the site up to levels suitable for construction.

This audit was relatively straightforward, and serves as a good illustration of the environmental auditing process. As 12 months had elapsed between the assessment (ESA) and the remediation works, the auditor required limited retesting of the groundwater at the time of remediation, because the Site Audit Report should reflect the environmental status of the site on the date that the audit is completed.

Case Study 2—Henley Oval Annex, South Australia

A former sports oval, located in an area known as 'The Reedbeds' near the River Torrens, was to be redeveloped as 16 residential allotments, together with a reserve and access roadway. Groundwater was encountered at about 3 m below ground surface. Investigations showed that for the southern half of the site, benzo(a)pyrene levels were above the nominated acceptance criterion of 2 mg/kg, in brown clays overlying a black clay layer. Remediation was accomplished by excavation of about 3000 m^3 of soil, removal and off-site disposal as Intermediate Landfill Cover (ILC), and replacement with clean imported fill after satisfactory validation of the excavated surfaces.

The main challenge in this audit was in the setting of a site-specific acceptance criterion for benzo(a)pyrene, based on health risk assessment principles, and in the delineation of the extent of remediation required. The auditor accepted that the northern half of the site did not require remediation, even though some test results showed benzo(a)pyrene levels above 2 mg/kg. The argument was that the high readings were in the 300–600 mm depth interval so that there was a minimum of 300 mm cover of less contaminated material. Furthermore, backhoe test pits dug along the demarcation line (between the northern and southern portions of the site) revealed that the PAHs appeared to be bound up in the brown clays overlying the black clay layer, and that availability of the benzo(a)pyrene and other PAHs could be expected to be limited.

Case Study 3—Former Norwood Council works depot, South Australia

This site was a former Council works depot with various buildings, underground storage tanks (USTs) and a 6 m deep waste pit. The site was to be redeveloped for medium- to high-density residential use, and the sale depended on the satisfactory completion of a Site Audit Report specifying that the site was suitable for 'unrestricted use'. PAHs (associated with the bitumen layer) and very high arsenic concentrations (source unknown) resulted in excavation to an average depth of 1.0 m over the whole site and replacement with clean imported materials following validation.

The site was originally valued at about $400 000. Remediation works cost in the order of $300 000. The site sold at auction for $1 m. There was clearly a commercial challenge in conducting this audit, both in terms of minimising remediation costs to achieve an 'unrestricted use' classification, and in terms of meeting the client's tight time schedule. The technical challenge was in setting acceptance criteria (and thus 'remediation goals') for the contaminants of concern: arsenic, lead, PAHs and petroleum hydrocarbons.

Case Study 4—Centre for Performing and Visual Arts, South Australia

This site, on Light Square, had been occupied since 1847. Past uses included a metal forge, a bakery, a book-binding factory, and a mechanical workshop. During the course of site remediation, seven USTs were removed and the tank pits were validated. Fill materials contaminated with PAHs and Lead were stripped to 0.5–0.7 m depth, stockpiled on site, classified as either ILC or LLCS, and disposed of appropriately.

One of the challenges of this audit was to provide satisfactory evidence that groundwater underlying the site had not been affected adversely by on-site activities. Unfortunately it was impracticable to undertake direct testing of the groundwater at the appropriate time, due to restrictions on access to the site. The site was underlain by several metres of low-permeability Hindmarsh Clay, and excavation of the 4 m deep basement carpark showed no contamination at depth. Further evidence of uncontaminated conditions at depth was gleaned from the Contractor's and Geotechnical Consultant's piling records, which gave information on 8–10 metre deep piles drilled over the entire site, generally on a 5 × 5 m grid. The auditor concluded that site data, including site investigations and pier construction records, provided strong evidence that hydrocarbons and other contaminants from former on-site sources had not affected soils at depth, and that therefore there was negligible risk that groundwater underlying the site (greater than 11 m depth) had been adversely affected by on-site sources. However, given the existence of potentially contaminating sources in the vicinity of the site, there was the potential for groundwater contamination from off-site sources. Therefore the auditor recommended that groundwater usage at the site be prohibited.

The Site Audit Report concluded that the site was suitable for its proposed use as the new Roma Mitchell Centre for Performing and Visual Arts.

Case Study 5—Fuel depot, Alice Springs, Northern Territory

Fuel releases of about 100 000 L in 1997 resulted in phase-separated hydrocarbon (PSH) and dissolved-phase hydrocarbon plumes on groundwater underlying a fuel depot site in Alice Springs. Groundwater is at 7–8 m depth, and the plume is moving in a south easterly direction towards the Town Basin. The remediation strategy comprises in situ groundwater treatment—a combination of Soil Vapour Extraction (SVE) and PSH skimming. Commissioned in October 1999, the system is now operating successfully.

After 'clean-up to the extent practicable', the auditor's 'sign-off' is likely to be based on independently derived risk-based criteria for the residual dissolved phase contamination. Remediation of the residual groundwater contamination may be by natural attenuation. The auditor notes that according to recent advice given by EPA Victoria, where a natural attenuation solution is proposed:

- multiple lines of evidence are required to demonstrate that natural attenuation is occurring, including decreasing contaminant concentrations, associated changes in groundwater chemistry (e.g. decreasing dissolved oxygen), and supporting information on microbiological activity;
- where the beneficial uses of the groundwater are not protected due to pollution, pending degradation under a natural attenuation proposal, consideration should be given to the controls necessary to ensure that use of the groundwater is avoided; and
- a site should not be allowed to continue to pollute another segment of the environment (including continued pollution of groundwater off site) unless clean up to the extent practicable has been effected.

Case Study 6—Former sulfur stockpile, Humpty Doo, Northern Territory

The Humpty Doo site was used by Energy Resources of Australia (ERA) as a stockpile for the temporary storage of imported elemental sulfur between 1982 and 1997. On decommissioning, all remaining sulfur and a 5 cm thick surface layer was excavated and removed for safe disposal in a mined-out pit at Ranger Uranium Mine. Soil contamination with sulfur, and subsequent oxidation, has led to accelerated weathering and low pH in both surface water and groundwater. In situ soil treatment has involved the application of crushed limestone (based on core leaching trials) to ameliorate residual acidity, thus stopping the acid drainage process.

Based on an appreciation of the regional hydrogeological setting, the data suggested that there were two aquifer systems underlying the site: a shallow, unconfined (watertable) aquifer, with depth to watertable varying seasonally, from near surface at the end of the wet season to 8–10 m below ground surface at the end of the dry season; and a confined aquifer extending from about 40 m to 70 m below ground surface, whose potentiometric surface is typically 10–20 m below ground surface. There was likely to be a significant upward hydraulic pressure associated with this confined aquifer. Contamination of the shallow unconfined aquifer (depressed pH, elevated heavy metals) was detected over several sampling events at one location immediately downgradient of the former sulfur stockpile. At time of writing, the confined aquifer had not been tested.

There are existing beneficial uses associated with the confined aquifer; for example, in the vicinity of the site, the groundwater is realised and used for potable and irrigation uses. There are also beneficial uses associated with the shallow unconfined aquifer. Existing uses include the maintenance of ecosystems (for example, discharge to surface water bodies such as creeks and wetlands); potential likely beneficial uses might include irrigation and protection of buildings and structures (defined as a 'beneficial use' by the State Environment Protection Policy (SEPP) Groundwaters of Victoria).

The auditor has required continued surface water and groundwater monitoring to ensure that there are no unacceptable health or environmental risks and no off-site impacts. The constraints (set out as conditions in the Site Audit Report) also include specifications for construction in the former stockpile area (for example, the use of sulfate-resistant concrete) and constraints on groundwater use in the vicinity of the contaminated bore.

CONCLUSIONS

Auditing of site contamination is a statutory process in Victoria and New South Wales. It is practised as a non-statutory process in South Australia and Northern Territory. The National Environment Protection (Assessment of Site Contamination) Measure (NEPM) provides guidelines on the competencies and acceptance of environmental auditors, th us fore-shadowing acceptance of the environmental auditing system at a national level. The auditor effectively carries out a delegated function of the environmental regulatory authority, on a user-pays basis.

Independence, technical competence, and an ability to make cost-effective decisions quickly and with minimum uncertainty would appear to be the key benefits of the Australian environmental auditing system.

ACKNOWLEDGEMENTS

In preparing this paper, the assistance of EPA Victoria, South Australian EPA and Tonkin Consulting is gratefully acknowledged.

NOTE

Adrian Hall now works at URS Australia Pty Ltd, 25 North Terrace, Hackney, SA 5069.

REFERENCES

Contaminated Land Management Act 1997 (New South Wales)

Environment Protection Act 1970 (Victoria)

Environment Protection Act 1993 (South Australia)

NEPM (1999), National Environment Protection (Assessment of Contaminated Sites) Measure. National Environment Protection Council

State Environment Protection Policy (SEPP): Groundwaters of Victoria, 1997

4

Heavy Metal Contamination of Soil with Domestic Sewage Sludge

M.B. Kirkham

INTRODUCTION

The disposal of urban sewage sludge from wastewater-treatment plants is the most difficult and increasingly financially draining problem confronting sanitary district staffs in countries throughout the world. Sludge management authorities have had to accept dramatic increases in disposal costs. For example, in the US in the 1970s, costs were generally less than $US110 per metric ton. By the late 1980s, however, estimates often had jumped to $US550 per metric ton or more. Some short-term contracts have run as high as $US880 per metric ton (Bastian, 1997).

The traditional means of sludge handling and disposal are: (a) ocean disposal through barging or pipeline transport; (b) dewatering and drying with disposal in a landfill; (c) incineration; and (d) lagooning (Girovich, 1996a). Ocean disposal was banned in the US in 1992. Dried sludge sometimes can be sold or given away as a fertilizer, but drying is expensive. Land application of wet sludge is a method of disposal that is both economical and solves the disposal problem in a beneficial way, because the organic matter, nutrients, and water in the sludge are recycled back to the land (Kirkham, 1979a).

Recycling the water is of special urgency. Available fresh water amounts to less than one half of one percent of all the water on earth (Barlow, 1999). Global consumption of water is doubling every 20 years, more than twice the rate of human population growth. According to the United Nations, more than one billion people on earth already lack access to fresh drinking water. If this current trends persist, by 2025, the demand for fresh water is expected to rise by 56% more than is currently available (Barlow, 1999). To help ensure that water supplies will not dwindle, polluted water, including water from sludge, must be reclaimed.

Department of Agronomy, 2004 Throckmorton Hall, Kansas State University, Manhattan, Kansas (KS) 66506-5501, USA; Tel: 785-532-5731; Fax: 785-532-6094; E-mail: mbk@ksu.edu

The world's population has recently reached six billion (Pearlstine, 1999), and it is predicted to reach 11 bn by 2050 (Swerdlow, 1998). Disposal of domestic sewage sludge is a problem that will increase with the growing population. Increase in wastewater volumes and treatment levels projects a corresponding increase in the volume of residuals . From 1972 to 1990, the volume of sewage sludge produced annually nearly doubled, because of new treatment plants coming on-line and existing plants upgrading to secondary or higher levels of treatment (Bastian, 1997). In the US, with a population of 276 m (Pearlstine, 1999), current sludge production is estimated to be some 6.9 m dry English tons (6.2 tons) per year, of which 56% of the sludge is land applied, 15% incinerated, 22% disposed of on the surface of the soil (e.g. lagooning), and the rest managed in some other way (e.g. long-term storage; landfill) (Bastian, 1997). Some of the sludge that has been stored for a long time (10–15 years) is eventually used on land (Fondahl, 1999).

As in the US, sludge disposal to agricultural land is widely practised in Europe. In the United Kingdom, 42% of the estimated 1.1 m dry tons of sludge produced annually is applied to agricultural land (Wilson and Jones, 1999). This amount is expected to increase as a result of the ban by the European Union (EU) on disposal of sludge to sea in 1998 and the EU Directive to improve wastewater quality. In Japan, with a population of 127 m (Pearlstine, 1999), approximately 1.7 m tons per year of sludge are generated, of which about 60% is incinerated; only 15% of the sludge is put to beneficial use (Goldstein, 1999). But disposal methods in Japan are changing, because construction and operating costs related to incineration-based treatment are rising due to measures needed to reduce pollutants such as dioxins (Goldstein, 1999).

In the past, three major constraints have limited land application of sewage sludge: nitrogen in excess of crop needs; pathogens; and heavy metals. Recent concerns include a fourth category: endocrine disruptors (Desbrow et al., 1998). Reports of adverse effects (vitellogenin synthesis in male fish) following exposure to extremely low concentrations of endocrine-disrupting contaminants have focused the scientific community's attention on these formerly-ignored contaminants (Sedlak et al., 2000). In 1992, the United States Environmental Protection Agency (EPA) promulgated a *Code of Federal Regulations* (CFR) that governs the disposal of sewage sludge (United States Environmental Protection Agency, 1992). The regulations address the first three constraints (nitrogen, pathogens, heavy metals), but not the fourth (endocrine disruptors). Nitrogen management of sludge applied to land, following the CFR regulations, is followed by all publicly-owned wastewater treatment plants applying sludge to land, using the procedures listed in the published literature (Crohn, 1996; Forste, 1996; Gilmour and Skinner, 1999).

In the US, sewage sludge cannot be applied to land unless it has been stabilized so as to reduce the pathogens. The regulations contain two classes of pathogen reduction: Class A and Class B (United States Environmental Protection Agency, 1994). Class A pathogen reduction renders the sludge virtually pathogen-free after treatment, and pathogens are below levels detectable by specified analytical methods (Bastian, 1997). Class B pathogen reduction significantly reduces, but does not eliminate, all pathogens. Landholders who use sewage sludge certified by the preparer as Class A have no requirements with respect to pathogens. If the sewage sludge is Class B, site restrictions must be imposed in order to allow time for natural processes to reduce pathogen levels further than treatment has done. Most intestinal pathogenic bacteria are either destroyed or their populations greatly reduced by anaerobic digestion of sewage solids (Kirkham, 1979a). For example, digestors operated with a mean residence time, or 'sludge age' of at least 15 days at 35°C typically will meet the CFR requirements for Class B sewage sludge (Brinkman and Voss, 1999). The longer the sludge is allowed to age, the greater the pathogen reduction (Butler et al., 1999). The fate of viruses during anaerobic digestion is less certain. A Pathogen Risk Assessment computer model has been developed to assess the risk of human infection from enteric viral pathogens in municipal sludge applied to land (Wilson et al., 1992). However, more information is needed to improve the usefulness of the model and arrive at conclusions. In this context, bacteria (*Escherichia coli* and *Pseudomonas fluorescens*), with genes encoding bacterial luminescence inserted into them, are being added to the soil to monitor heavy element toxicity from sludge additions (Chaudri et al., 1999).

Other concerns about land application of sludge are the vector attraction and synthetic organic compounds, like dioxins and polychlorinated biphenyls. Vectors are rodents, flies, and birds that might be attracted to sewage sludge and, therefore, could transmit pathogenic organisms, if present, to humans. The EPA regulations require that vectors must be reduced either through treatment or by the use of barriers (subsurface injection or incorporation into the soil within six hours after application to the land) (United States Environmental Protection Agency, 1994). No synthetic organic contaminants are currently regulated under Part 503 of the CFR (Harrison et al., 1999), but they are to be addressed in the future by the EPA (Bastian, 1997).

This chapter focuses on heavy metals in urban sewage sludge applied to land. It does not deal with rural sewage, which is often treated by septic tanks (Sauer and Tyler, 1996). Literature since the last conference on Contaminants and Soil Environment in the Australasia-Pacific Region (Naidu et al., 1996) is the focus of this review. Note that now the term 'biosolids' is often used instead of sewage sludge, and it is defined as 'solid organic matter recovered from a sewage treatment process and used especially as fertilizer' (Christen, 1998). However, in this chapter, the term sewage sludge will be used, because the sludge under consideration is liquid and digested with 3–5% solids. It looks like crude oil and is not solid.

HEAVY METALS IN SLUDGE

A heavy metal has a density greater than 5.0 g/mL (Page, 1974). Land application of liquid sewage sludge results in heavy metal accumulation in surface soils, because sludge contains an abundance of them. The classic and pioneering work of Page considers 17 trace elements (As, Ba, B, Cd, Cr, Co, Cu, Pb, Mn, Mo, Hg, Ni, Se, Ag, Sn, V, and Zn). All of these elements are heavy elements except for Ba, B, and Se. Only B, Cu, Fe, Mn, Mo, and Zn are essential in trace amounts for plant growth. Hence, they are the 'trace elements' (Sauchelli, 1969), and there are seven of them, Cl being the seventh. The essentiality for some plants of other trace elements—for example, V for algae (Sauchelli, 1969, p. 219) and Ni for legumes (Eskew et al., 1983)—has been suggested, but it has not been proved for plants in general. The cogent review by Leeper (1978) considers 12 heavy metals (Cd, Cr, Co, Cu, Fe, Pb, Mn, Hg, Mo, Ni, Sn, and Zn). In the EPA regulations, concentration limits are set for 10 elements in sewage sludge (As, Cd, Cr, Cu, Pb, Hg, Mo, Ni, Se, and Zn) and their cumulative loading limits (Table 4.1) (Bastian, 1997). All of them are heavy metals except Se. Harrison et al. (1999) suggest that Be, B, F, Fe, Ag, and Sb should be added to the EPA list of elements to be analyzed in sludge for land application. The most popular toothpaste in the US, made by Proctor and Gamble (Cincinnati, Ohio), contains titanium dioxide. Even though

Table 4.1 Maximum concentrations of elements in sewage sludge allowed by the US Environmental Protection Agency regulations. Cumulative loadings allowed on soil also are given (Data from Bastian, 1997)

Element	*Maximum concentration (mg/kg)*	*Cumulative loading (kg/ha)*
Arsenic	75	41
Cadmium	85	39
Chromium	3000	3000
Copper	4,300	1,500
Lead	840	300
Mercury	57	17
Molybdenum	75	18
Nickel	420	420
Selenium	100	100
Zinc	7500	2800

Ti is considered to be physiologically inert (Weast, 1964, p. B-142), the metal might be analyzed in sludges to determine its concentration and whether it is increasing in sludge.

A number of assumptions in the US EPA 503 risk assessment for phytotoxic metals have been challenged as scientifically unsound (Schmidt, 1997; McBride, 1998b; Harrison et al., 1999). The regulations are defended, even though it is accepted that more research is needed (Vance and Pierzynski, 1999). Pretreatment and pollution prevention efforts in place in the US are resulting in decreasing concentrations of metals in some sludges, e.g. sludges in publicly owned wastewater treatment facilities in Massachusetts, US (Isaac and Boothroyd, 1996.)

Of the heavy elements in sludge, Cd, Cu, Ni, and Zn (Leeper, 1978) are of concern. Of these, Cd is the element of most concern, because it poses the greatest threat to human health. Food obtained from normal-looking plants grown on sludge-treated soil might contain concentrations of Cd toxic to man and animals (Kirkham, 1979a). Breeding of cultivars low in Cd offers a method to reduce human health concerns (Clarke et al., 1997).

SURFACE AND GROUNDWATER POLLUTION OF HEAVY METALS IN SLUDGE

The heavy metals in sludge can run off and pollute surface waters or move to groundwater through cracks or holes left by roots and worms. McBride et al. (1997) reported on the potential for groundwater and surface water contamination, and Camobreco et al. (1996) showed that preferential transport can accelerate metal leaching through soils. Steenhuis et al. (1999) found that, under field conditions, a substantial portion of less strongly-adsorbed metals, like Cd and Sr, can leach out of the zone of incorporation of sludge. Laboratory studies performed on homogenous soils may underestimate metal mobility in the field. McBride et al. (1999) analyzed groundwater collected at a field site in New York that had been heavily loaded with sewage sludge more than 15 years earlier. The soil was near-neutral and fine-textured, described in detail by Richards et al. (1998). The soil series was a Hudson silty clay loam (fine, illitic, mesic, Glossaquic Hapludalf). The pH on the sludge plot ranged from 5.07 to 7.64 and the pH on the control plot ranged from 4.76 to 7.84 (Richards et al., 1998). The groundwater had elevated concentrations of Cu, Zn, Sr, Rb, Mo, Cd, As, Cr, Ni, Sb, W, Ag, Hg, and Sn compared to the control site. For most of the heavy metals, the increased leaching was in response to the high metal loadings (McBride et al., 1999).

Sludge additions change both fine microporosity (diameter <50 μm) and coarse microporosity (>50 μm) in soil (Sort and Alcañiz, 1999). The morphology of fine micropores is not as much affected as coarse micropores. Coarse micropores tend to be longer than fine ones, and they usually present a vertical orientation which may enhance downward movement of metals.

Muddy water (sludge contains 3–5% solids) infiltrates more slowly into the soil than clean water (Wang et al., 1999). Therefore, one would expect more runoff when the sludge infiltrates into the soil rather than when rainwater or irrigation water infiltrates into the soil. The continuing change in management practices over the past 25 years, moving from disposal toward reuse (i.e., recycling to land), was noted by Bastian (1997) of the EPA; this is encouraging environmentally sound disposal methods (Kuchenrither and Hite, 1999). As wastewater plants are encouraged by the US government to apply more sludge from growing populations to the land, the burden of heavy metals added to soil builds up. Total concentrations of heavy elements in the 0–30 cm depth of a silt loam soil treated with sludge for 35 years at Dayton, Ohio, USA, were increased by the following factors: Cd, 35; Cu, 16.5; Fe, 1.1; Mn, 1.2; Ni, 2; Pb, 16.5, Zn, 13 (Kirkham, 1975). Long-term applications of sewage sludge have resulted in elevated levels of heavy metals at other locations (Bell et al., 1991; Dowdy et al., 1991; Juste and Mench, 1992; Barbarik et al., 1998; Berti and Jacobs, 1998; Brown et al., 1998; Hun, 1998; Sloan et al., 1998; Baveye et al., 1999).

IMMOBILIZATION OF HEAVY METALS IN SLUDGE

Immobilization of heavy metals in the soil has been a common practice to limiting the uptake (Czupyrna et al., 1989). Three methods are often used: liming, addition of P, and increasing the clay content. Liming is one of the most effective methods to reduce plant availability of metals (Chaney, 1973; Kirkham, 1977; Derome and Saarsalmi, 1999). Raising the pH of soil increases the adsorption of many heavy metals (Mo being the major exception). At sludge disposal sites, adding enough lime to keep the soil pH above 6.5 is recommended (Leeper, 1978, pp. 8, 100). Towers and Paterson (1997) found that some metals, for example Cd and Zn, in Scottish soils are potentially more mobile than others, such as Pb, but that most soils displayed a strong binding capacity. However, they also noted that this pattern could be sustained only if the soil pH values were maintained. A fall of one pH unit marked a dramatic shift towards weak binding. Due to the nitrification reaction and the microbial production of carbon dioxide, sewage sludge usually lowers the pH of soil (Kirkham, 1979a), necessitating the liming of sludge sites.

The application of alkaline sludge products to agricultural soils is now commonplace (McBride, 1998a). In particular, the patented "N-Viro Process" is one of the alkaline-stabilization technologies that is being used. This process involves mixing of partially dewatered sewage sludge with an alkaline material. Solid content of the sludge is 15–40%, and a number of industrial byproducts are used as alkaline reagents, including cement kiln dust, lime kiln dust, lime (CaO), limestone, alkaline fly ash, flue gas desulfurization ash, other coal-burning ashes, and wood ash. During preparation, the sludge mixture is kept at a pH greater than 12, with temperatures between 52 and 62°C and solids content greater than 50% for 12 hours. The pH then must remain greater than 12 for at least three days. The procedure destroys pathogens by a combination of high pH, heat, and drying. At this point, the final product has achieved the EPA's classification for complete pathogen destruction (Logan and Harrison, 1995).

Givorich (1996b) reported that the high pH of alkaline-stabilized sludge immobilizes heavy metals, and they become largely unavailable for plant uptake. However, McBride (1998a) found that the high pH of N-Viro sludge, when applied to land, modified metal leachability, and increased mobility of some metals (e.g. Cu, Mo, and Ni), while decreasing the mobility of others (e.g. Zn and Pb). He said that high leachability of Cu and Ni may be due to the increased dissolved concentration of organic matter that has been shown to bring strongly complexing metals into solution. Others have also shown that the leachability of metals cations is due, at least in part, to complexation with dissolved organic matter (Temminghoff et al., 1998). McBride (1998a) reported that the high solubility of Cu and Ni was a result of complexation with dissolved amines or amino acids in the alkaline (pH 12) water extracts. Because of the high affinity of both Cu^{2+} and Ni^{2+} for N-ligands in amines, these metals were mobilized much more than Pb, Zn, or Cd. The results suggested potential for the transport of Cu and Ni from land-applied alkaline sludges into surface water and groundwater.

The phosphate content of soil is another factor that controls metal availability (Chaney, 1973). Phosphate amendments have been used to immobilize Pb in soil (Ma et al., 1994; Zhang et al., 1998). Phosphates are often concentrated in the sludge during the treatment process, so phosphate fertilizers are usually not needed at land disposal sites. Large concentrations of P build up in the soil from sludge additions (Kirkham, 1982). The amount of P applied to land with sludge is generally more than plants need when the application rate is based on the potentially available N (Sui et al., 1999). Despite the large amounts of P added with sludge, heavy metal toxicities still prevail. In the previously mentioned study done at Dayton, Ohio, larger-than-normal concentrations of Cd and Cu were found in leaves of corn grown on the soil that had received sludge for 35 years (Kirkham, 1975). The P concentration in the treated soil was also elevated (0.07% P in surface of control soil; 0.57% P in surface of sludge-treated soil). The P concentrations of the plant parts, however, were within normal concentration ranges (Kirkham, 1976).

Conversely, P additions to the soil with the aim of immobilizing metals in contaminated soil can make the essential trace elements unavailable and lead to deficiency, causing growth inhibition (Boisson et al., 1999a).

Phosphorus fertilizers can be contaminated with Cd, because the two elements occur together geologically (Leeper, 1978). In fact, applications of phosphate fertilizers are largely responsible for the increasing Cd load in agricultural soils worldwide (Hamon et al., 1998). Therefore, if phosphate fertilizers were added to sludge-treated land, the soil would become further loaded with Cd. This is a matter of great concern in New Zealand, where sheep graze on phosphate-fertilized soil (Robinson et al., 1999a, 1999b; Zanders et al., 1999). The Cd, which accumulates slowly and is excreted slowly in sheep (Beresford et al., 1999), may pose a risk to their health and that of humans eating the meat. Cadmium was found in calf liver from animals grazing on an eroded prairie restored with applications of sludge from Oklahoma City, Oklahoma, US. However, the amount of Cd (0.003 μg/g dry weight) was considered clinically insignificant (Kessler and Kirkham, 1985). Addition of triple superphosphate, rock phosphate, or KH_2PO_4 is being advocated to reduce Pb bioavailability to humans (Hettiarachchi and Pierzynski, 1998). However, this treatment might be deleterious, if the P is high in Cd.

Clay (montmorillonite) has been used either to immobilize Zn and Cd in soils (Lothenbach et al., 1998, 1999) or to decrease the hydraulic conductivity of the soil so as to prevent the movement of pollutants to the groundwater (Rakhshandehroo et al., 1998). Use of clay avoids the increase in pH that occurs with liming, which may cause reduced Mn and phosphate availability to plants (Lothenbach et al., 1999). Beringite (a modified aluminosilicate related to clay which is chiefly aluminum silicate) and steel shots have been added to soil to immobilize Cd and Ni (Boisson et al., 1999b). Beringite decreased transfer of Cd and Ni to corn plants. Steel shots decreased transfer of Ni to corn, but not that of Cd.

Immobilization leaves the metals in the surface of the soil, and some of the metals, like Cd, are linked to neoplastic disease (Jungmann et al., 1993). The high rates of cancer that we now see in the US could be related in some way to heavy-metal pollution of edible plants or drinking water from surface or groundwaters. According to the current administration in the US, 'new protections [are needed] to give all our children the gift of clean, safe water in the twenty-first century' (Perciasepe, 1998). To this soil should be added, since children frequently put their hands to their mouths (Schmidt, 1999), and children playing on gardens fertilized with sludge could ingest potentially-toxic quantities of metals in the surface soil.

Phytoremediation

Instead of immobilizing the polluting metals in soil, another approach might be to remove them using the process of phytoremediation, which is the use of green plants (higher plants) to remediate contaminated soil (Cunningham et al., 1996; Maywald and Weigel, 1997; Watanabe, 1997; Gleba et al., 1999; Terry and Bañuelos, 2000). Phytoremediation contrasts with bioremediation, which is the use of lower plants like bacteria to remove pollutants from soil (Bajpai and Zappi, 1997; Leeson and Alleman, 1999). The main advantages of phytoremediation are that the procedure is carried out in situ, without physical removal of the soil, and it is relatively inexpensive. Classical methods for in situ remediation of water-soluble pollutants can cost \$100,000-\$1,000,000 per hectare, and other procedures are more expensive. Phytoremediation is estimated to cost between \$200-\$10,000 per hectare (Brooks, 1998). By 2005, the phytoremediation market is predicted to be between \$214 and \$370 million (Glaze, 1998).

Many studies show that hyperaccumulators, or plants that accumulate heavy metals from the soil into shoots, can be used in phytoremediation (Cunningham and Ow, 1996; Robinson, 1997; Robinson et al., 1997a, 1997b, 1998; Lasat et al., 1998; Salt et al., 1999). The largest group of these metal hyperaccumulators is found in the genus *Alyssum*, in which Ni concentrations can reach 3% of leaf dry biomass (Krämer et al., 1996). Hyperaccumulators are native plants. Also, wetland plants, like water hyacinth (*Eichhornia crassipes*), an aquatic floating plant, are candidates for phytoremediation of wastewater polluted with Cd, Cr, Cu, and Se (de Souza et al., 1999; Zhu et al., 1999). Crop plants, not just hyper-accumulators, have been used in phytoremediation to remove metals from soil, and metal uptake has been enhanced by using chelates (Huang et al., 1997). Now, let us consider chelates in soil.

CHELATION AND PHYTOREMEDIATION

Except for Mo, the 12 heavy elements in sludge listed by Leeper are primarily cationic in soils (Leeper, 1978). All cations of heavy metals have chemical features that favor their retention in soils. They differ from Na^+ in being more hydrolyzed, while some of these complexes are dimerized and polymerized, as is well known for Al and Fe. Therefore, the heavy metals, as well as Al, are held more strongly by negative colloids than would be otherwise expected. At the same time, they form complexes with other anions, some of which are so stable as to move the metals into a negatively charged state. The most notable of such compounds are the *chelates*, which are organic compounds. Leeper defines *chelation* as a reaction with organic colloids that removes a metallic cation (Leeper,

1978). But the same kind of combination can have the opposite effect and so allow the metal to stay dissolved. The word chelate denotes the binding of a metal ion by two or more linked atoms so as to give a stable ring. It does not imply anything about solubility. Soluble chelates are formed between the heavy metals and low molecular weight organic molecules, of which citric acid is typical and the most familiar. These soluble chelates usually carry a negative charge at ambient soil pH; so they are not fixed on the generally negative surfaces of the soil. In fact, it seems likely that for any soil that is richly supplied with organic material and, correspondingly with its biological products such as citric acid or amino acids, the amount of any heavy metal in the drainage should be expected to reach a minimum value, in the anionic or uncharged form. However, as stated earlier, most heavy metals are fixed as cations. Only a minute proportion of the total is mobile, and it becomes mobile through its association with organic molecules which are themselves not plentiful (Leeper, 1978).

However, in recent years, the natural chelates in the soil have been augmented by adding synthetic chelates to enhance phytoremediation (Brooks, 1998). Chelates have long been used in water culture for the study of plant nutrition (Hewitt, 1966). Iron is an essential trace element, yet plants will become severely chlorotic if only an iron source, such as $FeSO_4 \cdot 7H_2O$, is added to nutrient culture (G.C. Gerloff, University of Wisconsin, Madison, 1968, personal communication). For this reason, ferric tartrate and ferric citrate have been used for many years in place of an inorganic iron salt. However, FeEDTA (iron ethylenediamine-tetraacetic acid) and FeEDDHA [iron ethylenediamine di(o-hydroxyphenyl) acetate] are probably the most satisfactory Fe sources in culture media. These synthetic chelating agents keep Fe from precipitating. As it is more readily available in a highly purified form, FeEDTA is recommended unless the pH of the nutrient solution will be above 7.0. If higher than recommended levels of chelate are used in nutrient culture, large amounts of Fe are taken up by plants with reduced growth (Kirkham, 1979b).

EDTA has not been shown to be toxic (Harmsen and van den Toorn, 1982). In fact, for humans, chelation therapy is of value in the treatment of major poisoning by heavy metals. Chelates like EDTA form compounds with heavy metals, such as Pb, As, Cu, and Ag, which then are excreted from the human body. Chelation reduces mortality from acute, childhood Pb poisoning (Cotter-Howells, 1998), and the United States Food and Drug Administration approves the use of chelation as a treatment for Pb poisoning (Thomas, 1996). Chelation therapy is also used to remove heavy-element radionuclides, such as ^{60}Co, ^{95}Nb, ^{212}Pb, ^{226}Ra, and ^{65}Zn (Kirkham and Corey, 1977), after accidental exposure at nuclear plants (J.C. Corey, Savannah River Laboratory, personal communication, 1976).

EDTA is a common industrial agent for complexing metal ions in water. Aqueous solutions of EDTA have been studied extensively (Klewicki and Morgan, 1998). EDTA is used in applications such as metal plating, water softening, pharmaceuticals, photography, textile and paper manufacture, cosmetics, food processing, water treatment, and industrial cleaning (Madden et al., 1997; Baron and Hering, 1998; Davis and Green, 1999). EDTA is also used as a substitute for phosphates in detergents (Harmsen and van den Toorn, 1982). Processes that involve the removal of metal oxide scale from heat transfer surfaces, such as in boilers or nuclear reactors, also use organic complexing agents like EDTA for their ability to dissolve metals (Bürgisser and Stone, 1997). Complexation with EDTA can significantly alter the behavior of toxic metals in waste streams as well as in surface- and groundwater. For example, complexation with EDTA has been identified as the reason for enhanced mobility of ^{60}Co and other radionuclides in the subsurface at nuclear waste sites (Baron and Hering, 1998; Gorby et al., 1998, McArdel et al., 1998). The chelate has been detected in environmental waters, but the concentrations of EDTA typically found in river systems are unlikely to contribute to solubilization of trace elements in sediments (Gonsior et al., 1997).

Many studies report the addition of synthetic chelates to soil in order to facilitate phytoremediation. For example, Indian mustard (*Brassica juncea*), which has been identified as a good plant to use in phytoremediation (Dushenkov et al., 1995; Nanda Kumar et al., 1995), was grown in a chelate-treated soil to increase uptake of Pb at a contaminated site in New Jersey. The site was formerly an industrial area occupied by manufacturers of Magic Marker pens and lead-acid batteries (Betts, 1997). In less than seven weeks of cultivation, Pb uptake exceeded 800 ppm on a plant dry-weight basis. Vassil et al. (1998) exposed Indian mustard (*B. juncea*) plants to Pb and EDTA in hydroponic solution. The plants were able to accumulate upto 55 mmol kg^{-1} Pb in dry shoot tissues (1.1%, w/w). This represents a 75-fold concentration of Pb in shoot tissue over that in solution. Epstein et al. (1999) showed that EDTA enhanced the uptake of Pb by *B. juncea* in lead-contaminated soil. Nanda Kumar et al. (1995) compared cultivars of Indian mustard (*B. juncea*) and found some that accumulated Pb more efficiently than others. Ebbs and Kochian (1998) reported that the addition of EDTA to soil increased Zn accumulation by *B. juncea*, but not by the crop barley (*Hordeum vulgare*). Nevertheless, barley accumulated more Zn than Indian mustard in the presence of EDTA.

Other amendments, such as ammonium salts, have been shown to increase the bioavailability of contaminants to crop plants (Lasat et al., 1997; Dushenkov et al., 1999). Huang et al. (1998) added organic acids (acetic acid, citric acid, and malic acid) to the soil to increase uranium phytoextraction. Citric acid was the most effective in enhancing U accumulation in plants. Shoot U concentrations of *Brassica juncea* and *Brassica chinensis* grown in an uranium-contaminated soil increased from less than 5 mg/kg to more than 5000 mg/kg in citric acid-treated soils. Francis and Dodge (1998) removed radionuclides and toxic metals from contaminated soils and wastes by extracting them with citric acid. Fischer et al. (1998) used organic chelating agents (molasses, blood meal, silage effluents) to leach metal-polluted soils, which were contaminated after decades of sewage sludge applications. Grass silage effluent removed about 75% of the Cd and more than 50% of the Cu and Zn at pH 4.4. Wu et al. (1999) demonstrated that N, N′-di (2-hydroxybenzyl) ethylenediamine N,N′-diacetic acid (HBED) resulted in a higher concentration of Pb in roots of corn (*Zea mays* L.) than did EDTA.

Vogeler et al. (2001) studied the movement of the heavy element Cu, which is used as a fungicide to preserve wood and is a pollutant at timber-treatment sites, as affected by EDTA. They investigated basic transport processes to understand the partitioning of Cu between the soil and uptake by poplar roots. The experiment showed that EDTA solubilized the Cu and made it available for root uptake by poplar. This, apparently, was the first experiment to show the impact of EDTA on water and chemical (Cu) transport through and beyond a root zone. The work was also the first to compare measured results with a new mechanistic model that joins water and solute movements through the root zone with uptakes of water and solutes by roots. In the model, Vogeler et al. (2001) assumed that EDTA decreased the adsorption of Cu by the soil. The solubilization was considered to result in a different isotherm, which expressed the weak adsorption of Cu in the presence of EDTA. A Langmuir equation was used to model Cu without EDTA, and a Freundlich equation was used to model Cu with EDTA. The agreement between the predicted and measured values for Cu distribution with depth at the end of the experiment was good (Fig. 4.1). The model successfully predicted the movement of Cu through the EDTA-treated soil. As a first approximation to the process of chelation-induced mobility, the model provided a reasonable rendition of the observations.

People studying phytoremediation, either with or without EDTA, have focussed on the plants and have not considered how EDTA changes the partitioning of a heavy metal between the solid and solution phases in the soil, i.e. how the adsorption isotherm is changed, which is the relationship between the adsorbed and dissolved concentrations of an element in the soil. Even though the distribution of heavy elements between the solid phase and the water of the soil has been studied (Allen et al., 1995; Elzinga et al., 1999), works by Vogeler et al. (2001) and by Kedziorek et al. (1998) have measured and modeled heavy-element

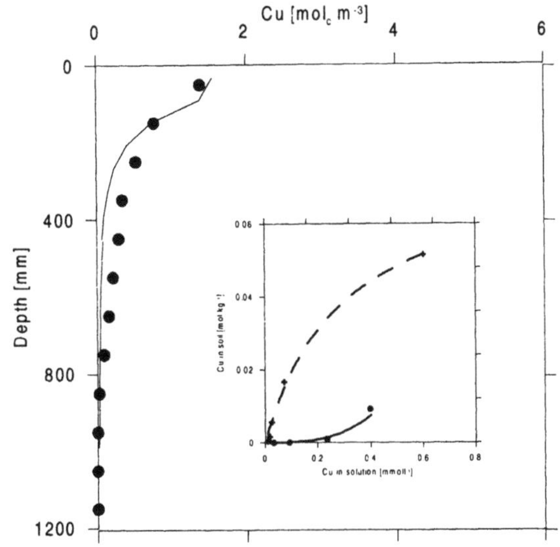

Fig. 4.1 Measured and predicted concentrations of copper in a Manawatu fine sandy loam soil at different depths in a lysimeter with a poplar tree. The inset shows the relationship between the adsorbed and the dissolved concentrations of copper (the adsorption isotherm) in the soil with or without the chelate EDTA. The crosses show the measured data from a batch experiment without EDTA, and the broken line was fit using a Langmuir equation. The solid circles show the measured data from a batch experiment with EDTA, and the solid line was fit using a Freundlich equation (From Vogeler et al., 2001, with permission).

transport as affected by EDTA (Fig. 4.1). However, Kedziorek et al. (1998) did not study plants.

CHELATE-FACILITATED REMOVAL OF HEAVY METALS IN SLUDGE-TREATED SOIL

Since apparently no work has been done to show whether or not EDTA might be used to remove heavy elements from sludge-treated soil, between 6–29 April 1999, Kirkham (2000) carried out a greenhouse experiment with sunflower (*Helianthus annuus* L.) grown on sludge-treated soil with EDTA added at a rate of 1 g per kg soil. The soil was a commercial potting mix, and the sludge came from the Manhattan, Kansas, Sewage Treatment Plant. At harvest, the plant and soil samples were analyzed (Table 4.2). The Cu, Fe, and Zn in the soil were determined by extracting with DTPA using the standard methods of the Soil Testing Laboratory at Kansas State University. With or without sludge, the plants grown with EDTA in the soil had a higher concentration of Cu, Fe, Zn, Cd, and Pb in their leaves than plants grown without EDTA, except for Zn and Cd in sludge-treated soil where the sludge-crust must have affected uptake in

Table 4.2 Concentration of metals (mg/kg) in sunflower and soil treated with sludge and the chelate EDTA

Controls without sludge and without EDTA also are given. Standard errors are shown with the mean (Data from Kirkham, 2000.)

Plant Treatment	Cu	Fe	Zn	Cd	Pb
Sludge+EDTA	45±10	282±13	152±15	1.5±0.5	3.8±0.8
Sludge+no EDTA	35±03	236±93	167±01	6.0±2.0	0.8±0.2
No sludge+EDTA	36±03	264±40	153±04	5.0±1.0	1.4±0.2
No sludge+no EDTA	24±01	99±01	103±01	4.0±2.0	0.1±0.1
			Soil		
Treatment	Cu	Fe	Zn	Cd†	Pb†
Sludge+EDTA	6.2±0.8	165±30	15±1.4	...†	...
Sludge+no EDTA	5.7±0.2	126±9	11±0.3	...	...
No sludge+EDTA	4.7±0.3	152±23	11±1.7	...	...
No sludge+no EDTA	4.0±0.0	132±10	9±1.0	...	...

† Below detection limits, which were 1.0 mg/kg for Cd and 10 mg/kg for Pb.

the EDTA-treated soil. The EDTA increased the extractable concentration of Cu, Fe, and Zn in the soil and apparently made these elements more available for uptake. The concentration of the elements was not reduced in the soil, but, with further harvests, the elements should be lowered. A 'fast-plant' technique, taught at the University of Wisconsin by the Wisconsin Fast Plants Program and described on the Internet at http://fastplants.cals.wisc.edu, might be appropriate at sludge-disposal sites. It is used to grow plants quickly in spaceships where astronauts need food for survival. Fast generations of plants at sludge disposal sites would remove metals expeditiously. The treatment with EDTA offers a new method to reduce metals in soil with sludge. Sludge can be recycled to the land to conserve its organic matter, nutrients, and water. If the metals become elevated in soil, then treatment with EDTA followed by plant removal and ashing of metal-contaminated plants could scour the soil of metals for further recycling of sludge.

When a synthetic chelate like EDTA is added to soil, it is probably absorbed by the plant roots. For example, work with [^{14}C]EDTA in hydroponic solution has shown that the Pb-EDTA complex is taken up rather than the free metal ion (Vassil et al., 1998; Epstein et al., 1999). The EDTA then can alter the transport of other metals inside the plant. For example, the EDTA in the solution for the Fe source probably affected the uptake of Cd by corn reported by Page et al. (1972). An unknown proportion of the Cd and other heavy metals, including the essential ones, was present in the chelated form. The EDTA absorbed by plants might have served to transport some Cd from roots to tops (Leeper, 1978, pp. 51, 57-58).

CONCLUDING COMMENTS

Land disposal of sludge, followed by EDTA treatment for removal of heavy metals, if necessary, may prove to be less polluting and cheaper than landfilling or lagooning of sludges and may also result in a better quality environment. Only future work, which includes field studies, will show if chelates like EDTA are useful in cleansing sludge-treated soils of metals. As pointed out by environmental engineers and scientists gathered in Monterey, California, in January, 1998, to participate in an Environmental Engineering Frontiers Workshop sponsored by the United States National Science Foundation and the Association of Environmental Engineering Professors, the solutions to environmental problems are complex (Logan and Rittmann, 1998). A comprehensive, integrated view of the environment is needed for sustainability. In this respect, this future work, which would seek to determine how heavy metals are partitioned between the soil and plant as they move through the root zone in the presence of chelates, would integrate the soil-plant system in order to find a solution to the problem of heavy metal accumulation on sludge-treated land.

REFERENCES

Allen, H.E., Chen, Y.T., Li, Y. and Huang, C.P., 1995, Soil partition coefficients for Cd by column desorption and comparison to batch adsorption measurements. *Environmental Science and Technology*. 29: 1887–1891.

Bajpai, R.K. and Zappi, M.E. (Eds), 1997, Bioremediation of Surface and Subsurface Contamination. *Annals of the New York Academy of Sciences*, Vol. 829. The New York Academy of Sciences, New York. 341 pp.

Barbarick, K.A., Ippolito, J.A., and Westfall, D.G., 1998, Extractable trace elements in the soil profile after years of biosolids application. *Journal of Environmental Quality* 27: 801–805.

Barlow, M., 1999, Blue gold. The global water crisis and the commodification of the world's water supply. *World Rivers Review* 14(5): 6–7, 13. (Published by International Rivers Network, 1847 Berkeley Way, Berkeley, California 94703, USA).

Baron, D. and Hering, J.G., 1998, Analysis of metal-EDTA complexes by electrospray mass spectrometry. *Journal of Environmental Quality* 27: 844–850.

Bastian, R.K., 1997, Biosolids management in the United States. *Water Environment and Technology* 9(5): 45–50.

Baveye, P., McBride, M.B., Bouldin, D., Hinesly, T.D., Dahdoh, M.S.A., and Abdel-sabour, M.F., 1999, Mass balance and distribution of sludge-borne trace elements in a silt loam soil following long-term applications of sewage sludge. *Science of the Total Environment* 227: 13–28.

Bell, P.F., James, B.R., and Chaney, R.L., 1991, Heavy metal extractability in long-term sewage sludge and metal salt-amended soils. *Journal of Environmental Quality* 20: 481–486.

Beresford, N.A., Mayes, R.W., Court, N.M.J., Maceachern, P.J., Dodd, B.A., Barnett, C.L., and Lamb, C.S., 1999, Transfer of cadmium and mercury to sheep tissues. *Environmental Science and Technology* 33: 2395–2402.

Berti, W.R. and Jacobs, L.W., 1998, Distribution of trace elements in soil from repeated sewage sludge applications. *Journal of Environmental Quality* 27: 1280–1286.

Betts, K.S., 1997, Phyto-cleanup of lead progressing at N.J. site. *Environmental Science and Technology* 31: 496A–497A.

Boisson, J., Ruttens, A., Mench, M., and Vangronsveld, J., 1999a, Evaluation of hydroxyapatite as a metal immobilizing soil additive for the remediation of polluted soils. I. Influence of hydroxyapatite on metal exchangeability in soil, plant growth and plant metal accumulation. *Environmental Pollution* 104: 225–233.

Boisson, J., Mench, M., Sappin-Didier, V., Solda, P., and Vangronsveld, J., 1999b, Short-term in situ immobilization of Cd and Ni by beringite and steel shots application to long-term sludged plots. *Agronomie* 18: 347–359.

Brinkman, D. and Voss, D., 1999, Egg-shaped digesters. *Water Environmental and Technology* 11(10): 28–33.

Brooks, R.R, 1998, General introduction. In: R.R. Brooks (Ed.), *Plants that Hyperaccumulate Heavy Metals*. CAB International, Wallingford, Oxon, United Kingdom: 1–14.

Brown, S.L., Chaney, R.L., Angle, J.S., and Ryan, J.A., 1998, The phytoavailability of cadmium to lettuce in long-term biosolids-amended soils. *Journal of Environmental Quality* 27: 1071–1078.

Bürgisser, C.S. and Stone, A.T., 1997, Determination of EDTA, NTA, and other amino carboxylic acids and their Co(II) and Co(III) complexes by capillary electrophoresis. *Environmental Science and Technology* 31: 2656–2664.

Butler, R., Clark, D., Cleveland, C., Finger, R., Newlands, D., and Newman, G., 1999, A matter of time. *Water Environmental and Technology* 11(10): 34–39.

Camobreco, V.J., Richards, B.K., Steenhuis, T.S., Peverly, J.H., and McBride, M.B., 1996, Movement of heavy metals through undisturbed and homogenized soil columns. *Soil Science* 161: 740–750.

Chaney, R.L., 1973, Crop and food chain effects of toxic elements in sludges and effluents. In: *Recycling Municipal Sludges and Effluents on Land*. Joint Conf. Proc. (Champaign, Illinois). National Association of State Universities and Land-Grant Colleges, Washington, DC: 129–141.

Chaudri, A.M., Knight, B.P., Barbosa-Jefferson, V.L., Preston, S., Paton, G.I., Killham, K., Coad N., Nicholson F.A., Chambers B.J., and McGrath S.P., 1999, Determination of acute Zn toxicity in pore water from soils previously treated with sewage sludge using bioluminescence assays. *Environmental and Science Technology* 33: 1880–1885.

Christen, K, 1998, Biosolids and beyond. 40 CFR 503 author, Alan Rubin, looks five years after the rule was promulgated. *Water Environment and Technology* 10(5): 55–57.

Clarke, J.M., Leisle, D., and Kopytko, G.L., 1997, Inheritance of cadmium concentration in five durum wheat crosses. *Crop Science* 37: 1722–1726.

Cotter-Howells, J. (Ed.), 1998, The prevention of lead poisoning in children—from Baltimore to the world initiatives. An appreciation of the work of J. Julian Chisolm, Jr., M.D. Report of a symposium held at The John Hopkins University School of Medicine, Baltimore, Maryland, June 1, 1998. *Newsletter of the Society for Environmental Geochemistry and Health*. Autumn/Winter, p. 1.

Crohn, D.M, 1996, Planning biosolids land application rates for agricultural systems. *Journal of Environmental Engineering (Amer. Soc. Civil Eng.)* 122: 1058–1066.

Cunningham, S.D. and Ow, D.W., 1996, Promises and prospects of phytoremediation. *Plant Physiology* 110: 715–719.

Cunningham, S.D., Anderson, T.A., Schwab, A.P., and Hsu, F.C., 1996, Phytoremediation of soils contaminated with organic pollutants. *Advances in Agronomy* 56: 55–114.

Czupyrna, G., Levy, R.D., MacLean, A.I., and Gold, H., 1989, *In situ Immobilization of Heavy-Metal-Contaminated Soils*. Noyes Data Corp., Park Ridge, New Jersey. p. 155.

Derome, J. and Saarsalmi, A., 1999, The effect of liming and correction fertilisation on heavy metal and macronutrient concentrations in soil solution in heavy-metal polluted Scots pine stands. *Environmental Pollution* 104: 249–259.

Desbrow, C., Routledge, E.J., Brighty, G.C., Sumpter, J.P., and Waldock, M., 1998, Idenfication of estrogenic chemicals in STW [sewage treatment works] effluent. 1. Chemical fractionation and in vitro biological screening. *Environmental Science and Technology* 32: 1549–1558.

de Souza, M.P., Huang, C.P.A., Chee, N., and Terry, N,. 1999, Rhizosphere bacteria enhance the accumulation of selenium and mercury in wetland plants. *Planta* 209: 259–263.

Dowdy, R.H., Latterell, J.J., Hinesly, T.D., Grossman, R.B., and Sullivan, D.L., 1991, Trace metal movement in an Aeric Ochraqualf following 14 years of annual sludge applications. *Journal of Environmental Quality* 20: 119–123.

Dushenkov, V., Nanda Kumar, P.B.A., Motto, H., and Raskin, I., 1995, Rhizofiltration: The use of plants to remove heavy metals from aqueous streams. *Environmental Science and Technology* 29: 1239–1245.

Dushenkov, S., Mikheev, A., Prokhnevsky, A., Ruchko, M., and Sorochinsky, B., 1999, Phytoremediation of radiocesium-contaminated soil in the vincinity of Chernobyl, Ukraine. *Environmental Science and Technology* 33: 469–475.

Ebbs, S.D. and Kochian, L.V., 1998, Phytoextraction of zinc by oat (*Avena sativa*), barley (*Hordeum vulgare*), and Indian mustard (*Brassica juncea*). *Environmental Science and Technology* 32: 802–806.

Elzinga, E.J., van Grinsven, J.J.M., and Swartjes, F.A., 1999, General purpose Freundlich isotherms for cadmium, copper and zinc in soils. *European Journal of Soil Science* 50: 139–149.

Epstein, A.L., Gussman, C.D., Blaylock, M.J., Yermiyahu, U., Huang J.W., Kapulnik Y., and Orser C.S., 1999, EDTA and Pb-EDTA accumulation in *Brassica juncea* grown in Pb-amended soil. *Plant and Soil* 208: 87–94.

Eskew, D.L., Welch, R.M., and Cary, E.E., 1983, Nickel: An essential micronutrient for legumes and possibly all higher plants. *Science* 222: 621– 623.

Fischer, K., Bipp, H.P., Riemschneider, P., Leidmann, P., Bieniek, D., and Kettrup, A., 1998, Utilization of biomass residues for the remediation of metal-polluted soils. *Environmental Science and Technology* 32: 2154–2161.

Fondahl, L, 1999, Biosolids management in the western region. *BioCycle* 40(7): 70–74.

Forste, J.B, 1996, Land application. In: Girovich, M.J., (Ed.), *Biosolids Treatment and Management. Processes for Beneficial Use*. Marcel Dekker, New York: 389–448.

Francis, A.J. and Dodge, C.J., 1998, Remediation of soils and wastes contaminated with uranium and toxic metals. *Environmental Science and Technology* 32: 3993–3998.

Gilmour, J.T. and Skinner, V., 1999, Predicting plant available nitrogen in land-applied biosolids. *Journal of Environmental Quality* 28: 1122–1126.

Girovich, M.J., 1996a, Biosolids characterization, treatment and use: An overview. In: Girovich, M.J. (Ed.), *Biosolids Treatment and Management. Processes for Beneficial Use*. Marcel Dekker, New York: 1–45.

Girovich, M.J., 1996b, Alkaline stabilization. In: Girovich, M.J. (Ed.), *Biosolids Treatment and Management. Processes for Beneficial Use*. Marcel Dekker, New York: 343–388.

Glaze, W.H. (Ed.), 1998, News Briefs. By 2005, the phytoremediation market will swell to between $214 and $370 million. *Environmental Science and Technology* 32: 399A.

Gleba, D., Borisjuk, N.V., Borisjuk, L.G., Kneer, R., Poulev, A., Sarzhinskaya, M., Dushenkov, S., Logendra, S., Gleba, Y.Y., Raskin, I., 1999, Use of plant roots for phytoremediation and molecular farming. *Proceedings of the National Academy of Sciences (Washington, D.C.)* 96: 5973–5977.

Goldstein, J. (Ed.), 1999, Japanese researchers seek better ways to manage sludge. *BioCycle* 40(8): 8.

Gonsior, S.J., Sorci, J.J., Zoellner, M.J., and Landenberger, B.D., 1997, The effects of EDTA on metal solubilization in river sediment/water systems. *Journal of Environmental Quality* 26: 957–966.

Gorby, Y.A., Caccavo, F., Jr., and Bolton, H., Jr, 1998, Microbial reduction of cobalt^{III}EDTA^{-} in the presence and absence of manganese (IV) oxide. *Environmental Science and Technology* 32: 244–250.

Hamon, R.E., McLaughlin, M.J., Naidu, R., and Correll, R., 1998, Long-term changes in cadmium bioavailability in soil. *Environmental Science and Technology* 32: 3699–3703.

Harmsen, J. and van den Toorn, A., 1982, Determination of EDTA in water by high-performance liquid chromatography. *Journal of Chromatography* 249: 379–384.

Harrison, E.Z., McBride, M.B., and Bouldin, D.R., 1999, Land application of sewage sludges: An appraisal of the US regulations. *International Journal of Environmental Pollution* 11: 1–36.

Hettiarachchi, G.M. and Pierzynski, G.M., 1998, Effects of phosphorus and other soil amendments on lead bioavailability. In: 1998 *Annual Meeting Abstracts, Baltimore, Maryland, October 18-22, 1998*. Amer. Soc. Agron., Crop Sci. Soc. Amer., Soil Sci. Soc. Amer., Madison, Wisconsin: 333.

Hewitt, E.J., 1966, *Sand and Water Culture Methods Used in the Study of Plant Nutrition*. Tech. Commun. No. 22 (Revised 2nd Ed.) Commonwealth Agricultural Bureaux, Farnham Royal, Bucks., England, p. 547.

Huang, J.W., Chen, J., Berti, W.R., and Cunningham, S.D., 1997, Phytoremediation of lead-contaminated soils: Role of synthetic chelates in lead phytoextraction. *Environmental Science and Technology* 31: 800–805.

Huang, J.W., Blaylock, M.J., Kapulnik, Y., and Ensley, B.D., 1998, Phytoremediation of uranium-contaminated soils: role of organic acids in triggering uranium hyperaccumulation in plants. *Environmental Science and Technology* 32: 2004–2008.

Hun, T., 1998, Biosolids. The results are in; long-term study confirms agricultural value. *Water Environmental Technology* 10(9): 38, 40, 42 (3 pages).

Isaac, R.A. and Boothroyd, U., 1996, Beneficial use of biosolids: progress in controlling metals. *Water Science and Technology* 34: 493–497.

Jungmann, J., Reins, H.A., Schobert, C., and Jentsch, S., 1993, Resistance to cadmium mediated by ubiquitin-dependent proteolysis. *Nature* 361: 369–371.

Juste, C. and Mench, M., 1992, Long-term application of sewage sludge and its effects on metal uptake by crops. In: Adriano, D.C. (Ed.), *Biogeochemistry of Trace Metals*. Lewis Publishers, Boca Raton, Florida: 159–193.

Kedziorek, M.A.M., Dupuy, A., Bourg, A.C.M., and Compère, F., 1998, Leaching of Cd and Pb from a polluted soil during the percolation of EDTA: Laboratory column experiments modelled with a non-equilibrium solubilization step. *Environmental Science and Technology* 32: 1609–1614.

Kessler, E. and Kirkham, M.B., 1985, Restoration of eroded prairie with digested sludge. *Proceedings of the Oklahoma Academy Sciences* 65: 25–34.

Kirkham, M.B., 1975, Trace elements in corn grown on long-term sludge disposal site. *Environmental Science and Technology* 9: 765–768.

Kirkham, M.B., 1976, Correspondence on the Note, Trace Elements in Corn Grown on Long-Term Sludge Disposal Site. *Environmental Science and Technology* 10: 285.

Kirkham, M.B., 1977, Trace elements in sludge on land: Effect on plants, soils, and ground water. In: Loehr, R.C. (Ed.), *Land as a Waste Management Alternative*. Ann Arbor Science, Ann Arbor, Michigan: 209–247.

Kirkham, M.B., 1979a, Sludge disposal. In: R.W. Fairbridge and C.W. Finkl, Jr. (eds) *Encyclopedia of Earth Sciences Series, Vol. 12. The Encyclopedia of Soil Science, Part 1. Physics, Chemistry, Biology, Fertility, and Technology*. Dowden, Hutchinson and Ross, Stroudsburg, Pennsylvania: 429–33.

Kirkham, M.B., 1979b, Effect of FeEDDHA on the water relations of wheat. *Journal of Plant Nutrition* 1: 417–424.

Kirkham, M.B., 1982, Agricultural use of phosphorus in sewage sludge. *Advances in Agronomy* 35: 129–163.

Kirkham, M.B., 2000, EDTA-facilitated phytoremediation of soil with heavy metals from sewage sludge. *International Journal of Phytoremediation* 2:159–172.

Kirkham, M.B. and Corey, J.C., 1977, Pollen as indicator of radionuclide pollution. *Journal of Nuclear Agriculture and Biology* 6: 71–74.

Klewicki, J.K. and Morgan, J.J., 1998, Kinetic behavior of Mn(III) complexes of pyrophosphate, EDTA, and citrate. *Environmental Science and Technology* 32: 2916–2922.

Krämer, U., Cotter-Nowells, J.D., Charnock, J.M., Baker A.J.M., and Smith, J.A.C., 1996, Free histidine as a metal chelator in plants that accumulate nickel. *Nature* 379: 635–638.

Kuchenrither, R. and Hite, R., 1999, The national biosolids partnership invests in sustainability. *Water Environment and Technology* 11(5): 6 & 8.

Lasat, M.M., Norvell, W.A., and Kochian, L.V., 1997, Potential for phytoextraction of ^{137}Cs from a contaminated soil. *Plant Soil* 195: 99–106.

Lasat, M.M., Baker, A.J.M., and Kochian, L.V., 1998, Altered Zn compartmentation in the root symplasm and stimulated Zn absorption into the leaf as mechanisms involved in Zn hyperaccumulation in *Thlaspi caerulescens*. *Plant Physiology* 118: 875–883.

Leeper, G.W., 1978, *Managing the Heavy Metals on the Land*. Marcel Dekker, New York. p. 121.

Leeson, A. and Alleman, B.C. (Eds), 1999, *Bioremediation of Metals and Inorganic Compounds*. Battelle Press, Columbus, Ohio, and Richland, Washington. p. 190.

Logan, T.J. and Harrison, B.J., 1995, Physical characteristics of alkaline stabilized sewage sludge (N-viro soil) and their effects on soil physical properties. *Journal of Environmental Quality* 24: 153–164.

Logan, B.E. and Rittmann, B.E., 1998, Finding solutions for tough environmental problems. *Environmental Science and Technology* 32: 502A–507A.

Lothenbach, B., Krebs, R., Furrer, G., Gupta, S.K., and Schulin, R., 1998, Immobilization of cadmium and zinc in soil by Al- montmorillonite and gravel sludge. *European Journal of Environmental Quality* 49: 141–148.

Lothenbach, B., Furrer, G., Schärli, H., and Schulin, R., 1999, Immobilization of zinc and cadmium by montmorillonite compounds: Effects of aging and subsequent acificiation. *Environmental Science and* Technology 33: 2945–2952.

Ma, Q.Y., Traina, S.J., Logan T.J., and Ryan, J.A., 1994, Effects of aqueous Al, Cd, Cu, Fe(II), Ni, and Zn on Pb immobilization by hydroxyapatite. *Environmental Science and Technology* 28: 1219–1228.

Madden, T.H., Datye, A.K., Fulton, M., Prairie, M.R., Majumdar, S.A., and Stange, B.M., 1997, Oxidation of metal-EDTA complexes by TiO_2 photocatalysis. *Environmental Science and Technology* 31: 3475–3481.

Maywald, F. and Weigel, H.J., 1997, Zur Biochemie und Molekularbiologie der Schwermetallaufnahme und -speicherung bei höheren Pflanzen. (Biochemistry and molecular biology of heavy metal accumulation in higher plants). *Landbauforschung Völkenrode* 47(3): 103–126. (In German with English abstract).

McArdel, C.S., Stone, A.T., and Tian, J., 1998, Reaction of EDTA and related aminocarboxylate chelating agents with $Co^{III}OOH$ (heterogenite) and $Mn^{III}OOH$ (manganite). *Environmental Science and Technology* 32: 2923–2930.

McBride, M.B., 1998a, Soluble trace elements in alkaline stabilized sludge products. *Journal of Environmental Quality* 27: 578–584.

McBride, M.B., 1998b, Growing food crops on sludge-amended soils: problems with the US Environmental Protection Agency method of estimating toxic metal transfer. *Environmental Toxicology and Chemistry* 17: 2274–2281.

McBride, M.B., Richards, B.K., Steenhuis, T., Russo, J.J., and Sauvé, S., 1997, Mobility and solubility of toxic metals and nutrients in soil fifteen years after sludge application. *Soil Science* 162: 487–500.

McBride, M.B., Richards, B.K., Steenhuis, T., and Spiers, G., 1999, Long-term leaching of trace elements in a heavily sludge-amended silty clay loam soil. *Soil Science* 164: 613–623.

Naidu, R., Kookuna, R.S., Oliver, D.P., Rogers, S., McLaughlin, M.J. (Eds), 1996, *Contaminants and the Soil Environment in the Australasia-Pacific Region*. Kluwer Academic Publishers, Dordrecht. p. 717.

Nanda Kumar, P.B.A., Dushenkov, V., Motto, H., and Raskin, I., 1995, Phytoextraction: The use of plants to remove heavy metals from soils. *Environmental Science and Technology* 29: 1232–1238.

Page, A.L., 1974, Fate and effects of trace elements in sewage sludge when applied to agricultural lands. EPA-670/2-74-005. U.S. Environmental Protection Agency, Cincinnati, Ohio. p. 98.

Page, A.L., Bingham, F.T., and Nelson, C., 1972, Cadmium absorption and growth of various plant species as influenced by solution cadmium concentration. *Journal of Environmental Quality* 1: 288–291.

Pearlstine, N., 1999, Six billion . . . and counting. *Time*. 18 October 1999, p. 67.

Perciasepe, R., 1998, Restoring and protecting US waters: An action plan with promise. *Water Environment and Technology* 10(5): 37–42.

Rakhshandehroo, G.R., Wallace, R.B., Boyd, S.A., and Voice, T.C., 1998, Hydraulic characteristics of organomodified soils for use in sorptive zone applications. *Soil Science Society of America Journal* 62: 5–12.

Richards, B.K., Steenhuis, T.S., Peverly, J.H., and McBride, M.B., 1998, Metal mobility at an old, heavily loaded sludge application site. *Environmental Pollution* 99: 365–377.

Robinson, B.H., 1997, The phytoextraction of heavy metals from metalliferous soils. Ph.D. thesis. Massey University, Palmerston North, New Zealand. p. 145.

Robinson, B.H., Chiarucci, A., Brooks, R.R., Petit, D., Kirkman, J.H., Gregg, P.E.H., and Dominicis, V. De., 1997a, The nickel hyperaccumulator plant *Alyssum bertolonii* as a potential agent for phytoremediation and phytomining of nickel. *Journal of Geochemical Exploration* 59: 75–86.

Robinson, B.H., Brooks, R.R., Howes, A.W., Kirkman, J.H., and Gregg, P.E.H., 1997b, The potential of the high-biomass nickel hyperaccumulator *Berkheya coddii* for phytoremediation and phytomining. *Journal of Geochemical Exploration* 60: 115–126.

Robinson, B.H., Leblanc, M., Petit, D., Brooks, R.R., Kirkman, J.H., and Gregg, P.E.H., 1998, The potential of *Thlaspi caerulescens* for phytoremediation of contaminated soils. *Plant Soil* 203: 47–56.

Robinson, B.H., Mills, T.M., Petit, D., Fung, L., Green, S.R., and Clothier, B.E., 1999a, Phytoremediation of cadmium-contaminated soils using poplar and willows. In: Wang, H. and Tomer, M. (Eds), *Function and Management of Plants in Land Treatment Systems*. NZ Land Treatment Collective, Tech. Session No. 19. Forest Research, Private Bag 3020, Rotorua, New Zealand, pp. 3–10.

Robinson, B.H., Mills, T.M., and Clothier, B.E., 1999b, Phytoremediation of heavy-metal contaminated soils. In: B.E. Clothier (Ed) *WISPAS. A Newsletter about Water in the Soil-*

Plant-Atmosphere System, No. 72. HortResearch, A Crown Research Institute, Palmerston North, New Zealand: 1–2.

Salt, D.E., Prince, R.C., Baker, A.J.M., Raskin, I., and Pickering, I.J., 1999, Zinc ligands in the metal hyperaccumulator *Thlaspi caerulescens* as determined using X-ray absorption spectroscopy. *Environmental Science and Technology* 33: 713–717.

Sauchelli, V., 1969, *Trace Elements in Agriculture*. Van Nostrand Reinhold Co., New York. p. 248.

Sauer, P.A. and Tyler, E.J., 1996, Heavy metal and volatile organic chemical removal and treatment in on-site wastewater infiltration systems. I. Catch basins and septic tanks. *Water, Air, and Soil Pollution* 89: 221–232.

Schmidt, C.W., 1999, A closer look at chemical exposures in children. *Environmental Science and Technology* 33: 72A–75A.

Schmidt, J.P., 1997, Understanding phytotoxicity thresholds for trace elements in land-applied sewage sludge. *Journal of Environmental Quality* 26: 4–10.

Sedlak, D.L., Gray, J.L., and Pinkston, K.E., 2000, Understanding micro contaminants in recycled water. *Environmental Science and Technology* 34: 509A–515A.

Sloan, J.J., Dowdy, R.H., and Dolan, M.S., 1998, Recovery of biosolids-applied heavy metals sixteen years after application. *Journal of Environmental Quality* 27: 1312–1317.

Sort, X. and Alcañiz, J.M., 1999, Modification of soil porosity after application of sewage sludge. *Soil and Tillage Research* 49: 337–345.

Steenhuis, T.S., McBride, M.B., Richards, B.K., and Harrison, E., 1999, Trace metal retention in the incorporation zone of land-applied sludge. *Environmental Science and Technology* 33: 1171–1174.

Sui, Y., Thompson, M.L., and Mize, C.W., 1999, Redistribution of biosolids-derived total phosphorus applied to a Mollisol. *Journal Environmental Quality* 28: 1068–1074.

Swerdlow, J.L., 1998, Population. *National Geographic* 194(4): 2–5 (October issue).

Temminghoff, E.J.M., van der Zee S.E.A.T.M., and de Haan F.A.M., 1998, Effects of dissolved organic matter on the mobility of copper in a contaminated sandy soil. *European Journal Soil Science* 49: 617–628.

Terry, N. and Bañuelos, G. (Eds), 2000, *Phytoremediation of Contaminated Soil and Water*. Lewis Publishers, Boca Raton, Florida. p. 389.

Thomas, P. (Ed.), 1996, By the way, doctor . . . *Harvard Health Letter* 21(11): 3. (Harvard Medical School Health Publications Group, 164 Longwood Ave., Boston, Massachusetts 02115, USA).

Towers, W. and Paterson, E., 1997, Sewage sludge application to land—A preliminary assessment of the sensitivity of Scottish soils to heavy metal inputs. *Soil Use and Management* 13: 149–155.

United States Environmental Protection Agency, 1992, Standards for the use and disposal of sewage sludge. 40 CFR Part 503. WWBKRG35. Final. 25 November, 1992. US Environmental Protection Agency, Washington, D.C. p. 69. (Available from National Small Flows Clearinghouse, West Virginia University, P.O. Box 6064, Morgantown, West Virginia 26506–6064).

United States Environmental Protection Agency, 1994, Land application of sewage sludge. EPA/831-B-93-002b. Office of Enforcement and Compliance Assurance. US Environmental Protection Agency, Washington, D.C. 4 chapters + 6 appendices (pages not numbered consecutively).

Vance, G.F. and Pierzynski, G.M., 1999, Bioavailability and fate of trace elements in long-term, residual-amended soil studies. In: W.W. Wenzel, D.C. Adriano, B. Alloway, H.E. Doner, C. Keller, N.W. Lepp, M. Mench, R. Naidu, and G.M. Pierzynski, (Eds) Proceedings of the Extended Abstracts from the 5th International Conference on the Biogeochemistry of Trace Elements, July 11-15, 1999, Vienna, Austria. International Society for Trace Element Research, PO Box 81, Vienna, Austria: 116–117.

Vassil, A.D., Kapulnik, Y., Raskin, I., and Salt, D.E., 1998, The role of EDTA in lead transport and accumulation by Indian mustard. *Plant Physiology* 117: 447–453.

Vogeler, I., Green, S.R., Clothier, B.E., Kirkham, M.B., and Robinson, B.H., 2001, In: Iskandar, I.K. and Kirkham, M.B. (Eds), *Trace Elements in Soil. Bioavailability, Flux, and Transfer*. Lewis Publishers, A Division of CRC Press, Boca Raton, Florida: 175–197.

Wang, Q., Shao, M., and Horton, R., 1999, Modified Green and Ampt models for layered soil infiltration and muddy water infiltration. *Soil Science*. 164: 445–453.

Watanabe, M.E., 1997, Phytoremediation on the brink of commercialization. *Environmental Science and Technology* 31: 182A–186A.

Weast, R.C., 1964, *Handbook of Chemistry and Physics*. The Chemical Rubber Co., Cleveland, Ohio. (Now the CRC Press, Boca Raton, Florida). 7 sections (pages not numbered sequentially).

Wilson, M., Hadden, C.T., and Gibson, M.C., 1992, Preliminary risk assessment for viruses in municipal sewage sludge applied to land. EPA/600/R-92/064. Environmental Criteria and Assessment Office. Office of Health and Environmental Assessment. Office of Research and Development. US Environmental Protection Agency, Cincinnati, Ohio. 9 chapters + appendix (pages not numbered consecutively).

Wilson, S.C. and Jones, K.C., 1999, Volatile organic compound losses from sewage sludge-amended soils. *Journal of Environmental Quality* 28: 1145–1153.

Wu, J., Hsu, F.C., and Cunningham, S.D., 1999, Chelate-assisted Pb phytoextraction: Pb availability, uptake, and translocation constraints. *Environmental Science and Technology* 33: 1898–1904.

Zanders, J.M., Hedley, M.J., Palmer, A.S., Tillman, R.W., and Lee, J., 1999, The source and distribution of cadmium in soils on a regularly fertilised hill-country farm. *Australian Journal of Soil Research* 37: 667–678.

Zhang, P., Ryan, J.A., and Yang, J., 1998, In vitro soil Pb solubility in the presence of hydroxyapatite. *Environmental Science and Technology* 32: 2763–2768.

Zhu, Y.L., Zayed, A.M., Qian, J.H., de Souza, M., and Terry, N., 1999, Phytoaccumulation of trace elements by wetland plants: II. Water hyacinth. *Journal Environmental Quality* 28: 339–344.

5

Heavy Metal Accumulation and Mobility in Soils Affected by Irrigating Agricultural Crops with Sewage Water

O. Wiger[1], *J. Hamedi*[1], *R.P. Narwal*[2], *B.R. Singh*[1*] and *M.S. Kuhad*[2]

INTRODUCTION

In different parts of the world, the menace of a rapidly-increasing population, the wanton growth of industries and increasing urbanization has created major problems with the disposal of sewage and industrial effluents. The disposal of wastes is a matter of serious concern because along with some essential plant nutrients, wastes also contain potentially toxic heavy metals such as Pb, Ni, Cd, and Cr. The subsequent entry of these metals into the food chain, through the crops grown on soils irrigated with sewage water, may cause disorders in animals and humans. The extent of metal uptake in agricultural crops is, in addition to the actual metal concentrations in the soil, also dependent on various physical and chemical soil properties.

Accumulation of heavy metals in sewage-irrigated soils, transferring into the agricultural crops grown on these soils, is well documented. A field experiment in the central region of Spain has shown that long-term application of solidified sewage sludge significantly increased the concentrations of Cu, Cr, Ni, Pb, Zn and Cd, both in the soil and in the agricultural crops (Walter et al., 1990).

In India, industrial effluents, which usually contain heavy metals and other harmful ingredients, are discharged into the sewer system (Azad et al., 1987: Gupta et al., 1994). Due to a shortage of sewage treatment plants, raw sewage, solid wastes, and industrial effluents are generally disposed of in nearby water-bodies, dumped in the open on roadsides, or applied to agricultural land. For the farmers, sewage water is a valuable source of irrigation and plant nutrients. Continuous application of

[1]Department of soil and water sciences, Agricultural University of Norway, P.O. Box 5028, N-1432 As, Norway.
[2]Department of soil science, CCS HAU, Hisar 125 004, India.
*Corresponding author

untreated sewage effluent and solid sludge on agricultural fields is shown to cause heavy metal accumulation in soils (Narwal et al.,1990; Kuhad et al., 1990).

Irrigation with sewage water often leads to considerable organic matter accumulation in the soil (Gupta et al., 1994; Azad et al., 1987). The varieties of soil in Haryana, India are naturally very low in organic carbon, and this accumulation may be valuable for the soil structure as also for resistance against erosion. However, high contents of suspended solids in the irrigation water may, in some instances, choke the soil micropores and thereby impede water infiltration. Hydraulic conductivity has been reported to approach zero after continuous application of sewage water (Narwal et al., 1988). Sewage waters are also carriers of different salts, and consequently, sewage-irrigated soils may have salinity and sodicity problems.

Haryana state generates about 165×10^6 L per day of sewer water. These waters are used for irrigating the crops around the cities. Thus, it has become imperative to study the effect of sewage irrigation on soil properties, and heavy metal accumulation and mobility.

The specific objectives of this study were to:

- investigate the heavy metal accumulation in sewage-irrigated soils around some of the major industrial cities in Haryana, and compare these soils with the soils irrigated by tubewell or canal water; and
- study the vertical movement of metals in polluted soils, and estimate the risk of groundwater contamination.

MATERIALS AND METHODS

Site Description

Soil samples were collected from nine different locations in the Haryana state of India: Sirsa, Kaithal, Panchkula, Rohtak, Gurgaon, Faridabad, Sonepat, Karnal and Jind. All the sites are located in the alluvial plain of the Indo-Gangetic region (27.5° to 30.9° N latitude and 74.5° to 77.6° E longitude). The climate of Haryana is continental and monsoonic with very hot summers and cool winters. Annual average rainfall in the state ranges from about 300 mm in the southwest to about 1200 mm in the northeast, and the rain is usually concentrated in the months of July, August, and September. Annual potential evapotranspiration is about 2000 mm, with moisture conditions ranging from semi-arid to sub-humid (Goyal and Kuhad, 1998). According to the USDA Soil Taxonomy system (Soil Survey Staff, 1998), the soils of this study area are classified as Typic Ustochrepts except at Panchkula (Typic Ustorthents) and at Sonipat (Typic Ustifluvents).

Soil Sampling

At each location, soil samples were collected from a field with a known history of irrigating with sewage water. At each location, one field with sewage water irrigation and one with canal or well water irrigation was selected for this study. The two selected fields were adjoining in order to avoid variations in soil and climatic factors. From each field, cylinders of soils were taken out from five different holes from 0–15, 15–30, 30–60 and 60–90 cm depths and mixed later according to depth to make a representative sample of each depth and location. The samples were collected with a soil auger. The same holes for each location were used for all depths. The final sample, a mixture of five holes, contained about 1 kg of composite soil. The soil samples were air-dried, ground and passed through a 2-mm sieve.

Physio-chemical Properties

Soil pH of the samples was determined in 1:2 soil:water suspension. Electrical conductivity (EC) was measured by a conductivity bridge in the supernatant liquid above the settled soil of the same suspension. The content of organic carbon (OC) in the soil samples was determined by the wet digestion method of Walkley and Black (1947). Cation exchange capacity (CEC) was determined by sodium acetate and ammonium acetate, a method described by Schollenberger and Simon (1945). Textural composition of the soil samples was determined by the hydrometer method. Hydrogen peroxide was used for digestion of organic matter, while sodium hexametaphosphate served as the dispersing agent.

Heavy Metals

To determine total metal concentrations, the soil samples were first digested with HNO_3 overnight, and then with HNO_3 and $HClO_4$ on a hot plate for several hours. The residue was finally dissolved in 2N HNO_3. For available contents of the metals, the DTPA method of Lindsay and Norvell (1978) was used. The concentration of metals in the digested or extracted solutions was measured by atomic absorption spectrophotometer (GBC 902).

RESULTS AND DISCUSSION

Characteristics of Sewage Water

The dry-matter content in the sewage water samples ranged from 0.7 to 2.5 g L^{-1}, with a mean value of 1.5 g L^{-1}. Organic carbon ranged from 12 to 90 mg L^{-1}, with a mean value of 40 mg L^{-1}. There was a significant positive correlation (r = 0.66, p = 0.05) between organic carbon and dry matter contents. pH of water samples ranged from 6.80 to 8.15, with a mean value of 7.15. These values are consistent with the given pH-values of typical sewage waters (Kovda et al., 1973).

Electrical conductivity ranged from 0.88 to 2.9 mS/cm. According to the classification by the US Salinity Laboratory, the sewage waters of Sirsa, Fatehabad, Panchkula, Gurgaon, Karnal, Jind and Hisar are highly saline (0.75–2.25 mS/cm), and the waters of Narwana, Kaithal, Rohtak, Faridabad and Sonepat are very highly saline (2.25–5.0 mS/cm).

Physico-chemical Properties

With a few exceptions, the soils were of sandy clay loam texture (Soil Taxonomy) with varying chemical properties (Table 5.1). The continuous use of sewage water as a source of irrigation has influenced the soil characteristics. The irrigation with sewage water had affected the soil pH at all locations, but the extent of change varied with the composition of sewage water and the duration of irrigation. Sewage water irrigation increased the soil pH at Kaithal, Panchkula, Gurgaon and Karnal, but decreased at other locations. The highest increase of soil pH in the surface was observed at Gurgaon (from 7.4 to 9.0), and maximum reduction was at Sirsa (from 8.5 to 7.2) after 20 years of sewage irrigation (Table 5.2). The effect of sewage water irrigation on pH was also observed in the subsurface horizons of all profiles. No significant differences between depths were found, but at most places, pH slightly increased downward in the profile.

In general, the differences in the salt content (EC) of soils at all depths were not statistically significant (Table 5.1). At Gurgaon, EC of the surface soil increased by ten times (0.19 to 1.90 mS/cm) as a result of sewage irrigation for 20 years (Table 5.2). Due to high salt accumulation, these once productive soils have become almost unproductive. These results suggest that high EC of sewage water may limit its disposal on agricultural land.

The content of organic matter was not significantly increased by sewage irrigation. However, at Sirsa, 20 years of sewage water irrigation has increased the surface content of organic carbon from 0.64 to 2.54 g/100 g soil. At Sonepat, Jind, Rohtak and Karnal, organic matter had accumulated to some extent, while the inverse relationship was evident at the other sites.

Zinc

In the surface layer, total concentrations of Zn were 70–370 mg/kg in control soils, and 70–1000 mg/kg in soils irrigated with sewage water

Table 5.1: Effect of sewage irrigation on some physico-chemical properties of the soils (ranges and mean values for all profiles)

Profile depth (cm)	*pH*				*EC (mS cm^{-1})*				*OC (g/100 g)*				*CEC (cmol$_c$ kg^{-1})*				*Clay (%)*			
	Control soils		*Sewage irr. soils*		*Control soils*		*Sewage irr. soils*		*Control soils*		*Sewage irr. soils*		*Control soils*		*Sewage irr. soils*		*Control soils*		*Sewage irr. soils*	
	Range	*Mean*	*Range*	*Mean*	*Range*	*Mean*	*Range*	*Mean*	*Range*	*Mean*	*Range*	*Mean*	*Range*	*Mean*	*Range*	*Mean*	*Range*	*Mean*	*Range*	*Mean*
0–15	7.4–8.7	8.2	7.2–9.0	8.1	0.14–1.6	0.50	0.16–1.9	0.66	0.60–1.27	0.88	0.61–2.54	1.16	6.1–15.3	10.2	9.3–14.0	11.7	16–31	23	19–28	24
15–30	7.7–8.9	8.3	7.3–9.2	8.2	0.08–1.3	0.36	0.15–2.1	0.66	0.40–1.09	0.61	0.26–2.09	0.75	7.0–18.3	9.5	8.5–15.3	11.3	17–35	24	19–39	25
30–60	8.2–9.2	8.5	7.4–9.1	8.4	0.12–1.3	0.35	0.12–2.4	0.69	0.18–0.41	0.27	0.20–1.26	0.38	7.3–16.6	10.7	7.3–17.6	10.5	21–32	27	21–51	29
60–90	8.0–9.2	8.4	7.8–9.1	8.3	0.12–1.0	0.41	0.13–3.8	0.80	0.13–0.33	0.18	0.12–0.51	0.24	7.5–17.9	12.1	7.8–17.0	11.1	21–42	30	19–61	30

Table 5.2: Physico-chemical properties of control and sewage water irrigated surface soils (0–15 cm) at different sites

Location	*Control/Duration of sewage irr.*	*pH*	*EC ($mS\ cm^{-1}$)*	*OC (g/100 g)*	*CEC ($cmol_c\ kg^{-1}$)*	*Clay (%)*
Sirsa	Control	8.5	0.25	0.64	7.0	23
	20 years	7.2	0.95	2.54	14.0	22
Kaithal	Control	8.2	1.3	1.09	12.5	29
	20 years	9.0	0.89	0.88	12.3	26
Panchkula	Control	7.8	0.23	1.27	13.4	31
	10 years	8.0	0.16	0.67	12.3	24
Rohtak	Control	8.7	1.6	0.60	6.1	21
	18 years	8.0	0.42	0.98	9.8	28
Gurgaon	Control	7.4	0.19	0.77	15.3	29
	20 years	9.0	1.9	0.65	11.0	25
Faridabad	Control	8.4	0.29	0.67	8.0	16
	7 years	8.5	0.40	0.61	9.3	24
Sonepat	Control	8.1	0.45	0.94	9.8	17
	20 years	7.5	0.67	1.59	12.0	19
Karnal	Control	8.2	0.14	1.01	9.5	19
	15 years	8.5	0.21	1.17	11.5	20
Jind	Control	8.2	0.22	0.94	9.8	21
	10 years	7.9	0.32	1.37	13.0	27

(Table 5.3). The corresponding ranges of DTPA-extractable concentrations were 2.0–60 mg/kg and 3.8–200 mg/kg, in control and sewage-irrigated soils, respectively. The control levels of total Zn are substantially higher than the mean background concentration of 26 mg/kg, as reported by Kuhad et al. (1990). The DTPA-extractable concentrations also exceed the values obtained earlier by Narwal et al. (1988) in both control and sewage-irrigated soils. Singh and Kansal (1983) and Sharma and Kansal (1986) have also reported that the use of sewage water for irrigation substantially increased the accumulation of Zn, Cu, Pb and Cd in soils of different cities of Punjab. The accumulation was higher in soils receiving sewage water from industrial cities. Sharma and Kansal (1986) found the concentration of Zn, Fe, Cu, Mn and Cd to be 13.4, 75.3, 24.4, 33.2 and 0.119 mg/kg soil in Budha Nulla (this *nulla* carries sewer water and other wastes of Ludhiana city) sewage-irrigated soils as compared to 4.2, 38.0, 7.8, 28.4 and 0.074 mg/kg, respectively, in soils irrigated with normal water.

At Sonepat, Sirsa and Jind, both total and DTPA-extractable concentrations had increased distinctly due to sewage water irrigation (Table 5.3 and Figure 5.1). The situation was especially severe at Sonepat, where the total concentration of Zn had reached 1000 mg/kg in the sewage irrigated soil (Table 5.3). But statistically, the concentrations of neither total nor DTPA-extractable Zn were significantly higher in the sewage treated

Table 5.3: Effect of sewage irrigation on total soil concentrations of Zn, Cu and Ni

Profile depth (cm)	*Total Zn*		*Total Cu*		*Total Ni*	
	Control	*Sewage irr.*	*Control*	*Sewage irr.*	*Control*	*Sewage irr.*
			mg/kg			
			Sonepat			
0–15	370	1000	118	220	59	74
15–30	245	700	70	225	20	99
30–60	95	700	33	130	20	121
60–90	70	295	15	63	15	63
			Sirsa			
0–15	95	320	18	118	21	23
15–30	95	320	13	85	24	17
30–60	95	95	18	43	18	17
60–90	95	95	18	25	15	35
			Jind			
0–15	70	270	13	118	21	11
15–30	95	95	21	28	32	14
30–60	95	70	20	18	25	33
60–90	95	70	22	18	26	27
			Rohtak			
0–15	95	120	20	37	14	25
15–30	120	95	21	23	12	14
30–60	70	95	21	23	19	18
60–90	70	95	21	26	23	8
			Kaithal			
0–15	95	195	20	53	12	15
15–30	120	120	20	35	8	28
30–60	95	145	18	35	10	31
60–90	95	145	20	35	14	28
			Karnal			
0–15	120	145	43	43	18	13
15–30	120	120	48	40	24	8
30–60	95	95	30	28	17	8
60–90	95	95	15	28	13	30
			Gurgaon			
0–15	95	95	20	19	26	31
15–30	95	95	20	21	31	28
30–60	70	95	19	22	26	22
60–90	70	70	17	18	17	22
			Faridabad			
0–15	120	120	26	25	11	9
15–30	95	95	26	18	12	13
30–60	70	70	20	17	12	8
60–90	70	95	19	17	14	19
			Panchkula			
0–15	245	145	23	10	20	17
15–30	245	95	23	18	31	20
30–60	95	95	17	8	38	9
60–90	95	70	16	15	29	26

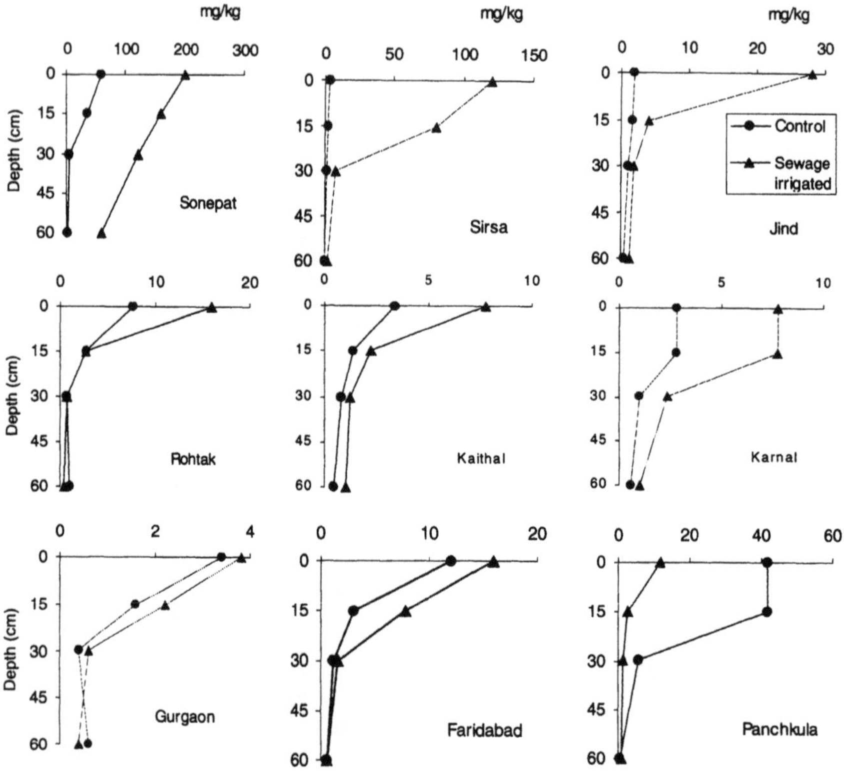

Fig. 5.1: Distribution of DTPA-extractable Zn in the soil profile at different locations

soils than in the corresponding control soils. At Panchkula and Faridabad, the sewage water irrigated fields contained less Zn than the control fields (Table 5.3 and Figure 5.1). Earlier applications of sewage water, unknown to the present farmer, may be an explanation for the latter observations.

The ratio of DTPA-extractable Zn to total concentrations generally increased with increasing total levels in the soil. Though not statistically significant, this correlation suggests that zinc, when added through sewage water, or perhaps other external sources, occurs in a more soluble and mobile form than the native Zn in soils. The availability to plants of the metals in sewage water depends on several soil factors apart from organic matter content. It has been shown that in alkaline soils, organic matter may enhance the intrinsically poor availability of metals. This is in contrast to the effect of organic matter in acid soils (Brummer and Herms, 1983), where organic matter is known to immobilize the metals. Decomposition of complex organic substances into simpler molecules with chelating properties may also gradually increase the metal availability.

Both in control and sewage water irrigated soils, total ($p = 0.05$, LSD = 34 mg/kg) and DTPA-extractable ($p = 0.05$, LSD = 12 mg/kg) concentrations significantly decreased from the surface to the layers 30–60 cm and 60–90 cm. The accumulation of Zn in the surface of control soils can be explained by regular application of inorganic Zn as fertilizer. However, these observations contradict the conclusion of Kuhad et al. (1990) who stated that zinc concentrations were lower in the Ap-horizon than the C-horizon.

Copper

Total Cu concentrations in the surface layer were 13–118 mg/kg in control soils, and 10–220 mg/kg in sewage-irrigated soils (Table 5.3). DTPA-extractable concentrations were 1.0–35 mg/kg, and 2.7–114 mg/kg, in control soils, and sewage-irrigated soils, respectively. In contrast to Zn, all these values are lower compared to the results of Kuhad et al. (1990).

Total Cu concentrations were significantly increased ($p = 0.05$, LSD = 9.0 mg/kg) by sewage water irrigation. Also, DTPA-extractable Cu concentrations were significantly higher ($p = 0.05$, LSD = 3.0 mg/kg) in the surface of soils receiving sewage water. The highest concentrations were found at Sonepat; but also at Sirsa and Jind, both DTPA extractable and total concentrations of Cu had increased sharply after sewage irrigation (Figure 5.2 and Table 5.3). The soils of Kaithal and Rohtak were moderately affected, while minor or no differences were found at the remaining sites.

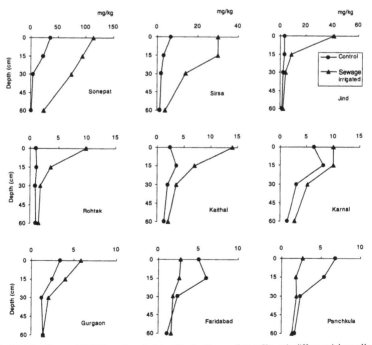

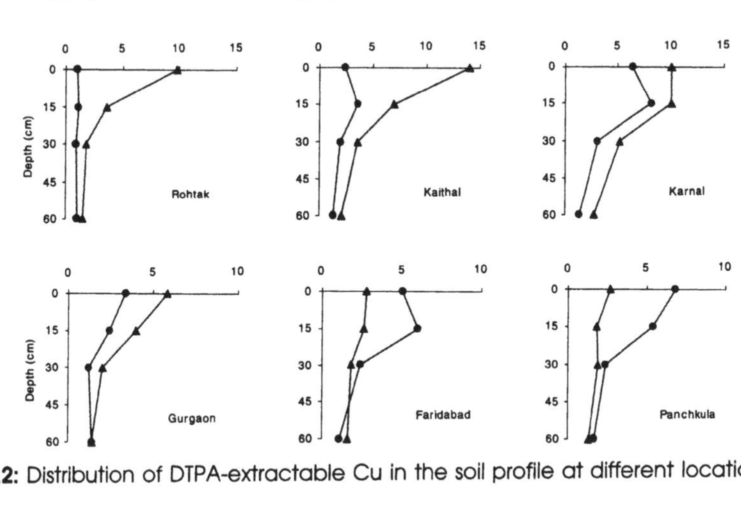

Fig. 5.2: Distribution of DTPA-extractable Cu in the soil profile at different locations

Also for Cu, the DTPA-extractable/total ratio tends to be higher in the most polluted soils, but the relationship is not as distinct as for zinc. However, in the sewage-irrigated soil of Sonepat, about half of the total Cu is extractable by DTPA.

Total Cu is evenly distributed throughout the profile of control soils. In sewage-treated soils, Cu seems to have accumulated more in the surface than in deeper layers, but the differences between depths are not statistically significant. DTPA-extractable Cu is, however, significantly higher in the surface than in the two lower depths ($p = 0.05$, LSD = 4.4 mg/kg). This is most evident in sewage-irrigated profiles, as well as in control soils.

Lead

The results obtained for total concentrations of Pb were not considered reliable. Therefore, only the values for the DTPA-extractable fraction are presented.

Surface concentrations of DTPA-extractable Pb were 1.6–12.4 mg/kg in control soils, and 2.7–14.1 mg/kg in sewage water-irrigated soils. The increase of DTPA-extractable Pb as a result of sewage water irrigation was statistically significant ($p = 0.05$, LSD = 1.0 mg/kg). As for Zn and Cu, the highest Pb concentration was found in the sewage-irrigated profile of Sonepat (Figure 5.3). Also at Sirsa and Kaithal, the increase of DTPA-extractable Pb due to sewage water irrigation was very distinct. At Gurgaon, Karnal and Jind, the increase was less pronounced, while no major differences were observed at Panchkula, Rohtak and Faridabad (Figure 5.3).

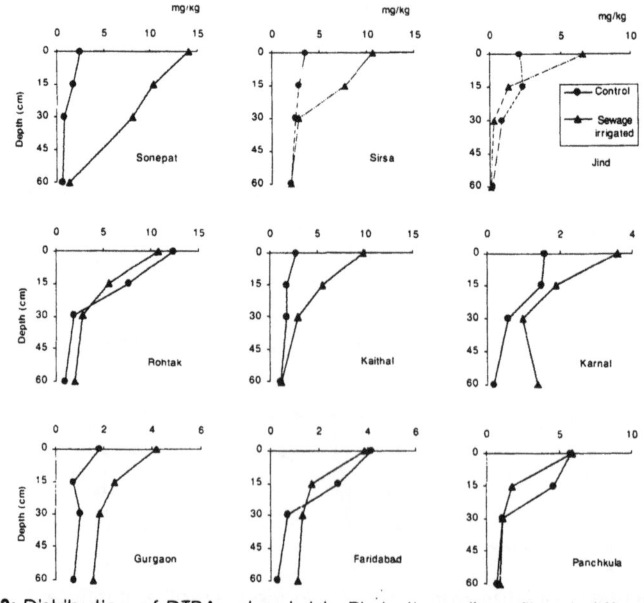

Fig. 5.3: Distribution of DTPA-extractable Pb in the soil profile at different locations

The DTPA-extractable concentrations of Pb sharply decreased downwards in all the profiles, of both sewage water irrigated and control soils (Figure 5.3). The differences were significant between all pairs of depths from the surface to the 30–60 cm layer ($p = 0.05$, LSD = 1.4 mg/kg). The clear surface accumulation of mobile lead in control profiles also may be caused by direct deposition from vehicles. Many of the investigated sites were close to roads with various traffic loads. The control sites of Panchkula and Rohtak, where high extractable Pb concentrations were found, had adjacent roads with considerable traffic.

Other Metals

Sewage water irrigation has affected the level of total Ni only at Sonepat (Table 5.3), where the Ni concentration increased, especially in the lower depths. The concentrations of DTPA-extractable Ni were also higher at Sonepat compared to the other sites, in the sewage-irrigated soil as well as in the control field. However, in most of the profiles, DTPA-extractable Ni was close to the the detection limit.

Soil concentrations of total Fe were not influenced by sewage water irrigation at any site. But the DTPA-extractable concentrations had markedly increased at Sirsa, Rohtak and Karnal, while the highest extractable level was found in the control field at Gurgaon. Only minor effects were found at the other sites.

Irrigation with sewage water did not influence the soil concentrations of Mn and Co. Neither total nor DTPA-extractable levels of these two metals were increased at any of the sites.

Total and DTPA-extractable concentrations of Cd were below the detection limit in almost all the soil samples.

Metal Accumulation and Leaching

The sewage-irrigated soil at Sonepat had by far the highest level of Zn, Cu and Ni of all the studied profiles (Table 5.3). However, metal concentrations were also high in the control profile, which may reflect high background values at this site. On the contrary, high proportions of DTPA-extractable metals in the control profile may indicate former applications of sewage water to that field, unknown to the present farmer. High metal levels throughout the whole profile are evidences of extensive leaching. This leaching was probably enhanced by the coarse-textured sandy loam soil. High mobile concentrations of Zn (Figure 5.1), Cu (Figure 5.2) as well as Ni in the 60–90 cm layer indicate a significant risk of groundwater contamination.

The most distinct effect of sewage water irrigation was found at Sirsa. Total concentrations (Table 5.3), and especially DTPA-extractable concentrations of Zn, Cu and Pb, increased substantially at this site (Figures

5.1, 5.2 and 5.3). Leaching of metals from the surface layer appears to be limited to about 30 cm depth. This is evident from the distribution of DTPA-extractable Zn (Figure 5.1) and Cu (Figure 5.2). The light textured soil (75% sand) in Sirsa may explain the leaching tendencies, but the risk of groundwater contamination is probably still low, because the mobility of metals was limited only to the upper 30 cm soil depth.

At Jind, Zn, Cu and Pb have accumulated in the surface of the sewage water irrigated profile, but the vertical mobility seems to be restricted (Figures 5.1, 5.2 and 5.3). A reasonable explanation may be the relatively high clay content (27%) found in this profile. Small amounts of extractable metals in the deepest layer indicate that the risk of groundwater contamination at Jind is low.

The sewage-irrigated field at Rohtak showed slight increases of Zn and Cu (Figures 5.1 and 5.2), while DTPA-extractable Pb was high in both profiles (Figure 5.3). Downward mobility of metals seems to be quite restricted inspite of very low CEC values in this soil.

At Kaithal and Karnal, the DTPA-extractable concentrations of several metals were slightly increased in the sewage-irrigated field. Total concentrations were not affected, and at these two sites, the leaching of the mobile metal fractions was generally restricted to the upper 30 cm.

At Gurgaon and Faridabad, sewage water irrigation had obviously not affected soil concentrations of heavy metals. Only at Gurgaon were the DTPA-extractable Cu and Pb slightly increased (Figures 5.2 and 5.3). At Panchkula, metal concentrations were generally lower in the sewage-irrigated profile than in the control field.

Relationship between Soil Properties and Metal Concentrations

All soil samples were grouped according to textural class (USDA) and clay content. Analyses of variance showed that total concentrations of Fe, Mn and Co were significantly higher ($p = 0.05$) in the textural classes clay, clay loam and sandy clay than in the classes sandy clay loam and sandy loam. Likewise, the same relationships were evident for samples with more than 30% clay compared to samples with less than 30% clay. Compared to sand and silt, the parent material of clays consist of more easily-weathering basic minerals with generally high metal concentrations. In sandy soils, stronger leaching also contributes to lower metal concentrations. However, the variations of Zn, Cu, and Ni could not be explained by texture alone if the heavily polluted and light-textured profile of Sonepat was excluded.

The value of this analysis is obviously limited by the variable number of samples in the different textural groups. Sandy clay loam is the dominating soil type, while heavier textural classes like sandy clay, clay loam and clay are poorly represented.

High concentrations of all metal fractions, except Mn and total Fe, were generally associated with high contents of organic carbon. The correlations between metal concentrations and organic carbon were significant ($p = 0.05$) for total Zn ($r = 0.52$), DTPA-extractable Zn ($r = 0.68$), total Cu ($r = 0.64$), DTPA-extractable Cu ($r = 0.56$), DTPA-extractable Fe ($r = 0.38$), DTPA-extractable Co ($r = 0.37$) and DTPA-extractable Pb ($r = 0.47$). The strong correlations between total Zn, Cu and organic carbon reflect the fact that organic matter was added along with these metals at the polluted sites. High contents of organic matter in the soil itself cannot be the direct cause of increased total metal concentrations. Total concentrations of Pb, Fe and Co were not correlated with organic carbon, in contrast to the DTPA-extractable concentrations of the same metals. This indicates that organic substances may have increased the solubility of Pb, Fe and Co.

Regression analyses also showed that both total and DTPA-extractable metal concentrations generally increased as pH decreased. Concentrations of DTPA-extractable Fe ($r = -0.64$), DTPA-extractable Zn ($r = -0.48$), and total Cu ($r = -0.37$) were significantly negatively correlated ($p = 0.05$) with the pH-value. The increased extractability at lower pH is generally expected because of less fixation to oxides and dissolution of hydroxides. However, the relationship between high total metal concentrations and low soil pH indicates that sewage water has lowered the soil pH in the most polluted profiles. Breakdown of organic matter applied with the sewage water creates acidity by release of carbon dioxide, organic acids, and strong inorganic acids like H_2SO_4 and HNO_3 (Brady and Weil, 1996). The lowered pH observed in metal polluted soils is, therefore, probably an effect of extended organic matter decomposition, rather than the direct cause of high metal levels.

Soil pH is normally one of the most important factors determining the availability and plant uptake of metals in the field. The DTPA-solution is buffered at pH 7.3, and during soil extraction, soil pH will change to this value. Hence, the original pH in the soil will have only minor influence on the extracted amounts. This serious modification of the soil environment clearly restricts the value of the results from a DTPA-extraction. An unbuffered extracting solution like NH_4NO_3 does not alter the soil pH (Mellum et al., 1997), and may in some cases be a better predictor of plant uptake under field conditions.

Inter-relationships Among Metal Fractions

Inter-relationships among the different metals and metal fractions were analysed by linear regression. The analysis was based on all soil samples, irrespective of depth or irrigation practice. The results are shown in Table 5.4.

Table 5.4: Correlation matrix with regression coefficients (r) for all pairs of metal combinations.

	t. Fe	*t. Mn*	*t. Zn*	*t. Cu*	*t. Co*	*t. Ni*	*e. Fe*	*e. Mn*	*e. Zn*	*e. Cu*	*e. Co*	*e. Pb*
t. Mn	0.43*											
t. Zn	0.01	-0.11										
t. Cu	-0.02	-0.13	0.93*									
t. Co	0.02	0.07	0.03	0.03								
t. Ni	0.06	-0.05	0.77*	0.70*	0.14							
e. Fe	-0.04	-0.03	0.48*	0.37*	-0.05	0.35*						
e. Mn	-0.18	-0.04	-0.04	-0.09	-0.17	-0.01	0.22*					
e. Zn	-0.07	-0.12	0.96*	0.92*	0.02	0.72*	0.48*	-0.01				
e. Cu	-0.05	-0.15	0.97*	0.96*	0.02	0.75*	0.45*	-0.04	0.94*			
e. Co	-0.22*	-0.03	0.08	0.12	0.16	0.10	0.19	0.37*	0.18	0.09		
e. Pb	0.03	-0.07	0.65*	0.63*	-0.13	0.35*	0.40*	0.05	0.67*	0.63*	0.12	
e. Ni	0.01	-0.06	0.71*	0.76*	0.04	0.76*	0.18	-0.05	0.72*	0.76*	0.02	0.38*

t denotes total concentrations, e denotes DTPA-extractable concentrations. An asterisk (*) denotes that the correlation is significant at $p = 0.05$.

Total and DTPA-extractable concentrations of Zn, Cu and Ni were highly correlated (Table 5.4). This means that high pollution levels of these metals were normally associated with increased mobile fractions. For Fe, Mn and Co, no such significant correlations were found between total and DTPA-extractable concentrations (Table 5.4). The extractable concentrations of these metals seemed to be more dependent on other soil properties rather than on total contents.

There were significant positive correlations for all of the paired combinations of both total and DTPA-extractable Zn, Cu and Ni (Table 5.4). The concentrations of DTPA-extractable Fe and Pb were also generally correlated with the concentrations of Zn, Cu and Ni. This obviously means that contaminations of Zn, Cu, Ni, Pb and Fe are closely interconnected in the investigated profiles. It can, therefore, be concluded that these metals have not been added separately to the soil, but rather have been added together as a mixture of metals.

The DTPA-extractable concentrations of Mn and Co were significantly correlated, but the occurrence of these metals was not correlated with the concentrations of Zn, Cu, Ni, Pb and Fe (Table 5.4). The correlation between extractable Mn and Co can be explained by the strong adsorption of Co on Mn-oxides. Poor extractability of Mn generally indicates that a large fraction of this metal occur as oxides. Consequently, a high fraction of Co is adsorbed and made immobile.

The effect of sewage water on the concentrations of Mn and Co in soils was found to be minor or completely absent. This correlation analysis also shows that increased soil concentrations of Zn, Cu, Ni and Pb were generally not accompanied by corresponding increase of Mn and Co.

CONCLUSIONS

This study clarifies that nothing general can be said about the status of sewage-irrigated soils in Haryana. Impacts of sewage water on metal accumulation and different physico-chemical soil parameters are highly variable from site to site. In addition to heavy metal loadings, sewage water may have other degrading effects on soil fertility. This was clearly demonstrated at Gurgaon, where good agricultural fields had been transformed to infertile land.

Copper, Zn, Pb, and at a few places Ni, were the metals with the highest concentrations in polluted soils relative to background levels. It was evident that these metals occurred together in contaminated soils. The vertical mobility of metals was generally restricted at most places. However, a clear exception was seen in the sewage-irrigated profile at Sonipat, where extensive leaching may have occurred.

The best way of sewage disposal is application on agricultural fields. Sewage water is a valuable resource for the farmer, both as a carrier of macronutrients and as a source of irrigation water. For many farmers, sewage water is the only source of irrigation, if alternative canal water or suitable tube well water is not available. The benefits of sewage water should not be destroyed by heavy metal hazards. Therefore, the metal emissions should be reduced at the initial point inside the factories. Eventually, new technology should be developed in order to extract metals from the effluents and recycle them in the production process.

REFERENCES

Azad, A.S., Arora, B.R., Singh, B. and Sekhon, G.S., 1987, Effect of sewage wastewaters on some soil properties. *Indian Journal of Ecology* 14: 7–13.

Brady, N.C. and Weil, R.R., 1996, *The Nature and Properties of Soils* (11th ed). Prentice-Hall, New Jersey, US: 307–326.

Brummer, G. and Herms, U., 1983, Influence of soil reaction and organic matter on the solubility of heavy metals in soils. In: Ulrich, B. and Pankrath, J. (Eds), *Effects of Accumulation of Air Pollutants in Forest Ecosystems*. D. Reidel Publishing Co, Dordrecht: 233–343.

Goyal, V.P. and Kuhad, M.S., 1998, Land resources and their use in Haryana. National Seminar on Developments in Soil Science & 63rd Annual Convention of the Indian Society of Soil Science, 16–19 November 1998: 28–34.

Gupta, A.P., Narwal, R.P., Antil, R.S., Singh, A. and Poonia, S.R., 1994, Impact of sewage water irrigation on soil health. In: Behl, A. and Tauro, P. (Eds), *Impact of Modern Agriculture on Environment*. CCS HAU Hisar & Max Mueller Bhawan, New Delhi: 109–117.

Kovda, V.A., van den Berg, C. and Hagan, R.M. (Eds), 1973, *Irrigation, Drainage and Salinity. An International Source Book*. FAO/UNESCO. Unesco, Paris and Hutchinson & Co, London: 177–205

Kuhad, M.S., Malik, R.S., Singh, A. and Karwasra, S.P.S., 1990, Background levels of heavy metals and extent of contamination in agricultural soils of Haryana. *Proceeding Environmental Pollution*: 101–107.

Lindsay, W.L. and Norvell, W.A., 1978, Development of a DTPA soil test for zinc, iron, manganese and copper in soils. *Soil Science Soceity of America Journal* 42: 421–428.

Mellum, H.K., Arnesen, A.K. and Singh, B.R., 1997, Extractability and plant up take of heavy metals is Alum Shale Soils. *Communications in soil science and plant Analysis* 29: 1183–1198.

Narwal, R.P., Gupta, A.P., Singh, A. and Karwasra, S.P.S., 1990, Pollution potential of some sewer waters of Haryana. In: *Recent Advances in Environmental Pollution and Management*. Haryana Pollution Control Board and Haryana Agricultural University, Hisar: 121–126.

Narwal, R.P., Singh, M. and, Gupta, A.P., 1988, Effect of different sources of irrigation on the physico-chemical properties of soil. *Indian Journal of Environment and Agriculture*. 3: 27–34.

Schollenberger, C.J. and Simon, R.H., 1945. Determination of exchange capacity and exchangable bases in soil-ammonium acetate method. *Soil Sciences* 59: 13–24.

Sharma, V.K. and Kansal, B.D., 1986, Heavy metal contamination of soils and plants with sewage irrigation. *Pollution Research*. 4: 86–91.

Singh, J. and Kansal, B.D., 1983, Accumulation of heavy metals in soils receiving municipal wastewater and effect of soil properties on their availability. In: CPC Consultants Ltd. *Heavy Metals in the Environment*. Edinburgh, UK: 409–412.

Soil Survey Staff, 1998, *Keys to Soil Taxonomy*. (8th ed). United States Department of Agriculture, Natural Resources Conservation Service, 326 pp.

Walkley, A. and Black, I.A., 1947, Chromic acid titration method for determination of soil organic matter. *Soil Sciences*. 63: 251–255.

Walter, I., Miralles, R., Funes, E., Gorospe, M.J. and Bigeriego, M., 1990, Effect of sewage sludge used as fertilizer in the central region of Spain. In: L'Hermite, P. (ed.), *Treatment and Use of Sewage Sludge and Liquid Agricultural Wastes*. Proceedings of a symposium, Athens, October 1990. Elsevier Applied Science, London and New York: 304–309.

6

Heavy Metals Availability at Industrially Contaminated Soils in NSW, Australia

Ya Lang Tam and *Balwant Singh*

INTRODUCTION

In the last few decades of the previous century, public awareness of the potential harmful effects of heavy metals to humans and the environment has increased (Mellor and Bevan, 1999). This is no more apparent than in the Southeast Asia region, where many incidences of poisoning from inadvertent ingestion of metals have been recorded. Metals enter the environment through several anthropogenic activities. In the urban areas, industrial processes are one of the major sources of metals. Several studies have demonstrated metal contamination from various industrial activities such as mining, smelting and manufacturing (Beavington, 1975; DeKoning, 1974; Cartwright et al., 1976; Tiller, 1992). There is pressure on the utilization of contaminated areas in the urban environment. Similarly, with increasing industrialization, it is important to determine the possible environmental threats posed by contaminants. Soil decontamination processes are often very expensive but may be suitable for high commercial value sites.

Contaminated areas often support typical plant species that may have adapted to elevated levels of metals through various mechanisms such as internal inactivation, cellular/sub-cellular compartmentalization, partial exclusion and reduced root-to-shoot translocation (Vogel-Lange and Wagner, 1990; Paliouris and Hutchinson, 1991). Over the last decade, there has been increasing interest and research in developing plant-based technology to remediate heavy metal contaminated soils. Several plant species have been shown to accumulate very high metal concentrations (Reeves and Brooks, 1983; McGrath et al., 1993). While the results from these and several other studies have been promising, some researchers have suggested that the slow growth and the small size of many plant

Faculty of Agriculture, Food and Natural Resources, The University of Sydney, Sydney, NSW 2006, Australia species may limit their utility for phytoremediation (Brown et al., 1995). Furthermore, many hyperaccumulators are yet to be identified and there is a need to gather more knowledge about such plant species (Kumar et al., 1995). In this chapter, we present data on the extent of heavy metal contamination at industrially-contaminated sites, soil-plant transfer coefficients for various heavy metals, and heavy metal accumulating plant species at various contaminated sites.

MATERIAL AND METHODS

Sites

Soil and plant samples were taken from three sites contaminated as a result of industrial activities in New South Wales, Australia. Two of the sites, Rhodes and Chiswick, are located in the metropolitan area of Sydney, whereas the third site, Boolaroo, is about 15 km southwest of Newcastle. Contamination at these sites has resulted from different industrial processes and all the sites have about 100 years of industrial history.

The Chiswick site is a former steelworks factory that produced wires, wire products and galvanizing products. Possible sources of contamination were from zinc residues burnt with coke in zinc oxides from the manufacturing processes of rubber and paint; and zinc dross, i.e. a byproduct of the galvanizing processes consisting of 95% Zn and 5% Fe and Zn pellets. Other contaminants may include lead oxides, carbonates, and hydroxides and copper sulphate from the wire-drawing processes. The Rhodes site is a former wastepaper mill and paint factory. Lead, from paint-based products, is believed to be the major metal contaminant. Anthropogenic waste such as broken clay crucible, brick, pigment material, charcoal and ashes from the paper mill buried in the subsoil were common at this site. The Boolaroo site is currently used for refining and smelting of metals including Pb, Zn, Cu and Cd. These metals are considered to be the major contaminants at this site.

Soil and Plant Sampling

A total of 90 soil samples and 83 plant samples were collected from the three sites. The soil samples were taken using a random stratified sampling technique, whereby, at each site, four plots were selected in a 10 m × 10 m measured area. Four random sampling points were selected in each plot and two soil samples were collected from each point, at depths of 0–30 cm (topsoil) and 30–60 cm (subsoil), making a total of eight samples from each measured plot. From the same plot, above-ground parts of all major plant species were also sampled.

Chemical Analysis

Soil samples were dried at 60°C, ground and passed through a 2 mm sieve before analysis for various soil properties and total metal contents. Soil pH and electrical conductivity (EC) were measured in 1:5 soil water suspensions. The particle size analysis was performed by wet sieving and the hydrometer method (Day, 1965). The organic carbon content of the soil was determined by wet oxidation using the modified Walkley and Black method (McCleod, 1975). Soil samples were digested in concentrated nitric acid, hydrogen peroxide and hydrochloric acid based on the US EPA method 3050B for the analysis of heavy metals (US Environmental Protection Agency, 1996).

The plant samples were put through a washing sequence in order to remove any surface contamination, which involved agitating and rinsing first in 0.1% teepol for 15 seconds, followed by 0.1% HCl for 15 seconds, then three separate washes in deionized water (Reuter et al., 1988). The clean plant samples were blotted dry and placed in an oven at 70°C for 48 hours. The dried samples were then ground, using a stainless steel coffee grinder and passed through a 1 mm stainless steel sieve. The ground plant samples were oven-dried at 70°C overnight to remove any absorbed moisture prior to the plant digestion. The ground plant samples were digested in a mixture of concentrated nitric and perchloric acids for metal analysis (Miller, 1998).

The digested plant and soil samples were analysed for heavy metals using a Vista Varian Inductively Coupled Plasma Atomic Emission Spectrometer. Three soil and two plant standards were digested in a similar manner to the contaminated samples and were then included in every batch during ICP analysis for quality control.

Statistical Analysis

The statistical analysis was performed using a statistical package JMP version 3.2.5 (SAS Institute, 1995). The biplot technique was used to determine an approximation to a set of both variables and points in one graph. In the principal component analysis, the biplot shows inter-unit distances and indicates the clustering of units as well as display variances and correlation of the variables (Gabriel, 1971).

RESULTS AND DISCUSSION

Soil Properties

Soils at the Rhodes and Chiswick sites were mostly alkaline with median pH (1:5 water) values of 7.96 and 7.27, respectively, whereas at the Boolaroo site, most soil samples were acidic (median of 6.08). For all the three sites, electrical conductivity (EC) values were low and ranged between 17–2560 μS cm^{-1} (Table 6.1). The organic carbon content was generally low for all samples and the median values were similar (around 1.5%) for the three sites. For the Rhodes and Boolaroo sites, clay contents were similar with a median value of about 18%. At the Chiswick site, however, the soils were relatively lighter textured with a median value for clay content of 9%.

Table 6.1: A summary of some physico-chemical properties of soils at the three sites

Soil property	*Rhodes*		*Chiswick*		*Boolaroo*	
	Range	*Median*	*Range*	*Median*	*Range*	*Median*
pH (1:5 H_2O)	5.30–10.88	7.96	3.83–10.05	7.60	5.15–7.81	6.08
EC (μS cm^{-1})	52–603	136	33–2560	175	17–90	51
Organic carbon (%)	0.14–6.27	1.45	0.36–3.54	1.62	0.33–5.80	1.41
Coarse sand % (200-2000 μm)	1.0–84.7	46.3	43.4–75.8	56.9	5.4–59.6	30.1
Fine sand % (20-200 μm)	9.3–48.6	21.1	14.0–38.1	27.5	14.1–53.9	27.7
Silt % (2-20 μm)	1.0–25.3	13.3	0.5–4.6	7.7	5.5–29.0	16.5
Clay % (< 2 μm)	1.9–60.8	18.0	3.9–16.8	9.3	9.8–56.5	18.4

Total Element Analysis of Soil

The heavy metal content is an important factor in determining the extent of contamination in soil. National standards were used to provide consistent guidelines in the assessment and management of contaminated sites (ANZECC and NHMRC, 1992). The development of these environmental investigation guidelines was based on threshold levels for phytotoxicity and the uptake of contaminants that can result in the impairment of plant growth or reproduction, or as unacceptable residue levels. The guideline values utilize overseas information and represent conservative values that should protect the environment. More recently, ecological investigation levels for urban landuse and a number of health-based investigation levels have been set by the National Environment Protection Council to accommodate the range of human and ecological exposures settings (NEPM, 1999). The ecological investigation levels (NEPM, 1999) for several heavy metals (for example As, Cd and Zn) are similar to ANZECC environment investigation values, and for some of the metals such as Pb and Cu, the values have been increased. The range and median values for the heavy metal content at the three sites along with the ANZECC environment investigation values are given in Table 6.2.

Cadmium

The concentrations of Cd in the soils at Rhodes were between 0.09–34.2 mg kg^{-1}. The mean value decreased with depth, from 2.75 mg kg^{-1} for the topsoils to 1.61 mg kg^{-1} for the subsoils. For the Chiswick site, Cd concentrations ranged between 0.05 to 64.6 mg kg^{-1} and the topsoil and

Table 6.2: Total heavy metal contents (mg kg^{-1}) in soils at the three sites

Heavy metal	*ANZECC environment investigation value*	*Rhodes*		*Chiswick*		*Boolaroo*	
		Range	*Median*	*Range*	*Median*	*Range*	*Median*
Cd	3	0.09–34.2	0.15	0.05–64.6	7.7	0.03–55.4	4.1
Co	50*	0.4–104.8	8.8	0.4–142.1	14.0	1.0–33.1	5.4
Cr	50	11.9–528.1	38.8	10.2–136.7	48.6	5.6–34.5	12.1
Cu	60	3.0–285.6	35.0	22.4–760.9	99.2	5.1–278.5	21.4
Fe	—	12 862–114 321	38 926	3 720–166 390	73 888	7 824–77 360	18 928
Mn	500	54–2004	315	79–989	429	25–982	180
Ni	60	4.6–273.7	19.3	4.0–710.3	54.9	0.7–76.3	8.3
Pb	300	3–46 293	689	339–76 931	412	20–3 609	172
Zn	200	1–15 962	167	641–590 003	89 543	20–2 613	226

*Dutch B investigation value is an indicative value set by the Dutch authorities for the investigation of a site for possible rehabilitation or restriction of land use.

subsoil mean concentrations were 9.2 and 13.9 mg kg^{-1}, respectively. Cadmium concentrations at the Boolaroo site varied between 0.03-55.4 mg kg^{-1}. The topsoil and subsoil mean concentrations were 2.2 and 17.2 mg kg^{-1}, respectively. Most of the soil samples at the three sites exceeded the ANZECC environment investigation value of 3 mg kg^{-1}. Most Cd values were also greater than the range reported for Australian soils (0.032–0.21 mg kg^{-1}) (Williams and David, 1973) and the normal range (0.001–0.7 mg kg^{-1}) reported for the world soils (Ure and Berrow, 1982). Many soil Cd values were also greater than the range (0.04–4.3) reported for soils from Adelaide city (Barzi et al., 1996).

Cobalt

At the Rhodes site, Co concentrations were between 0.4–104.8 mg kg^{-1} and the mean values in topsoils and subsoils were 21.4 and 10.0 mg kg^{-1}, respectively. The concentrations of Co at the Chiswick site occurred between 0.4–142.1 mg kg^{-1}. The mean values for the topsoils and subsoils at Chiswick were 31.0 and 20.9 mg kg^{-1}, respectively. Cobalt concentration at the Boolaroo site ranged between 1.0–33.1 mg kg^{-1}. The mean Co concentrations in the topsoils and subsoils were 6.8 and 8.2 mg kg^{-1}, respectively. ANZECC has set no environment investigation value for Co. However, most samples had Co contents below the Dutch B investigation value of 50 mg kg^{-1}, with a few exceptions at Rhodes and Chiswick sites. The Co values also occurred within the range reported for Australian soils (<2–43 mg kg^{-1}) and world soils (0.5–65 mg kg^{-1}) (Bowen, 1979; McKenzie, 1960).

Chromium

The Cr concentrations at the Rhodes site were between 11.9–528.1 mg kg^{-1} and mean concentration decreased with depth from 96.5 mg kg^{-1} in the topsoil to 55.1 mg kg^{-1} in the subsoil. At the Chiswick site, Cr concentrations were 10.2–136.7 mg kg^{-1}. There was an increase in concentration with depth with mean concentrations in topsoils and subsoils of 39.6 and 53.0 mg kg^{-1}, respectively. For the Boolaroo site, Cr concentrations ranged between 5.6–34.5 mg kg^{-1}. The mean concentration in topsoils and subsoils were 14.5 and 11.0 mg kg^{-1}, respectively. Worldwide soil Cr concentrations range between 0.9 and 1500 mg/kg (Ure and Berrow, 1982; Bowen, 1979). The Cr concentrations observed in most samples in this study were towards the lower end of the range reported in the literature, and 30% of the samples at Rhodes and Chiswick exceeded the ANZECC environment investigation value of 50 mg kg^{-1}.

Copper

At the Rhodes site, Cu concentrations varied between 3.0–285.6 mg kg^{-1} and the mean concentrations in the topsoils and subsoils were 56.5 and 43.7 mg kg^{-1}, respectively. The concentration range of Cu at Chiswick occurred between 22.4–760.9 mg kg^{-1}. There was a small increase in Cu concentration with depth with the mean concentrations in the topsoils and subsoils of 116.8 and 166.6 mg kg^{-1}, respectively. For the Boolaroo site, Cu concentrations ranged between 5.1–278.5 mg kg^{-1}, and the mean concentration in the topsoils and subsoils were 18.8 and 63.8 mg kg^{-1}, respectively. Normal Cu concentration in soils ranges between 2 and 250 mg kg^{-1} (Swaine, 1955; Alloway, 1990) and majority of the Cu values at the three sites occured within the range but exceeded the ANZECC environment investigation value (60 mg kg^{-1}).

Iron

The concentration range of Fe at the Rhodes site was between 12,862 to 114,321 mg kg^{-1} and there was no change in Fe values with depth. For the Chiswick site, Fe content ranged between 3720–166390 mg kg^{-1}, and concentrations in topsoils and subsoils were 73,867 and 68,213 mg kg^{-1}, respectively. The concentration range of Fe at Boolaroo site was from 7824 to 77,360 mg kg^{-1} and the mean concentration values in the topsoil and subsoil were 22,732 and 20,283 mg kg^{-1}, respectively. Ure and Berrow (1982) stated the worldwide concentration range as 0.01–21%, whereas Bowen (1979) reported the range for Fe value as 0.2–55%. The concentration range observed at the three sites occurred within the reported ranges of Fe.

Manganese

At the Rhodes site, Mn concentrations were between 54–2004 mg kg^{-1} and the topsoil and subsoil mean concentrations were 614 and 382 mg kg^{-1}, respectively. For the Chiswick site, Mn concentrations were between 79–989 mg kg^{-1}. There was a small decrease in soil Mn content with depth, with the topsoil and subsoil mean concentrations of 448 and 372 mg kg^{-1}, respectively. The concentration range of Mn for soil samples at Boolaroo site was between 25–982 mg kg^{-1}, and the topsoil and subsoil mean concentrations were 163 and 309 mg kg^{-1}, respectively. The concentrations of Mn at these sites fall within the worldwide concentration range of <1–18 300 mg kg^{-1} (Ure and Berrow, 1982) and the range of 39–1000 mg kg^{-1}, as reported by McKenzie (1960) for Australian soils. About 20% of the total samples exceeded the ANZECC environment investigation value of 500 mg kg^{-1} at these sites.

Nickel

For the Rhodes site, Ni concentrations were between 4.6–273.7 mg kg^{-1}. The mean topsoil and subsoil concentrations were 49.1 and 24.8 mg kg^{-1}, respectively. The concentration range of Ni for the Chiswick site was between 4.0 and 710.3 mg kg^{-1}. The topsoil and subsoil concentrations were similar, with values of 106.5 and 92.7 mg kg^{-1}, respectively. At the Boolaroo site, Ni ranged from 0.7 to 76.3 mg kg^{-1} and the mean top and subsoil concentrations were 6.3 and 13.0 mg kg^{-1}, respectively. Most Ni values at these sites occurred near the lower end of the worldwide range of 0.1–1523 mg kg^{-1} (Ure and Berrow, 1982), but Ni values exceeded the ANZECC environment investigation value (60 mg kg^{-1}) in 19% of the total samples.

Lead

The concentration of Pb in the soils at the Rhodes site ranged between 3–46 293 mg kg^{-1} and the mean topsoil and subsoil concentration were 7 158 and 5 046 mg kg^{-1}, respectively. At the Chiswick site, Pb concentrations were between 339–76 931 mg kg^{-1} and topsoil contained twice as much of Pb as in the subsoil (11847 and 5719 mg kg^{-1}, respectively). The concentration range of Pb in soils at Boolaroo was 20–3 609 mg kg^{-1}. There was a significant increase in Pb concentration with depth, with the mean topsoil and subsoil concentrations of 130 and 941 mg kg^{-1}, respectively. Bowen (1979) reported Pb content in common soils between 2–300 mg kg^{-1}. All the samples at the Chiswick site had Pb content greater than the upper limit, and nearly half of the samples at Rhodes and Boolaroo sites had Pb content exceeding the ANZECC environment investigation value of 300 mg kg^{-1}.

Zinc

Zinc content in soils at Rhodes was between 1–15,962 mg kg^{-1}. There was a significant decrease in Zn concentration with depth, with the mean top and sub-soil concentrations of 1361 and 741 mg kg^{-1}, respectively. The concentration range of Zn in the soil for the Chiswick site was between 641–590,003 mg/kg, and the topsoil and subsoil mean concentrations were 105,290 and 185,129 mg/kg^{-1}, respectively. At the Boolaroo site soil Zn content was between 20–2,613 mg kg^{-1}. The mean concentrations for topsoils and subsoils were 148 and 835 mg kg^{-1}, respectively. McKenzie (1960) reported Zn values between 11–86 mg kg^{-1} for Australian soils, whereas the worldwide range for Zn was reported to be between 1.5 and 200 mg kg^{-1} (Ure and Berrow; 1982). All the sites had Zn concentrations above these reported ranges and also the ANZECC environment investigation value (200 mg kg^{-1}). The Chiswick site had the highest Zn concentration of 590,003 mg/kg.

Metal Distribution in Relation to Depth

A general trend occurred at Rhodes with a majority of the heavy metal concentration decreasing with depth, and the site was considered contaminated with Cd, Cr, Mn, Ni, Pb and Zn. according to the ANZECC guidelines (Table 6.2).

At the Chiswick site, there was no definite trend with depth. For Cd, Cr, Cu and Zn, the concentration increased with depth, whereas Co, Fe, Mn, Ni, and Pb concentrations decreased with depth. The Chiswick site was contaminated with Cd, Cr, Cu, Mn, Ni, Pb and Zn according to ANZECC guidelines.

For most of the heavy metals (Cd, Cu, Mn, Ni, Pb and Zn), concentrations increased with depth at the Boolaroo site. The exceptions were Co and Fe, where the mean concentration decreased with depth. According to the ANZECC guidelines, Boolaroo was considered contaminated with Cd, Cu, Mn, Ni, Pb and Zn.

Heavy metal content of the soils at the three sites exceeded the investigation threshold limits. Based on the ANZECC guidelines, these soils need further investigation in order to assess the bioavailability and mobility of metals. Samples with high metal bioavailability will need remediation.

It must be emphasized that the distribution of metals at the studied sites was not normal and caution is needed in the interpretation of data based on the average values of heavy metals in the studied soils. The extent of metal contamination was, therefore, examined on the log transformed values in the following section.

Soil Classes in Relation to Heavy Metal Concentration

To further test the extent and distribution of contamination at these sites, the samples were classified into three categories based on the following criteria:

- *Class 1:* The soils in this group were considered *uncontaminated* as the mean concentration of the metal was significantly below the ANZECC investigation value.

- *Class 2:* When the mean concentration of the metal was greater than the ANZECC investigation value but the difference was not statistically significant, the soil was considered *potentially contaminated.*
- *Class 3:* The soils in this group were considered *contaminated,* as the mean concentration of the metal was significantly greater than the ANZECC investigation value.

Since the metal concentrations were not normally distributed, the log transformed data were also classified into the above described contamination classes by comparing with the log transformed ANZECC investigation values. Tables 6.3 and 6.4 show both forms of data for the topsoil and subsoil, respectively.

Table 6.3: Mean values (mg kg^{-1}) and the soil classes in topsoil of some environmentally-significant heavy metals

	Original data				*Log_{10} transformed data*			
Heavy metal	*ANZECC environment investigation value*	*Boolaroo*	*Chiswick*	*Rhodes*	*ANZECC environment investigation value*	*Boolaroo*	*Chiswick*	*Rhodes*
Cd	3	2.20	9.15	2.75	0.48	−0.06	0.60	−0.46
	Class	2	2	2		1	1	1
Co	50	6.3	28.7	19.6	1.7	0.6	1.2	1.0
	Class	1	1	1		1	1	1
Cr	50	14.5	36.6	96.5	1.7	−1.9	−2.1	−2.5
	Class	2	2	2		1	1	1
Cu	60	16.8	116.8	56.52	1.8	1.2	1.9	1.5
	Class	2	2	2		1	2	2
Mn	500	233.9	410.2	501.4	2.7	2.2	2.6	2.5
	Class	1	2	2		1	2	2
Ni	60	6.3	106.5	49.1	1.8	0.7	1.7	1.4
	Class	2	2	2		1	2	1
Pb	300	130	11846	7158	2.5	1.9	3.6	2.8
	Class	2	3	2		2	3	2
Zn	200	149	96189	1361	2.3	2	4.5	2.4
	Class	2	3	2		2	3	2

Figure 6.1 (a, b) shows examples of the graphical representation of various classes for Zn in soils at the three sites; the significant ($P \leq 0.05$) differences have been indicated by showing the site name in italics and filled circles. If the comparison circles for the sites did not overlap and were located away from the ANZECC circle, or there was at least an angle of intersection greater than 90°, then the mean values were considered to be significantly different (SAS Institute, 1995). The mean Zn concentration at each site was compared with the ANZECC investigation limit of 200 mg kg^{-1} (Figure 6.1a). Boolaroo and Rhodes were not significantly different to the investigation limit so the sites were

Table 6.4: Mean values (mg kg^{-1}) and the soil classes in subsoil of some environmentally-significant heavy metals

	Original data				Log$_{10}$ transformed data			
Heavy metal	*ANZECC environment investigation value*	*Boolaroo*	*Chiswick*	*Rhodes*	*ANZECC environment investigation value*	*Boolaroo*	*Chiswick*	*Rhodes*
Cd	3	17.22	13.85	1.61	0.48	1.09	0.85	−0.49
	Class	3	3	2	—	1	1	1
Co	50	7.3	19.0	9.7	1.7	0.7	1.1	0.9
	Class	1	1	1	—	1	1	1
Cr	50	11.0	53.0	55.1	1.7	1.1	1.7	1.6
	Class	1	2	2	—	1	1	1
Cu	60	63.8	166.6	43.7	1.8	1.6	2.1	1.5
	Class	2	3	2	—	2	1	3
Mn	500	309	372	382	2.7	2.4	2.5	2.4
	Class	2	2	2	—	1	2	1
Ni	60	13.0	92.7	24.8	1.8	0.9	1.8	1.3
	Class	1	2	2	—	1	2	1
Pb	300	941	5719	5067	2.5	2.7	3.6	2.9
	Class	2	2	2	—	2	3	2
Zn	200	835	253946	741	2.3	2.7	5.0	2.2
	Class	2	3	2	—	2	3	2

indicated in non-italic text and the ANZECC value and comparison circles shown as overlapped (unfilled). The value for the Chiswick site (shown in italics) was significantly greater than the ANZECC circle and the comparison circle was filled.

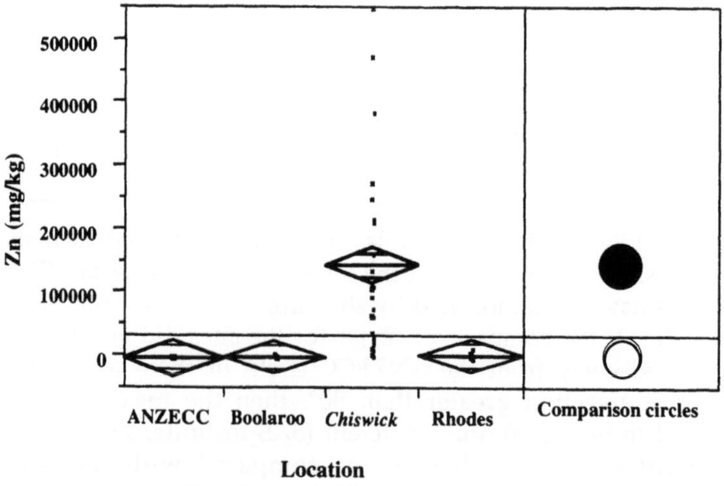

Fig. 6.1a: Original zinc concentrations in soil samples at the three sites tested against the ANZECC guideline value.

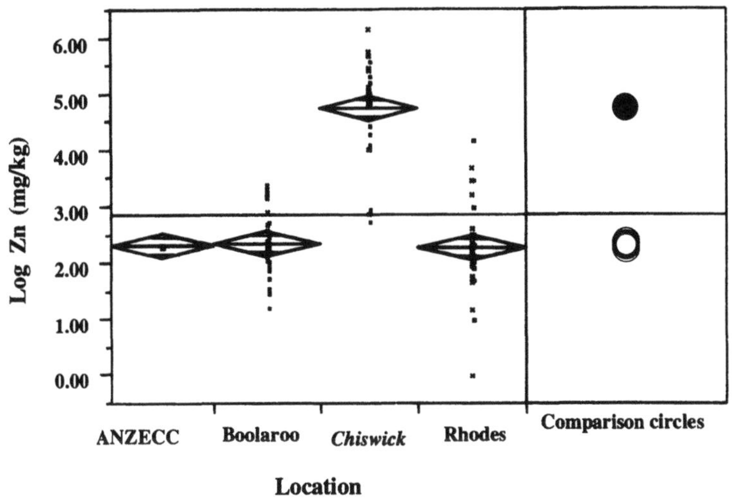

Fig. 6.1b: Log-transformed soil zinc concentrations at the three sites tested against the log-transformed ANZECC guideline value.

Figure 6.1(b) shows the contamination classes for the log transformed soil Zn data. Both forms of Zn data, however, showed the same results. Boolaroo and Rhodes were possibly contaminated with Zn (Class 2), whereas Chiswick was definitely contaminated with Zn (Class 3).

There were differences between classes in the topsoil and subsoil (Tables 6.3 and 6.4). Generally, greater contamination occurred in the subsoil from various contaminants buried in the subsoil. The transformation of the original data to log data showed similar trends for both topsoil and subsoil. In both cases, the extent of contamination for heavy metals such as Cd, Co, Cr was reduced and the metals considered uncontaminated (Class 1) in the log transformed data. The log transformation normalized the data to provide evenly-distributed values, implying that certain data points increased the apparent extent of contamination in the original data. The original and log-transformed Mn data showed that the Boolaroo site was not contaminated, whereas Chiswick and Rhodes sites might be contaminated. Both forms of data at Chiswick showed that the site was contaminated with Pb and Zn, whereas Boolaroo and Rhodes sites were possibly contaminated with these two metals. The original data for subsoil showed that the Boolaroo and Chiswick sites were contaminated with Cd, and the Rhodes site was possibly contaminated. The log-transformed data, however, showed that the three sites were uncontaminated with respect to Cd. At the Chiswick and Rhodes sites, original data indicates that possible Cr contamination had occurred, and in the log-transformed data, the extent of contamination reduced to being classified as uncontaminated (Class 1) for both sites.

On the basis of the original subsoil data, Cu was possibly a contaminant at Rhodes and Boolaroo; on the log-transformed data, Rhodes site was designated as contaminated. In both data sets, the status of the Boolaroo site remained the same. At Chiswick, the extent of Cu contamination reduced from Class 3 on the original data to Class 1 for the log-transformed data.

At the three sites, Mn was a possible subsoil contaminant on the original data, whereas on the log-transformed data, there was no contamination at the Boolaroo and Rhodes sites. The extent of contamination also reduced for Ni in the log-transformed data. For Pb, the three sites were possibly contaminated, based on the original data and Chiswick is contaminated on the basis of log transformed data. The same trend occurred for Zn at the three sites, where Boolaroo and Rhodes sites were possibly contaminated and Chiswick certainly contaminated.

Total Element Analysis of Plants

The range and median values of the original plant concentration at the three sites are given in Table 6.5, along with critical values. The plant metal concentration data for individual metal is briefly discussed below.

Table 6.5: Total heavy metal contents (mg kg^{-1}) in plant samples at the three sites

Heavy metal	Critical value*	Rhodes		Chiswick		Boolaroo	
		Range	Median	Range	Median	Range	Median
Cd	0.1	0.03–8.3	0.1	0.003–1.4	0.2	0.06–52.3	5.4
Co	1.0	0.01–13.9	0.3	0.09–4.7	0.8	0.09–13.9	0.9
Cr	2.0	0.05–26.6	0.6	0.25–31.7	1.5	0.05–76.1	4.5
Cu	20.0	0.04–47.7	9.9	5.2–2324.0	13.1	2.8–99.9	11.1
Fe	500.0	0.1–953.3	134.0	94.9–2693.0	363.7	21.6–3802.6	239.3
Mn	300.0	0.16–333.1	26.0	11.1–223.3	33.3	9.9–337.3	61.7
Ni	5.0	0.12–12.9	1.6	0.43–78.0	2.5	0.24–32.3	0.3
Pb	20.0	0.17–129.9	10.6	11.1–327.7	101.6	2.4–1030.3	121.4
Zn	100.0	0.05–403.2	65.2	94.6–4247.6	354.9	72.8–2318.6	328.8

*Critical value is the upper limit of normal values for a given metal in plants (Kabata-Pendias and Pendias, 1992; Miller, 1998)

Cadmium

Cadmium is not essential for plant growth and is highly toxic to plants, and such toxicity has been reported in plants on severely polluted sites (Alloway, 1990). Cadmium concentration in plants was highest at the Boolaroo site, ranging between 0.06 and 52.3 mg kg^{-1}. At the Rhodes and Chiswick sites, Cd concentrations were low with median values of 0.10 and 0.15 mg kg^{-1}, respectively. The accumulation of Cd at the Boolaroo site may be due to either greater availability under acidic conditions in the soil or the plant uptake of aerial deposition of Cd (Hovmand et al., 1983). Soil pH has been considered to be a particularly important factor in controlling Cd availability where the increase in soil pH has been found to decrease the solubility of Cd (William and David, 1973; Ure and Berrow, 1982).

Cobalt

The median values for Co showed a similar pattern to Cd with the minimum median value (0.03 mg kg^{-1}) occurring at the Rhodes site and the maximum median value (0.90 mg kg^{-1}) observed at the Boolaroo site. The concentration ranges of Co at the two sites are also similar.

Chromium

As for Cd and Co, the Boolaroo site had the highest median value (4.5 mg kg^{-1}) and concentration range (0.05–76.1 mg kg^{-1}) for Cr. In plants, Cr is not known to be an essential element and generally, concentrations are very low due to its low solubility in soils (Mertz, 1969). Alloway (1990) reported uncontaminated or background Cr concentrations of 0.23 mg kg^{-1}. Clearly many samples contained higher Cr concentrations than this value (Table 6.5).

Copper

The median values for Cu concentration at the three sites were similar. The Chiswick site had the greatest range in Cu concentrations, with values ranging between 5.2–2324 mg kg^{-1}. This site had plant species commonly known as fish fern with Cu concentration of 2324 mg kg^{-1}. Copper is an essential element for plant growth, but toxicity problems arise when the levels of Cu in the soil are high.

Iron

The Fe concentrations at the three sites overall range between 0.1–3803 mg kg^{-1}. The Boolaroo site contained plant species with the greatest Fe concentration of 21.6–3802.6 mg kg^{-1}. Kangaroo grass accumulated the highest Fe concentration (3803 mg kg^{-1}) among the plant species at the three sites. Iron is an essential element for the growth of plants. Deficiencies of Fe in plants are caused by soil factors such as high soil pH (Chaney et al., 1972). The values over 500 mg kg^{-1} in plant tissues are observed when availability of Fe is very high in soils.

Manganese

Manganese plant concentrations were similar at the Rhodes and Chiswick sites, whereas at the Boolaroo site, the concentration of Mn was nearly twice the value to the previous two sites (Table 6.5). Manganese is essential to the growth of plants, and its deficiency generally occurs in plants grown on neutral or alkaline soils. However, plants growing on strongly acidic soils can suffer from Mn toxicity owing to excessive uptake (Moraghan and Mascagni, 1991).

Nickel

The greatest Ni concentration range (0.43–78 mg kg^{-1}) and the highest median value (33.3 mg kg^{-1}) in plants occurred at the Chiswick site. Cotoneaster accumulated the highest Ni concentration (78 mg/kg). Such concentrations are observed in plants growing on soils formed on mafic rocks, such as serpentine, and plants growing on highly contaminated soils (Alloway, 1990).

Lead

Boolaroo had the greatest concentration range (2.4–1030 mg kg^{-1}) and the highest median value (121.4 mg kg^{-1}) for Pb in plants. Lead is not an essential element and is considered highly toxic to plants (Underwood, 1977). Lead is not readily transported to the shoots as it is accumulated in the roots, so high Pb concentrations in plants at the Boolaroo site may suggest aerial deposition and subsequent uptake by plants. The Pb content in soil was highest at the Chiswick site and lowest at the Boolaroo site. Tjell et al. (1979) using ^{210}Pb tracer, showed that the uptake of Pb into grasses from the soil was very low, and concluded that 90–99% of Pb in the leaf material was due to foliar uptake.

Zinc

The Chiswick and Boolaroo sites had a very high Zn concentration in plants, with median values of 354.9 and 328.8 mg kg^{-1}, respectively, compared to the Rhodes site (median of 65.2 mg kg^{-1}). Zinc is essential for plant growth; but in the presence of excessive amounts in the soil, Zn toxicity can occur to plants (Ure and Berrow, 1982). Many plants accumulated Zn concentration greater than the critical value of 100 mg kg^{-1}.

Plant Classes in Relation to Heavy Metal Concentrations

The mean heavy metal data for the plants were compared with critical values similar to the soil classes. Table 6.6 reports the original and log-transformed mean heavy metal concentrations and the contaminated classes according to the following criteria.

- *Class 1:* The plants in this group had mean metal concentrations significantly ($P \leq 0.05$) less than the critical value and were considered

Table 6.6: Mean values (mg kg^{-1}) of heavy metals in plants and plant contamination classes for the original and log-transformed data

	Original data				*Log_{10} transformed data*			
Heavy metal	*Critical value**	*Boolaroo*	*Chiswick*	*Rhodes*	*Critical value**	*Boolaroo*	*Chiswick*	*Rhodes*
Cd	0.1	10.5	0.3	0.8	-1.0	0.6	-0.8	-0.8
	Class	3	2	2	Class	3	2	2
Co	1.0	2.6	1.1	2.3	0	-0.2	-0.1	-0.4
	Class	2	2	2	Class	2	2	3
Cr	2.0	10.7	3.6	2.5	0.3	0.5	0.3	-0.3
	Class	3	2	2	Class	2	2	1
Cu	20.0	28.1	130.1	11.3	2.3	1.2	1.3	0.7
	Class	2	2	2	Class	2	2	1
Fe	500.0	462.7	759.0	150.7	2.7	2.4	2.7	1.8
	Class	2	2	2	Class	2	2	1
Mn	300.0	104.7	43.4	47.9	2.5	1.8	1.5	1.3
	Class	1	1	1	Class	1	1	1
Ni	5.0	2.7	5.9	2.8	0.7	-0.3	0.4	0.1
	Class	2	2	2	Class	1	2	1
Pb	20.0	256.6	108.5	16.4	1.3	2.1	1.9	0.7
	Class	3	2	2	Class	3	3	1
Zn	100.0	624.1	829.2	78.5	2	2.6	2.6	1.5
	Class	3	3	2	Class	3	3	1

* Critical value is the upper limit of normal values for a given metal in plants (Kabata-Pendias and Pendias, 1992; Miller, 1998)

uncontaminated (in the case of non-essential heavy metals) or *deficient* (for essential elements).

- *Class 2*: Where the plant metal concentrations were greater than the critical value but the difference was statistically insignificant and they were considered *potentially toxic*.
- *Class 3*: The plant metal concentrations were considered toxic when significantly greater than the critical value and indicating accumulation of metals to *toxic* levels.

Cadmium accumulation in plants at the Boolaroo site reached toxic levels, as indicated by both original and log-transformed data. At the Chiswick and Rhodes sites, Cd concentrations were potentially toxic on the basis of both original and log-transformed data. Cobalt plant concentrations were at potentially toxic levels for both original and log-transformed data at all sites except for the log-transformed values at Rhodes, where Co concentrations were at toxic levels.

At the Boolaroo site, the original Cr concentrations in plants were present at toxic levels that changed to potentially toxic levels on log transformation. For the Chiswick site, both original and log-transformed data indicated potentially toxic levels of Cr in plants, whereas at Rhodes sites, original data showed potentially toxic levels of Cr in plants that changed to non-toxic class on log transformation.

Copper and Fe concentrations in plants were at potentially toxic levels at all sites on both original and log-transformed data, except the log transformed values at Rhodes site, which showed non-toxic levels of Cu. Manganese concentrations of plants at all sites were classified in the non-toxic category at all sites on both original and log-transformed data.

The original data for Ni concentrations in plants showed potentially toxic levels at all three sites, and after log transformation, the category changed to non-toxic levels for the Boolaroo and Rhodes sites.

Lead concentrations in plants were present at toxic levels at the Boolaroo site, as indicated by both the original and the log-transformed values. At the Chiswick site, the original data showed potentially toxic levels and the category changed to toxic class on log transformation. At the Rhodes site, plants were least contaminated with respect to Pb as the original data showed potentially toxic level in plants that changed to non-toxic on log transformation.

The concentration of Zn in plants were at toxic levels at the Boolaroo and Chiswick sites, whereas at Rhodes, the original data fell in the potentially toxic class and changed to non-toxic class on log transformation.

Previous research has shown that certain grasses accumulate greater concentrations of certain heavy metals (Hamon et al., 1997; Smith, 1994a, b; Bramley and Barrow, 1994; and Taylor et al., 1989). Peterson (1983) defines an accumulator as a plant species whose metal concentration exceeds the normal values of metal concentration in plants for a particular substrate. Accumulators are capable of concentrating metals in above-ground parts of a plant to various degrees. Three plant species, kangaroo grass (*Themeda australis*), kikuyu grass (*Pennisetum clandestinum*) and ryegrass (*Lolium perenne*), showed the ability to accumulate high concentration of several heavy metals simultaneously. Kangaroo grass, an Australian native, accumulated the highest concentrations of Cr, Fe and third highest concentration of Pb. Kikuyu grass accumulated the highest concentrations of Pb, second highest concentration of Cr and Zn and third highest concentration of Cd, Cu and Ni. Ryegrass accumulated the highest concentrations of Zn and third highest concentrations of Fe and Zn.

Three plant species, fish fern, kikuyu grass and prickly black teatree (*Leptospermum juniperinum*), were considered possible hyperaccumulators. Hyperaccumulators are plants that can accumulate metal concentrations greater than 1000 mg kg^{-1} on a dry weight basis for Cu, Co, Cr, Ni and Pb or concentrations greater than 10 000 mg kg^{-1} for Zn and Mn (Baker and Walker, 1990). The fish fern plants accumulated 2324 mg kg^{-1} of Cu and thus can be considered Cu hyperaccumulator. Both the kikuyu grass and prickly black teatree accumulated Pb concentration above 1000 mg kg^{-1} (1030 mg kg^{-1} and 1029 mg kg^{-1}, respectively) and so would be classified as Pb hyperaccumulators.

Soil-Plant Relationships

Biplot Analysis

Three biplots were constructed to summarize the relationships among soil properties and soil and plant metal concentrations at the three sites.

The length of the lines in the biplot represents an approximation to the proportionality of standard deviation of variables where the cosine of angles between the variables approximates any correlation. An angle less than 90° indicates positive corelation between variables, and an angle of 180° indicates a negative linear correlation; when the angles between the variables were exactly 90°, no correlation existed. The spread of the points in the biplot also represents variables with similar patterns (Gabriel, 1982).

The biplot in Figure 6.2 shows the spread values for relationship between soil properties, such as pH, clay content, silt content, course and fine sand content, EC, organic carbon; and the log-transformed values of heavy metals. There were three distinct groups within the biplot, the most distinct group being the heavy metals in the soil and organic carbon and electrical conductivity. The second group consists of pH (H_2O and $CaCl_2$)

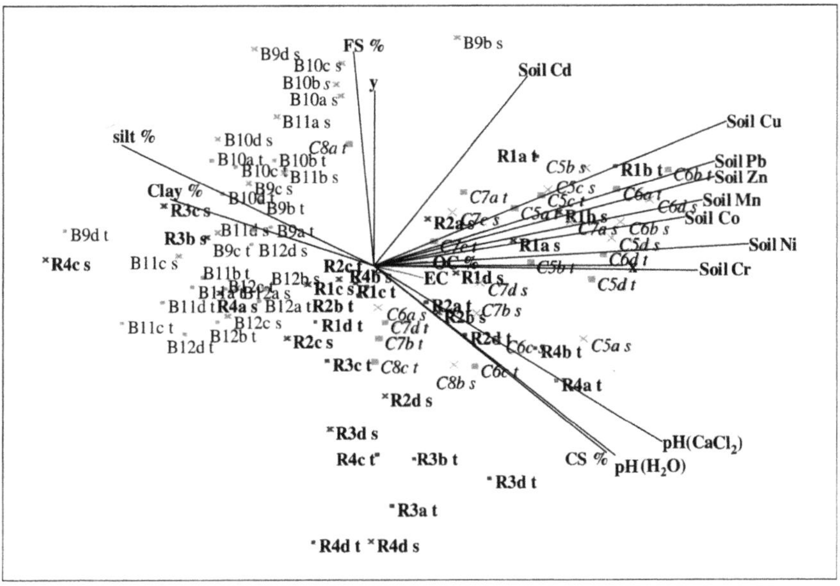

Fig. 6.2: Biplot for the soil properties and heavy metal contents at the three sites. Samples from the Rhodes site are labelled as 'R', from Chiswick site are labelled as 'C', from Boolaroo site labelled as 'B', topsoil samples labelled as 't' and subsoil samples labelled as 's'.

and coarse sand content. The third group consists of silt, clay and fine sand contents in the soils. These three groups suggest that correlation existed between all heavy metals and some soil properties such as organic carbon, electrical conductivity, coarse sand content and pH. The heavy metal content of the soil was not correlated with the silt, clay and fine sand contents.

Smith (1994 a, b) examined the effect of soil pH on the availability of metals in various crops and found that the main cause of Zn toxicity on yield reduction was low pH.

The spread and the extent of heavy metal contamination could also be observed in Figure 6.2. The biplot suggests that soils at Boolaroo sites were not contaminated with heavy metals as their values occurred outside the heavy metals lines. At Rhodes, only six samples were possibly contaminated: two samples were from the topsoil and four samples were from the subsoil. The trend at Chiswick was different as compared to Boolaroo and Rhodes, where about half of the samples were possibly contaminated. The extent of contamination was affected by the distance of the samples from the origin, so that the further away the samples were from the origin, the greater the possibility of heavy metal contamination at the site. At Rhodes, the topsoil for plots 1 and 2 had a greater possibility of being contaminated compared to the other plots. At the Chiswick site, the highest contamination occurred at plots 5 and 6. These results were consistent with the metal contamination observed on the basis of various contamination classes.

Figure 6.3(a) and (b) shows the biplots of log-transformed heavy metal data for soils and plants. The biplots are presented separately for the topsoils and subsoils so as to allow a comparison of their metal concentrations. Both biplots showed a similar trend in terms of relationship of heavy metals in the soils and plants, as there were two distinct groups with similar properties. Most of the data fell within these groups. The first group consisted of log-transformed soil heavy metals clustered together, whereas in the second group, log-transformed heavy metals in plants occurred together. Few exceptions include soil Co that correlated with plant Co in the topsoil and soil Cd with its plant concentration in the subsoil. The plant Ni showed a correlation with the Ni in both topsoil and subsoil samples. The biplots suggest that there is a little or no correlation between heavy metals in soil and plants, so the angles between the soil and plant heavy metal were close to 90°.

At Boolaroo, the spread of the log-transformed heavy metal content were similar as the spread were grouped together, with the exception of one plant species in the subsoil. At Rhodes, there were a greater spread of plant samples such that the variability of the plant species increased in relation to the heavy metal contents. At Chiswick, most plant samples

(a) Topsoil

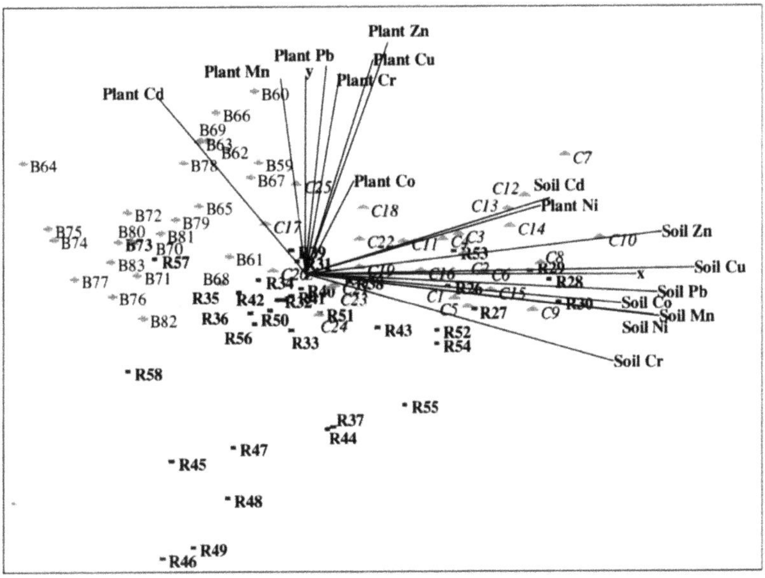

Fig. 6.3a: Biplot for heavy metal contents in the topsoil and plants at the three sites. The samples from the Rhodes site are as labelled 'R', from Chiswick site are labelled as 'C' and from Boolaroo site labelled as 'B'.

(b) Subsoil

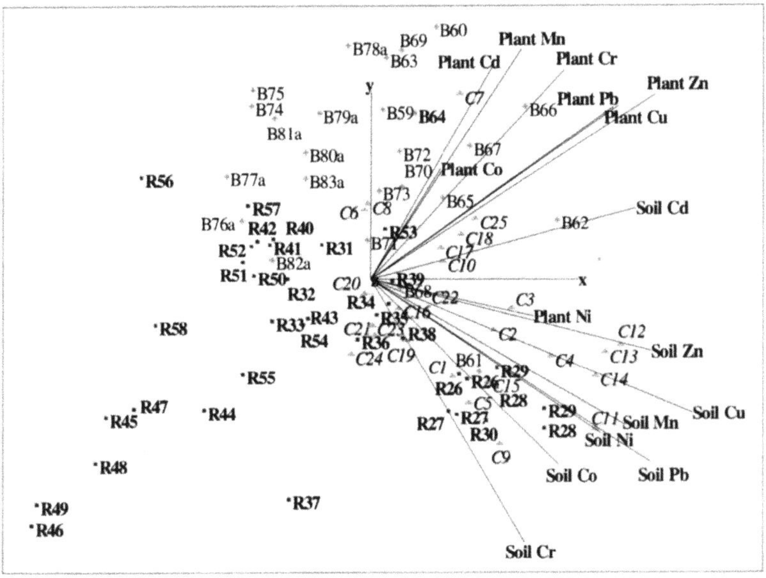

Fig. 6.3b: Biplot of heavy metal contents in the subsoil and plants at the three sites. The samples from the Rhodes site are as labelled 'R', from Chiswick site are labelled as 'C' and from Boolaroo site labelled as 'B'.

were confined to the soil heavy metal data, suggesting that there were similar plant species that absorb heavy metal and there was a greater possibility that the plant samples occurred where soil contamination existed.

Soil-Plant Transfer Coefficients

The transfer coefficient quantifies the relative differences in bioavailability of metals to plants or the efficiency of a plant species to accumulate a given metal. The coefficient was calculated by dividing the concentration of a metal in the aerial portion of the plant by the total metal concentration in the soil. Higher transfer coefficients reflect relatively poor retention in soils, whereas low coefficients reflect the strong sorption of metals to the soil colloids (Alloway and Ayres, 1997).

Table 6.7 outlines the range and median of the soil-plant transfer coefficients for the three sites. There was a definite trend with the median soil-plant transfer coefficients, where heavy metals Cd, Co, Cr, Cu, Mn, Pb and Zn had the highest median value at the Boolaroo site. The soil-plant transfer coefficients for the other two metals (Fe and Ni) were nearly equal for the three sites.

Table 6.7: The soil-plant transfer coefficients for heavy metals at the three sites

Heavy metal	*Rhodes*		*Chiswick*		*Boolaroo*	
	Range	*Median*	*Range*	*Median*	*Range*	*Median*
Cd	0.01–55.96	0.25	0.001–6.60	0.03	0.01–264.8	1.56
Co	0.001–6.07	0.02	0.004–11.10	0.06	0.01–6.34	0.07
Cr	0.0003–2.40	0.02	0.004–1.69	0.04	0.002–7.41	0.51
Cu	0.002–3.29	0.18	0.02–98.95	0.13	0.04–7.51	0.66
Fe	4.14–5.04	4.56	4.16–5.19	4.96	4.03–4.58	4.30
Mn	0.001–2.90	0.06	0.02–0.86	0.07	0.01–7.07	0.49
Ni	0.0001–0.81	0.07	0.004–1.85	0.05	0.004–4.02	0.05
Pb	0.0001–0.47	0.005	0.001–0.38	0.01	0.02–17.34	2.95
Zn	0.0001–87.48	0.23	0.0002–0.31	0.009	0.07–23.23	2.95

Kloke et al. (1984) suggested generalized transfer coefficient values of 0.01–0.1 for Co, Cr, Cu and Pb; 0.1–1 for Mn and 0–10 for Cd and Zn. Most soil-plant coefficients at these three sites occurred within these generalized ranges. At the Boolaroo site, the plants were able to take up more heavy metals as compared to the other sites. Kloke's transfer coefficients were based on the root uptake of metals, but plants can also accumulate relatively large amounts of metals by foliar absorption of atmospheric deposits on the plant leaves. Baker (1981) reported that most metals translocate from the roots to the shoots, such that the aerial portion of the plants generally had substantially higher concentration than the roots.

At the Boolaroo site, it was suspected that metal adsorption of aerially deposited metals from the presently operational smelter occurred through leaves. Tiller et al. (1987) reported that the soil could be contaminated with heavy metals by aerial deposition for up to hundreds of kilometres away from the source. The metals present in the air occur as aerosol particles of 5–20 nm diameter. Humans and animals can inhale metallic aerosols but the greater impact was through deposition under gravity, where the metals can be washed out onto vegetation, soil and waterbodies. The results from this study at the Boolaroo site showed that most of the contamination occurred by atmospheric deposition. Many of the metals were present at elevated concentrations in the plants despite very low concentrations in the soil. However, at the Boolaroo site, due to acidic soil pH (median = 4.93) compared to near neutral soil pH for the other two sites, metal availability to plants is also likely to be greater than the other two sites. The importance of soil pH in relation to metal uptake by plants is well known and in general plant availability of metals is high in the acidic pH range (Ure and Berrow, 1982).

CONCLUSIONS

The extent of contamination at the three sites depended on the industrial processes involved. Log transformation of the data for heavy metals generally indicated a reduced extent of contamination; even so, the results showed that the Chiswick, Rhodes and Boolaroo sites were contaminated with several heavy metals. Chiswick was the most contaminated site with Cd (71% of the samples), Cr (39%), Cu (75%), Mn (29%), Ni (43%), Pb (96%) and Zn (100%). Rhodes was the second most contaminated with Cd (16% of the samples), Cr (32%), Cu (23%), Mn (39%), Ni (13%), Pb (61%) and Zn (45%). At Boolaroo, contamination in soils was least, with Cd (55% of the samples), Cu (16%), Mn (13%), Ni (0.03%), Pb (36%) and Zn (52%).

The correlations in the biplots showed that the heavy metal concentrations in soil were governed by soil properties such as pH, organic carbon, electrical conductivity and coarse sand content. The spread and extent of contamination between the sites in the biplot confirmed that Chiswick is the most widely contaminated, where about half of the soil samples were considered possibly contaminated, followed by Rhodes and Boolaroo in that order. Biplots showed no significant relationship between soil and plant metal concentrations, suggesting that the metal availability to plants is much more complex and is not simply dependent on total soil metal levels.

In contrast to total metal levels in soils, it was observed that the plants at the Boolaroo site accumulated very high concentrations of heavy metals. We suggest that the accumulation of the metals by the plants is occurring through foliar absorption of atmospheric deposits on the plant leaves.

Three grass plants (kikuyu, kangaroo and rye) have the ability to accumulate high concentrations of several heavy metals simultaneously. Fish fern was identified as a Cu hyperaccumulator at the Chiswick site and, kikuyu grass and prickly black teatree were the two Pb hyperaccumulators at the Boolaroo site. These plants may have the potential for phytoremediation of contaminated sites. However, further research is required in order to assess the metal accumulation ability of the plants discussed.

REFERENCES

Alloway, B.J., 1990, *Heavy Metals in Soils*. Blackie and Sons Ltd., London.

Alloway, B.J. and Ayres, D.C., 1997, *Chemical Principles of Environmental Pollution*. Chapman and Hall, Great Britain.

ANZECC and NHMRC, 1992, Australian and New Zealand Guidelines for the Assessment of Contaminated Sites. Australian and New Zealand Environmental and Conservation Council. National Health and Medical Research Council.

Baker, A.J.M., 1981, Accumulators and excluders—Strategies in response of plants to heavy metals. *Journal of Plant Nutrition* 3: 643–653.

Baker, A.J.M. and Walker, P.L., 1990, Ecophysiology of metal uptake by tolerant plants. In: Shaw, A.J. (Ed), *Heavy Metal Tolerance in Plants*. CRC Press, Boca Raton, Florida: 155–178.

Barzi, F., Naidu, R. and McLaughlin, M.J., 1996, Contaminants and the Australian soil environment. In: Naidu, R., Kookana, R.S., Oliver, D.P, Rogers, S., McLaughlin, M.J. (Eds.), *Contaminants and the Soil Environment in the Australasia-Pacific Region*. Kluwer Academic Publishers, Dordrecht: 451–484.

Beavington, F., 1975, Heavy metal contamination of vegetables and soil in domestic gardens around a smelting complex. *Environmental Pollution* 9: 211–217.

Bowen, H.J.M., 1979, *Environmental Chemistry of Elements*. Academic Press, London.

Bramley, R.G.V. and Barrow, N.J., 1994, Differences in the cadmium content of some common Western Australian pasture plants grown in a soil amended with Cadmium: Describing the effects of level Cadmium supply. *Fertilizer Research* 39: 113–122.

Brown, S.L., Chaney, R.L., Angle, J.S. and Baker, A.J.M., 1995, Zinc and cadmium uptake by hyperaccumulator *Thlaspi caerulescens* grown in nutrient solution. *Soil Science Society of America Journal* 59: 125–133.

Cartwright, B., Merry R.H. and Tiller, K.G., 1976, Heavy metal contamination of soil around a lead smelter at Port Pirie, South Australia. *Australian Journal of Soil Research* 15: 69–81.

Chaney, R.L., Brown, J.C. and Tiffin L.O., 1972, Obligatory reduction of ferric chelates in iron uptake by soybeans. *Plant Physiology* 50: 208–213.

Day, P.R., 1965, Particle fractionation and particle-size analysis. In: Black, C.A. (Ed), *Methods of Soil Analysis*, Part I. Agronomy. 9, Amer. Soc. Agron. Inc. Madison, WI. : 545–567.

DeKoning, H.W., 1974, Lead and cadmium contamination in the area immediately surrounding a lead smelter. *Water, Air and Soil Pollution* 3: 63–70.

Gabriel, K.R., 1971, The biplot graphic display of matrices with application to principal component analysis. *Biometrika* 58: 453–467.

Gabriel, K. R., 1982, Biplot. In: Johnson, N.J. and Kotz S. (Eds.), *Encyclopedia of Statistical Sciences*. John Wiley and Sons, New York. 1: 263–271.

Hamon, R., Wundke, J., McLaughlin, M. and Naidu, R., 1997, Availability of zinc and cadmium to different plant species. *Australian Journal of Soil Research* 35: 1267–1277.

Hovmand, M.F., Tjell, J.C. and Mosbaek, H., 1983, Plant uptake of airborne cadmium. *Environmental Pollution* (Series A) 30: 27–38.

Kabata-Pendias, A. and Pendias, H., 1992, *Trace Elements in Soils and Plants*. CRC Press, Boca Raton, Florida.

Kloke, A., Sauerbeck, D.R. and Vetter, H., 1984, The contamination of plants and soils with heavy metals and the transport of metals in terrestrial food chain. In: Nriagu, J.O. (Ed), *Changing Metal Cycles and Human Health*. Springer-Verlag, Berlin: 113–141.

Kumar, N.P.B.A., Dushenkov, V., Motto, H. and Raskin, I., 1995, Phytoextraction: the use of plants to remove heavy metals from soils. *Environmental Science and Technology* 29: 1232–1238.

McCleod, S., 1975, Studies on wet oxidation procedures for the determination of organic carbon in soil. Notes on Soil Techniques. CSIRO Division of Soils: 73–79.

McGrath, S.P., Sidoli, C.M.D., Baker, A.J.M. and Reeves R.D., 1993, The potential for the use of metal-accumulating plants for the in situ decontamination of metal-polluted soils. In: Eijsackers, H.J.P. and Hamers, T. (Eds), *Integrated Soil and Sediment Research: A Basis for Proper Protection*. Kluwer Academic Publishers, Dordrecht: 673–676.

McKenzie, R.M., 1960, CSIRO (Australia) Division of Soils: Divisional Report. 6/60. 17 pp.

Mellor, A. and Bevan, J.R., 1999, Lead in the soils and stream sediments of an urban catchment in Tyneside, UK. *Water, Air and Soil Pollution* 112: 327–348.

Mertz, W., 1969, Chromium occurrence and biological systems. *Physiological Review* 49: 163–169.

Miller, R.O., 1998, Nitric-perchloric acid wet digestion in an open vessel. In: Karla, Y.P. (Ed), *Handbook of Reference Methods for Plant Analysis*. CRC Press, Boca Raton, Florida: 57–61.

Moraghan, J. T. and Mascagni, H.J., 1991, Environmental and soil factors affecting micronutrient deficiencies and toxicities. In: Mortvedt, J.J., Cox F.R., Shuman, L.M. and Welch R.M. (Eds), *Micronutrients in Agriculture*. Soil Science Society of America, Inc., Wisconsin, USA. 2: 371–425.

NEPM, 1999, National Environment Protection (Assessment of site contamination) Measure 1999. National Environment Protection Council, Adelaide, Australia.

Paliouris, G. and Hutchinson, T., 1991, Arsenic, cobalt and nickel tolerances in two populations of *Silene vulgaris* (Moench) Garcke from Ontario, Canada. *New Phytologist*. 117: 449–459.

Peterson, P.J., 1983, Adaptation to toxic metals. In: Robbs, D.A. and Pierpoint, W.S. (Eds), *Metals and Micronutrients: Uptake and Utilization by Plants*. Academic Press, London.

Reeves, R.D. and Brooks, R.R., 1983, European species of *Thlaspi* (Cruciferae) as indicators of nickel and zinc. *Journal of Geochemical Exploration*. 18: 275–283.

Reuter, D.J., Robinson, J.B., Peverill, K.I., and Price, G.H., 1988, Guidelines for collecting, handling and analysing plant material. In: Reuter, D.J. and Robinson, J.B. (Eds.), *Plant Analysis: Intrepretation Manual*. Inkata Press, Melbourne: 20–30.

SAS Institute, 1995, Statistics and Graphics Guide: Version 3.2.5 of JMP. SAS Institute Inc. Cary, NC, US.

Smith, S.R., 1994a, Effect of soil pH on availability to crops of metals in sewage sludge-treated soils. I. Nickel, copper and zinc uptake and toxicity to ryegrass. *Environmental Pollution* 85: 321–327.

Smith, S.R., 1994b, Effect of soil pH on availability to crops of metals in sewage sludge-treated soils. II. Cadmium uptake by crops and implications for human dietary intake. *Environmental Pollution* 86: 5–13.

Swaine, D.J., 1955, The Trace Element Content of Soils. Commonwealth Bureau of Soil Science, Technical Communication. Commonwealth Agricultural Bureau. Harpenden, England 48: 157.

Taylor, R.W., Ibeabuchi, I.O. and Shuford, J.W., 1989, Growth and metal accumulation of forage grasses at various clipping dates on acid mine spoils. *Journal of Environmental Science and Health*. A24 (2): 195–204.

Tiller, K.G., 1992, Urban soil contamination in Australia. *Australian Journal of Soil Research* 30: 937–957.

Tiller, K.G., Smith, L.H., Merry, R.H. and Clayton, P.M., 1987, The dispersal of automotive lead from the metropolitan Adelaide into adjacent rural areas. *Australian Journal of Soil Research* 25: 155–166.

US Environmental Protection Agency., 1996, Acid digestion of sediment, sludge and soils: Method 3050b. USEPA. Washington, DC.

Underwood, E.J., 1977, *Trace Elements in Human and Animal Nutrition*. Academic Press, London.

Ure, A.M. and Berrow, M.L., 1982, The elemental constituents of soils. In: Bowen, H.J.M., Berrow, M.J., Berrow, M.L., Cawse, P.A., Patterson, D.S.P., Statham, P.J. and Ure, A.M. (Eds.), *Environmental Chemistry: A Review of the Literature Published up to Mid*-1980. 2: 94–204.

Vogel-Lange, R. and Wagner, G.J., 1990, Subcellular location of cadmium and cadmium-binding peptides in tobacco leaves. Implication of a transport function for cadmium-binding peptides. *Plant Physiology* 92: 1086–1093.

Willams, C.H. and David, D.J., 1973, The effect of superphosphate on the Cadmium content of soil and plants. *Australian Journal of Soil Research* 11: 43–56.

7

Urban Solid Wastes and the Environment

K. Jeevan Rao[1] and *M.V. Shantaram*[2]

INTRODUCTION

Waste generation, either in solid or liquid form, is a gift of civilization. In large urban communities, huge quantities of solid wastes are generated as a by-product of normal human activity. The wastes generated pose serious problem of collection, transportation and ultimate disposal. Since the ecosystem does not have the capacity to absorb wastes, the accumulation of such wastes poses vast environmental problems. This is no more evident than in the Asia region, where land-based waste disposal has affected the stability of the biosphere. However, on reviewing waste reuse, Kirkham (2001) comments that safe and smart reuse of wastes could lead to significant benefits to the public given that many solid and liquid wastes retain valuable macro- and micro-nutrients, in addition to certain unwanted substances. In this chapter, we present an overview of urban solid wastes and their potential impact on the environment.

The Indian urban population is steadily increasing for various reasons, such as the concentration of industries and better employment opportunities in the urban areas. According to the census data, the number of cities having a population of more than a million has increased from 216 to 296 in the decade 1981–1991, with an increase of 46.87%. Consequently, municipal agencies face difficulties in providing various public services such as education, water supply and public sanitation, including collection and disposal of urban solid wastes (USW) to the desired level. Of these services, municipal solid waste management gets the lowest priority, mainly because disruptions and deficiencies in it do not directly and immediately affect public life and cause public reaction. About 255 million people are estimated to reside in Indian urban areas and about Rs 19,200 million (US$ 576 m) is spent every year on municipal solid waste management. Despite such a heavy expenditure, the present

[1] Associate Professor, Department of Soil Science and Agricultural Chemistry, College of Agriculture, ANGRAU, Rajendranagar, Hyderabad 500 030 A.P., India
[2] Dean of Post Graduate Studies, (Retd.) ANGRAU, Hyderabad 500 030 A.P., India

level of service in many urban areas is low and may be a potential threat to public health in particular and the environment in general (Shekdar, 1999).

CLASSIFICATION OF SOLID WASTES

In general, 'solid waste' is the term used internationally to describe non-liquid waste materials arising from domestic, trade, commercial, industrial, agricultural and mining activities and from the public services (Bhide and Sundaresan, 1983). The disposal of such residue on land creates many environmental and health problems in the cities. So, proper management and safe disposal of solid wastes in cities is needed to create a clean and better living condition for the urban people.

Classification of Solid Waste in Cities

Solid wastes in cities are classified into the following groups (NEERI, 1983).

Urban solid wastes—putrescible (decomposable). Examples are wastes from food slaughter houses, and canning and freezing industries.

Rubbish—non-putrescible wastes (either combustible or non-combustible). Combustible wastes include paper, wood, cloth, rubber, leather and garden wastes. Non-combustible wastes include metals, glass, ceramics, stones, dirt, masonry, paints and some chemicals.

Ashes and residues—Cinders and fly ash of the combustion of solid fuels for heating and cooking or from the incineration of solid wastes by municipal and apartment house incinerators.

Large wastes, demolition and construction rubble—Pipes, lumber, masonry, brick, plastic, roofing and insulating materials, automobiles, furniture, refrigerators and other home appliances, trees, tiles.

Dead animals—Household pets, birds, rodents, zoo animals; also anatomical and pathological wastes from hospitals.

Sewage treatment process solids—Screenings, settled solids, sludge (Varadarajan and Elangovan, 1992).

GENERATION OF URBAN SOLID WASTES

Municipal bodies have to manage the solid wastes arising from residential, commercial and institutional activities along with waste from street sweepings. Normally, the municipal bodies handle all the waste deposited in the community bins located at different places in the city. The municipal solid waste is transported to processing/disposal facilities.

Most municipalities do not weigh their solid waste vehicles, but estimate the quantities on the basis of the number of trips made by the vehicles. Since the density of waste is considerably less than the vehicles are designed to carry, such data on quantity cannot be relied upon. In a number of studies carried out by the National Environmental Engineering Research Institute (NEERI), the waste quantity was measured. The data indicates that the quantity varies between 0.2 and 0.4 kg per capita per day, depending upon the population of the urban centre. In metropolitan cities, quantities up to 0.5 kg/capita/day have been recorded (Table 7.1). The per capita waste quantity tends to increase with the passage of time due to various factors such as increased commercial activities and the standard of living (Table 7.2). Increase in per capita waste quantity is also known to occur at a slightly lesser rate than the increase in GDP/capita. This increase is estimated to occur in India at a rate of 1–1.33% per year. In most Indian cities, the per capita production of refuse ranges from 0.15 to 0.35 kg day^{-1}. The overall average garbage generation per capita is estimated at 0.33 kg day^{-1}. Per capita production of refuse quantity from other countries varies from 0.5 to 1.5 kg day^{-1}. The maximum production of 2.5 kg day^{-1} is reported from the US (Bhide, 1984). Normally, the per capita value does not change appreciably in locations with comparable climatic conditions and living and food habits. But the change in standard of living and density of population per unit area has a perceivable influence on the quantity and quality of garbage generated from urban areas (NEERI, 1983). In many Indian cities, the density of USW is 330–560 kg m^{-3} (Bhide, 1984).

Table 7.1: Per capita quantity of municipal solid wastes in Indian urban centres

Population range (million)	*Average per capita value (kg/capita/day)*
0.1–0.5	0.21
0.5–1.0	0.25
1.0–2.0	0.27
2.0–5.0	0.35
> 5.0	0.50

Source: (Shekdar, 1999)

Table 7.2: Solid waste generated in India's major cities

City	*Waste (tonnes)*
Mumbai	6050
Delhi	4000
Kolkata	3500
Bangalore	2000
Chennai	2000 (USW); 500 (debris)
Lucknow	1500
Hyderabad and Secunderabad	1300
Ahmedabad	1280
Surat	1000

Source: The Hindu, Survey of the Environment

Such a large quantity of waste requires appropriate measures of collection, transporation and disposal. For this, understanding of the composition of wastes and how they should be collected and disposed of is needed (Jeevan Rao and Shantaram, 1994e, 1998b).

CHARACTERISTICS OF URBAN SOLID WASTES

The increasing cost of chemical fertilizers and their limited availability, coupled with the concern for efficient utilization of energy and natural resources, have generated an interest in alternative uses and utilization of urban and industrial wastes. Golueke (1977) claimed that use of organic fertilizer instead of chemical fertilizer can result in a two-thirds energy saving.

The manurial value of compost and its use as a soil conditioner has been subjected to detailed study by several investigators (Gaur et al., 1984; Biswas et al., 1984) because of their advantages over inorganic fertilizers and its long-term beneficial effects. Many of these studies are, however, carried out with composted farm wastes. In contrast, studies involving town refuse (USW) or urban composts as an organic manure are comparatively meagre and of recent origin. Use of urban wastes requires detailed information on their chemical composition, including an estimate of toxic substances. Such information will enable regulatory authorities to develop criteria for dealing with them (NEERI, 1970; Bhide et al., 1972). Information on macro- and micro-nutrients indicate the fertility of USW, while NaCl and toxic metals are indicative of environmental pollutants.

The nature of urban waste understandably varies with country, city, suburb and season and can be determined precisely only by analysis in each particular case. The main sources of these solid wastes, for which a municipality normally assumes responsibility, are domestic premises, shops, offices, hotels, institutions and small factories, together with refuse swept from the streets. Domestic wastes and USW account for some 75% of total solid waste. Table 7.3 provides a comparison of wastes from an Indian city with that of a European city. As personal income rises, paper wastes increase (4% in the Indian city, and 27% in the North European city): kitchen wastes decline; metals and glass increase, total weight generated rises and the density of wastes declines.

Municipal solid waste composition is observed to vary demographically (Table 7.4). The organic content is high due to the common use of fresh vegetables and fruits in the food. The high organic content also necessitates frequent collection and removal of the waste; the paper, glass and plastic content is small. These materials are sold by the citizens to hawkers, who collect and sell them for reuse or recycling. Hence, it is only the fraction that does not have a resale value, is in a non-usable form, and remains in

Table 7.3: A comparison of waste from a typical Indian city and a typical North European city

Item	*Indian*	*N. European*
Paper	4%	27%
Vegetable/putrescible matter	75%	30%
Dust under 10 mm	12%	16%
Metals	0.4%	7%
Glass	0.4%	11%
Textiles	3%	3%
Plastics	0.7%	3%
Other (e.g. stones, ceramics)	7%	3%
Weight/person/day	414 g	845 g
Weight/dwelling/day	2.5 kg (6 persons)	2.5 kg (3 persons)
Density	570 (kg m^{-3})	132 (kg m^{-3})

Source: Trivedi and Gurdeep (1992 a,b)

the waste. The waste contains a high percentage of ash and fine earth. This may be attributed to the disposal of street sweepings in community bins. Similarly, in many cases, the surfaces adjoining the roads are uncovered and a large amount of surface soil is swept away and mixed with the waste materials. The calorific value of Indian solid waste is 800–1000 kcal/kg (on a dry inert-free basis), while the density is 300–500 kg m^{-3}.

In most Indian cities, the volatile or decomposable matter content was 10–30% (Bhide, 1984).

Physical Characteristics

The average physical constituents of USW generated in Indian cities with a population of more than 2 million is: paper 7.07%, rubber and leather 0.87%, plastics 0.86%, metals 1.03%, glass 0.76%, ash and fine earth 31.74%, total compostable matter 41.69%, and miscellaneous matter 15.98% by dry weight (moisture content 31.18%) (Bhide and Sundaresan, 1983). Olaniya and Saxena (1977) reported the physical characteristics of USW from different waste disposal grounds of Jaipur as 3.6–23.60% stone, 0.20–2.80% coal, 0.06–1.90% glass, 0.90–22.2% rags, 0.60–4.56% papers, 0.01–1.70% metals, 0.16–0.30% hairs, 0.30–3.30% leather and rubber, 0.05–3.0% plastics, 0.07–0.50% bones, 18.30–46.90% compostable matter and 34.0–64.50% soil. Similarly, Patel and Tiwari (1990) reported the characteristics of USW from industrial areas of Rourekela as 41.50% construction material, 47.70% dust and cinder, 3.40% scrap metal, 1.10% paper and cardboard, 0.80% glass, 1.20% broken china, 1.00% mica, 1.00% polythene and 2.30% miscellaneous. A similar variation in waste was reported by Jeevan Rao and Shantaram (1993a). The markedly different physical characteristics of wastes from these two cities demonstrates the lack of regulatory control on the nature and composition of material disposed in the city dumps.

Table 7.4: Physico-chemical characteristics of municipal solid waste in Indian cities

Population range (in million)	*Number of cities surveyed*	*Paper**	*Rubber* leather and synthetics*	*Glass**	*Metals**	*Total compostable matter*	*Inert* material*	*Nitrogen as total nitrogen n ** %*	*Phosphorus as P_2O_5** %*	*Potassium as K_2O** %*	*C/N ratio*	*Calorific value in kcal/kg*
0.1–0.5	12	2.91	0.78	0.56	0.33	44.57	43.59	0.71	0.63	0.83	30.94	1009.89
0.5–1.0	15	2.95	0.73	0.35	0.32	40.04	48.38	0.66	0.56	0.69	21.13	900.61
1.0–2.0	9	4.71	0.71	0.46	0.49	38.95	44.73	0.64	0.82	0.72	23.68	980.05
2.0–5.0	3	3.18	0.48	0.48	0.59	56.67	49.07	0.56	0.69	0.78	22.45	907.18
> 5	4	6.43	0.28	0.94	0.80	30.84	53.90	0.56	0.52	0.52	30.11	800.70

* All values are in per cent, and are calculated on wet weight basis.

** All values are in per cent, and are calculated on dry weight basis.

Source: Shekdar (1999)

Such an unscientific approach to dumping limits the potential for alternate uses of USW.

King et al. (1974) analysed the unsorted, shredded municipal refuse from St. Catherines, Ontario, Canada, which contained 71.2% paper, 5.20% plastic, 8.33% metal, 5.0% glass and dust, 1.30% miscellaneous and 48.6% moisture. Clearly, the wastes generated in urban centres in western countries contained more paper, plastic, metal and glass compared to Indian cities. Radia Khatib et al. (1990) studied the municipal waste from Karachi (Pakistan) and reported that it contained 89.5% (dry basis) total recyclable material with plastics 9.85%, paper 10.1%, glass 1.24%, bones 1.85%, metal 0.74%, leather and rubber 0.20% and organic vegetable matter 54.5% as the major components.

Physico-chemical Characteristics

pH

The USW samples were slightly alkaline to alkaline in reaction. The reported pH (1:50 waste:water ratio) in wastes was: Nagpur (refuse) 7.9–8.2 (Bhide and Muley, 1973); Jaipur (refuse) 7.20–8.2 (Olaniya and Saxena, 1977); Rourkela (waste) 6.80–7.40 (Patel and Tiwari, 1990); Kolkata (wastes) neutral to alkaline (Gupta et al., 1984) with its waste composts 7.50–8.10 (Gupta et al., 1986); and Mumbai (wastes) 7.70 to 8.0 (Ramachandran and D'Souza (1990). Bhide and Sunderesan (1983) reported that in Indian urban refuse, the average pH was 7.68. USW in general is usually alkaline in reaction due to the presence of CO_3, HCO_3, Na, K and other alkaline materials.

Electrical conductivity (EC)

The EC of Kolkata city waste composts ranged from 1.04 to 2.05 d Sm^{-1} in 1:2 waste:water extracts (Gupta et al., 1986); while for Hyderabad wastes, the mean values reported were 266 μ Sm^{-1} (Jeevan Rao and Shantaram, 1993b, 1994a).

Organic carbon (OC)

The OC content of Indian city refuse was 12.0–15.3% (Mutatkar, 1985). The reported OC contents for various cities were: Bangalore (waste compost) 9.4%, (Nandkishore, 1980); Jaipur (refuse) 18.90% (Olaniya and Saxena, 1977); Rourkela (waste) 19.8 to 33.8% (Patel and Tiwari, 1990); Delhi (compost) 14.5% (Talashilkar and Vimal, 1985); Nagpur (refuse) 18.9% (Bhide and Muley, 1973); Kolkata (wastes) 8.5 to 10.0% (Gupta et al., 1984); Mumbai (wastes); 10.4% (Ramachandran and D'Souza, 1990) and Hyderabad (wastes) 4.20% (Jeevan Rao and Shantaram, 1993b, 1994a).

These results contrast with the OC content of Ontario refuse, which was reported to be 37.0% (King et al., 1974). The widely different OC content reported for wastes from different cities again confirms the variations in the nature and composition of USW in Indian cities.

The OC content in USW is critical for the evaluation of its suitability for composting. The higher the OC content, the better is the waste for composting and subsequent use as manure.

Chemical oxygen demand (COD)

The COD largely depends on the organic matter content and its extent of decomposition. The COD estimates help evaluate the maturity of refuse during composting. The reported COD for Jaipur city waste was 528 mg/g (Olaniya and Saxena, 1977) and 278 mg/g for Bangalore waste compost (Nandkishore, 1980).

Total nitrogen (N)

In a similar manner to organic carbon content, total N in wastes varies markedly in USW from different cities (Table 7.5).

Table 7.5: Total nitrogen content of USW from selected Indian cities

City	*Total N (%)*	*Reference*
Jaipur	0.67	Olaniya and Saxena, 1977
Nagpur	0.70	Bhide and Muley, 1973
Bangalore	0.58	Nandkishore, 1980
Rourkela	0.49–0.87	Patel and Tiwari, 1990
Kolkata	0.63–0.69	Gupta et al., 1986
Hyderabad	0.24	Jeevan Rao and Shantaram, 1994a
Indian city wastes average	**0.56–0.62**	Mutatkar, 1985

This is not surprising, since N in wastes is largely contributed by organic matter and, therefore, directly proportional to organic matter in wastes.

Total phosphorus

The P content of Indian USW varies from 0.60 to 0.72% (Mutatkar, 1985). Examples are: Rourkela (wastes) 0.43–1.67% (Patel and Tiwari, 1990); Delhi (waste compost) 0.26% (Talashilkar and Vimal, 1985); Jaipur (refuse) 0.62% (Olaniya and Saxena, 1977); Nagpur (refuse) 0.867% (Bhide and Muley, 1973); and Kolkata (waste) 0.28–0.31% (Gupta et al., 1986). The main source of P in USW is the organic matter and other inorganic materials like fine earth, ash, metals, rubber and other components containing P to a certain extent.

Total potassium

The K content of Indian USW varied from 0.60 to 0.72% (Mutatkar, 1985). Examples are: Jaipur (wastes) 0.70% (Olaniya and Saxena, 1977); Kolkata (wastes) 1.67–2.79% (Gupta et al., 1986); and Delhi (waste compost) 0.70% (Talashilkar and Vimal, 1985). The main sources of K in wastes are organic and inorganic materials like fine earth, ash, vegetable matter, and metals.

Total sodium (Na)

The total Na content in USW was reported as 0.3–0.38% (Gupta et al., 1984). The sources for this element are organic matter, fine earth, ash and inorganic salts.

Water-soluble ions

Jaipur city waste contained water-soluble ions like sulphate and chloride, 6750 and 1800 mg/L, respectively (Olamiya and Saxena, 1977). USWs usually carry high amounts of soluble salts. Part of the salts in organic wastes is not initially soluble but dissolves during decomposition of wastes.

C/N ratio

The range of C/N ratio for Indian USW is 25–35 (Mutatkar, 1985). Examples are:

- Jaipur (waste) 28.2 (Olaniya and Saxena, 1977);
- Nagpur (waste) 27.02 (Bhide and Muley, 1973);
- Bangalore (waste composts) 16.24 (Nandkishore, 1980); and
- Rourkela (waste) 24.5–61.10 (Patel and Tiwari, 1990).

This ratio indicates the maturity of the wastes. In general, the C/N ratio is higher in fresh wastes rather than in stabilized wastes. Wastes with lower C/N ratio are preferable for agriculture.

Total trace elements

The total trace element content of wastes varies significantly in wastes from both Indian cities as well as other countries. Such differences in trace element composition have been related to the extent to which industrial wastes were disposed in city dumps. Some examples are:

- Talshilkar and Vimal (1985) reported total trace element content of Delhi city waste compost as Zn 400, Mn 77, Cu 210 and Fe 8500 mg/kg
- King et al. (1974) reported the total trace element content of Ontario (Canada) refuse as Mn 250, Cu 28, Zn 400, Pb 200, Cd 7.0 and Cr 31.0 mg/kg

- Cottrell (1975) reported total trace metal load of Washington city refuse as Fe 9250, Mn 286, Cu 334, Zn 615, Co 2.9 and Cr 17.0 mg/kg
- Ramachandran and D'Souza (1990) reported total Cr as 51.0 and Mn 622.0 mg/kg in Mumbai city waste composts
- The total Cd content of Kolkata city wastes was 4.0–10 mg/kg (Gupta et al., 1984)
- Jeevan Rao and Shantaram (1994a, 1995e) reported total trace metals of Hyderabad (Table 7.6) as Fe 4863 mg/kg (highest) and Cd 2.0 mg/kg (lowest).

Table 7.6: Mean (n=50) chemical characteristics of USW from Hyderabad city

Characteristic	*Unit*		*Value*
pH (1:50)			9.00
EC (1:50)	μ Sm^{-1}		266.00
Organic carbon	%		4.20
C/N			24.00
Total nitrogen	%		0.24
Total phosphorus	%		0.70
Total potassium	%		0.30
Total sodium	%		0.13
Trace metals	**Unit**	**Total**	**DTPA-extractable**
Fe	mg/kg	4863	51.0
Mn	mg/kg	379	15.0
Zn	mg/kg	234	30.5
Cu	mg/kg	113	31.5
Pb	mg/kg	135	29.0
Ni	mg/kg	12.0	2.3
Co	mg/kg	5.5	0.27
Cr	mg/kg	25.0	0.62
Cd	mg/kg	2.0	0.13
Mechanical composition			
(a) Waste > 2 mm			47.0%
(b) Waste < 2 mm			53.0%
(i) Sand			15.0%
(ii) Silt			15.0%
(iii) Clay			60.0%

Source: Jeevan Rao and Shantaram, 1994a

Trace elements in USW are less soluble and migrate more slowly than Na and K. Their utilization or disposal of USW on land adds to the trace element load in the ecosystem. In general, the total trace element load is higher in USW than in normal agricultural soils.

DTPA trace elements

In Kolkata city waste composts, DTPA-extractable trace elements ranged from Fe 84–90, Zn 144–150, Cu 2430 and Mn–38.4 mg/kg (Gupta et al.,

1986). In Mumbai city wastes, Mn and Cr were reported as 45 mg/kg and traces, respectively (Ramachandran and D'Souza, 1990). The DTPA-trace element load is higher in USW than in normal agricultural soils (Jeevan Rao and Shantaram, 1996a).

Cadmium/Zinc ratio

Chaney (1973) suggested that maintaining low (< 0.5%) Cd/Zn ratio in wastes to be applied to agricultural land would ensure against contamination of the food chain by either Zn or Cd. In this respect, USW is safer than sewage sludges for application on agricultural land.

The USWs analysed contained several plant nutrients, heavy metals and other soluble salts. There is a wide variability in several of these parameters. The characteristics of USW provide information on the possible effect of its application to land for agricultural use (Jeevan Rao and Shantaram, 1995e; 1996d). The USWs no doubt contain significant quantities of primary and secondary plant nutrients besides pollutant elements such as Pb, Ni, Co, Cr, Cd and soluble salts which may pose some problems in the long run. The ions like sodium, sulphate and chloride contribute towards salinity and so have a deleterious influence on plant growth; a high concentration of all these ions indicates the possibility of development of salinity and alkalinity hazards due to excessive applications of USW on land. Hence, an effective monitoring system that measures the concentration of a very large number of elements in USW is essential if they are to be used in food crop production (Jeevan Rao and Shantaram, 1993b, 1994a).

ENVIRONMENTAL POLLUTION DUE TO URBAN SOLID WASTES

Currently, most waste is disposed in low-lying areas without taking proper precautions. This has led to the breeding of rodents, flies and other domestic pests. It also tends to pollute both surface waters and groundwater. Hence, it is necessary to adopt sanitary landfilling techniques so that such problems can be avoided (Jeevan Rao and Shantaram, 1994f). Improper and unscientific disposal of urban solid wastes may pollute the air, water and soil in several ways (Liptak, 1974).

Atmospheric Pollution

One of the critical public health hazards from a sanitary landfill is the gasses emanating (CO_2, CO, CH_4 and H_2S) from the decomposition of organic matter under partially-anaerobic conditions. Both CH_4 and H_2 can be explosive. CO_2 can cause acidic reactions when dissolved in water and can be the cause for corrosive problems. Hydrogen sulphide is a known corrosive, malodorous and if produced in high quantity, can be irritating and toxic to both humans and animals. There are demonstrated examples of severe damage caused by CH_4 emissions from the landfills. Horizontal movement and distribution of the gases in dangerous concentrations beyond the confines of the fill up to 300 m away have been reported.

Water Pollution

The land disposal of solid wastes creates an important source of groundwater pollution. Most landfills throughout the world are refuse dumps, rather than sanitary landfills designed and constructed according to engineering specifications. Leachate from a landfill can pollute groundwater if water moves through the fill material (Liptak, 1974). The possible sources of water include precipitation, surface water infiltration, percolating water from adjacent land, and groundwater in contact with the fill.

Entry of pollutants into shallow aquifers occurs by percolation from ground surface, through wells, from surface waters, and by saline water intrusion. The extent of pollution in groundwater from a point source decreases due to natural attenuation, as pollutants move away from the source until a harmless or very low concentration level is reached. As the attenuation rate varies with the nature of pollutants, the distance that pollutants travel will vary correspondingly.

A major concern with groundwater pollution is the fact that it may persist underground for years, decades or even centuries. This is in marked contrast to surface water pollution. Reclaiming polluted groundwater is, therefore, much more difficult, time-consuming and expensive as compared to remediation of polluted surface water. Underground pollution control is achieved primarily by regulating the pollution source and secondarily by physically entrapping, and when feasible, removing the polluted water.

Soil Pollution

A major difficulty in predicting potential hazards associated with land disposal of USW is the inherent variability in the composition of wastes. The unleached composts may be high in available boron (B) and other toxic substances. For example, boron causes severe toxicity when shredded refuse is incorporated into land. Thus, potential phytotoxicity exists for B, Zn, Cu and Ni. Not only is refuse heterogeneous in total content of an element, but the chemical forms of the elements also vary.

WATER POLLUTION BY OPEN REFUSE DUMPS

The land disposal of solid wastes can be a major source of groundwater pollution unless precautions are taken to treat leachates emanating from these wastes. A landfill may be defined as any land area serving as a depository of urban or municipal solid waste (Liptak, 1974). As noted earlier, most landfills throughout the world are open refuse dumps rather than sanitary landfills.

As discussed earlier, leachate from landfill can pollute the groundwater if water moves through the fill material (Jeevan Rao and Shantaram, 1995j). However, the mobility of the pollutants in the soil largely depends on chemical, physical and biological reactions. Therefore, selection of sites with soils which have the capacity to remove contaminants from leachates is critical for an effective landfill sites. Movement of pollutants to groundwater under a landfill can be predicted from the local geohydrologic characteristics of the site (Sawhney and Raube, 1986). Urban refuse usually has a moisture content less than that of field capacity. Therefore, leachate from a landfill can be minimized if water is kept away from the landfill material. In a properly constructed sanitary landfill, any leachate generated can be controlled and prevented from polluting the groundwater (Liptak, 1974). Pollutants frequently found in leachate include BOD, COD, Fe, Mn, Cl, NO_3, hardness, and trace elements. Hardness, alkalinity and total dissolved solids are often elevated in groundwaters at landfill sites. Polluted surface water that contributes to groundwater recharge becomes a source of groundwater pollution.

During decomposition, organic wastes normally produce leachates that dissolve a wide range of materials as they pass through the landfill. Leachate may emerge from landfill with a BOD of over 20,000 mg/L, which is several times stronger than that of raw sewage (Liptak, 1974). Contamination of groundwater is common in areas where the rate of percolation through the soil is high.

Studies on groundwater pollution by open refuse dumps at Jaipur showed that the leachates from refuse had high concentrations of OC, Fe, and Mn and high concentrations of dissolved solids (Olaniya and Saxena, 1977). The leachate from tipped refuse was richer than sewage in N, P, Cl and COD. Where landfill leachates have contaminated the groundwater, the total hardness, TDS, Cl, Fe and COD in well waters decreased with increasing distance from refuse dumps. For instance, well water down to a distance of 50 m was highly polluted, but beyond a distance of about 200 m, the values remained virtually unchanged (Jeevan Rao and Shantaram, 1995d, 1995j), indicating significant migration of contaminants in groundwater.

Some Examples of Groundwater Pollution Around Landfill Sites

Many examples of groundwater pollution have been reported at poorly-designed landfill sites in countries throughout the world. For example Srinivasan (1977) investigated the contamination of groundwater by leachates from refuse dumps in Ontario, Canada. Analysis of surface water and groundwater samples revealed that excessive contamination of groundwater. However, inclusion of a barrier system above the aquifer of dry and silty soil with high capacity to adsorb contaminants reduced the extent of groundwater contamination. Murray et al. (1981) observed severe contamination of the groundwater near sanitary landfill operations at Missouri, US and concluded that the landfill was the main source of water pollution. Tester and Harker (1982) investigated the groundwater pollution at three domestic refuse sites and concluded that pollution plumes (zones) around domestic landfill are restricted and dilution is beneficial in ameliorating the effect of landfill leachate on groundwater quality.

Nicholson et al. (1983) studied groundwater pollution at a landfill in Ontario, Canada. The results revealed that Ca^{2+}, SO_4^{+} and HCO_3^{-} are the dominant ions with maximum concentrations of 400, 2000 and 1200 mg/L, respectively. Beneath the landfill, the most highly contaminated water in the aquifier had a TDS concentration of 4000 mg/L. Sulphate and Fe are the inorganic constituents that exceeded the recommended limits for drinking water. The plume contained above-background concentrations of dissolved Zn and Mn. No heavy metals or other hazardous inorganic trace elements accumulated above the maximum limits for drinking water. The plume contained total dissolved organic carbon (DOC) above background level, ranging from 30 mg/L beneath the landfill to 5 mg/L near the front of the plume.

Gopal et al. (1991) investigated the extent of groundwater pollution in Kanpur city by solid wastes based on various parameters such as pH, alkalinity, Cl (1800–50 mg/L), SO_4 (3000–600 mg/L), BOD (6000–100 mg/L), and hardness (3000–600 mg/L) as initial and after 2.5 months, respectively. The leachate contained excessive pollutants up to one month from the time the refuse was exposed to leaching action. After two months, the values of all other pollutants were within limits, except for hardness, which continued be high.

Jeevan Rao and Shantaram (1994c) studied the quality of runoff water from the Amberpet landfill site at Hyderabad: pH ranged from 8.3 to 8.8; EC from 0.15 to 0.18 μSm^{-1}; dissolved CO_2 from 0.3 to 0.72 mg/L; turbidity 1.4 to 2.1 NTU; and total solids ranged from 14.7 to 181.0 mg/L. These figures show the need to control the runoff.

Jeevan Rao and Shantaram (1995d, 1995j) studied the properties of groundwater in relation to the disposal of refuse at Amberpet and Golkonda landfill sites, Hyderabad. The results showed that the pH and EC and the total dissolved and suspended solids increased in the waters of the wells within 200 m of the fringe of the landfill site. The total hardness and alkalinity values exceeded the prescribed Indian Standard Institution (ISI, 1983) and WHO (1984) standards. The BOD and COD also increased beyond the prescribed limits in some wells. The investigators concluded that the groundwaters at both the landfill sites were polluted and deemed unfit for human consumption and domestic use but could be used for irrigation only. Further, they concluded that unless leachate is properly collected and treated, surface and groundwater may be polluted at all the landfill sites in India. Thus, the use of sanitary landfill technique is the way to prevent groundwater pollution at landfill sites. With the recent restriction and control of disposal of wastes, lowland disposal may increase in most urban centres.

Effect of Landfill Leachate on Soil Characteristics

Sanitary landfills are used widely to dispose of municipal solid wastes. Problems that arise even with apparently properly constructed sanitary landfill include leachate generation when the surface is permeable. The movement of water first through USW dumps and then through soil results in soil pollution.

Winant et al. (1981) studied the effect of sanitary landfill leachate on chemical properties of soil to a depth of 60 cm and reported that the leachate supplied more Ca and Mg than did lime. Appreciable amounts of Ca moved into the soil profile to 30 cm depth. The CEC increased from 21 to 30 C mol (p+) kg soil at 0–5 cm soil depth. Exchangeable Mg content also increased to 30 cm depth. On the other hand, the extractable K in soils was doubled. Exchangeable Mn, Fe and Zn increased in the soil profile. The total N, extractable P, ESP and soil pH increased considerably in the soil profile. Most elements were attenuated to some extent by movement through the soil.

Musmeci et al. (1985) studied the attenuation and movement of some metals such as Fe, Pb, Al, Cd and Cu in soil samples collected from different depths by drilling in two landfills in Italy. Reduction in metal concentrations in the soil layers and the differential release of metals from the soil was correlated to different soil characteristics. Hoertling (1989) reported that all trace metal concentrations increased considerably below the dump at the municipal waste disposal site at Pangui Tung, China.

Jeevan Rao and Shantaram (1994b, 1996c) studied the effect of long-term disposal of USW on soil properties at Amberpet landfill sites in Hyderabad. The results revealed that the soil pH is highly alkaline at some soil depths with substantial increase in electrical conductivity. The content of water-soluble salts and exchangeable Na and K also increased markedly in the soil profile at different depths. The ESP varied from 0.27 in control to 70.6. The total heavy metals and DTPA-extractable trace metals increased substantially. The soil under the USW dumps acts as a storehouse of high concentrations of soluble salts and heavy metal pollutants like Pb, Ni and Cr. It was concluded that the pollutants can cause damage to groundwater through constant and continuous leaching (Jeevan Rao and Shantaram, 1995f).

Effect of the Application of USW on the Soil-Plant System

In many countries treated, USW are commonly used as soil amendments. However, such usage are controlled by strict regulatory guidelines.

The physical, chemical and biological properties of USW produced by domestic and industrial activity are heterogeneous. Continuous use of such wastes on agricultural land for cultivation of crops may result in the pollution of soil. On the other hand, if problems of such pollution are taken care of, the USW can be considered as a very valuable resource for use as a sources of nutrients on agriculture land (Jeevan Rao and Shantaram, 1995k).

Soil Properties

USW has been reported to result in increased soil fertility and water retention in the soil along with decreased soil erosion and fertilizer requirement (Glaub and Golueke, 1989). Hortenstine and Rothwell (1969) reported that municipal compost applications increased P, K, Ca, Mg, soluble salts, Water Holding Capacity (WHC) and CEC in the soil. Srivastava et al. (1971) evaluated the urban compost and reported that compost having C/N ratio of 15:20 was satisfactory for soil application. Terman et al. (1973) reported that the compost made from USW contained N, P, K and Zn and had a considerable liming effect. However, its application caused Zn toxicity in plants (Mays et al., 1973). King et al. (1974) studied the effect of municipal refuse on crop production by applying pulverized refuse at two rates: 188.0 and 376.0 t/ha. Soil pH at 0–15 cm depth was not affected, but a slight decrease in pH occurred in the 15–30 cm layer wherever refuse had been applied. Increase in conductivity values were below the 500 μSm^{-1} level which is the upper limit for proper seed germination.

Haan (1981) reported results of municipal waste compost research conducted over more than 50 years at the Institute for Soil Fertility at Haren, Netherlands. He found high micro-elements and Ca and S levels in the city refuse composts and low contents of N, P and K. Biennial applications of upto 40 t/ha. of compost between 1948 and 1975 resulted in a large increase in micro-element concentrations in soils.

Kowald et al. (1982) reported that the applications of 40, 80 or 120 t/ha. refuse compost over a two-year period increased the soil pH from 5.1 to 5.9, 6.2 and 6.5, respectively. Carbon content of the soil increased from 1.25 to 1.59%. Phosphorus and K contents increased with increasing rates of application. Increase in soil pH, K_2O, P_2O_5 and N and of Ni, Cr, Cd and Pb in sandy soil were reported by Schossing (1983) on application of 156 or 311 tonnes dry matter of refuse compost per hectare to 35 cm depth.

However, Pelletin et al. (1986) did not find any accumulation of metallic elements in the soil on application of town refuse compost. They concluded that there was no indication that fresh composts should not be used.

Giusquiani et al. (1988) studied the influence of urban waste compost on soil fertility under laboratory conditions. No change of pH and CEC values were observed, whereas the compost addition increased the available phosphorus. Exchangeable potassium increased threefold and Zn markedly; available Mn also increased, but available Fe and Cu concentrations were not affected. It was concluded that the marked increase in Zn content could cause problems of phytotoxicity. Application of municipal refuse compost improves the physical structure of the soil and alkali-soluble substances (Hernando et al., 1989).

Murillo et al. (1989) reported that successive application of city waste compost on rye grass and tomato increased soil N, oxidizable organic matter, P and CEC in the top 0–10 cm soil and at a depth of 12–22 cm, following applications of 150 and 400 tonnes of compost per ha. However, a parallel significant increase of heavy metals (both total and DTPA-extractable fractions) was also observed at the two depths. Sodium content occasionally reached levels close to three per cent, which could cause problems, and it was suggested that compost quality control would be an important factor.

Petruzzelli et al. (1989) studied the heavy metal extractability of compost-treated soil, Copper concentration increased from 16.0 to 25.4, Zn 21.1 to 37.4, Pb 9.2 to 16.1, Cr 1.3 to 1.9, Ni 3.8 to 6.7 mg/kg in DTPA-extractable form. Soil pH increased by 0.50. The electrical conductivity increased by 0.04 S m^{-1} and 0.20 S m^{-1} on treatment of soil with compost at a low rate and the highest rate of application (30 t/ha. per year), respectively. Soil OC increased from 0.09 to 0.17% with the highest rate of compost application.

Jeevan Rao and Shantaram (1994a, 1995g, 1997a, 1998a) studied certain physico-chemical properties and micronutrient and heavy metal contents of soils affected by long-term application of USW in agricultural fields around Hyderabad. Results revealed that the soil pH increased from 6.1 to 8.5, EC from 0.045 to 1.85 dS m^{-1}, OC 0.48 to 1.85%, CEC 8.90 to 31.55′c mol (p+) kg soil (Table 7.6). The total and DTPA Zn, Cu, Mn, Pb, Ni, Co, Cr and Cd concentrations increased in the soil. It was suggested that the buildup of DTPA trace metals in high amounts in soil (Tables 7.7 to 7.12) may cause phytotoxicity in the long run.

Table 7.7: Physico-chemical characteristics of soils treated with urban solid wastes

Location	Depth (cm)											
	pH (1:2)			EC (dS m^{-1}) (1:2)			Organic carbon (%)			CEC (c mol (p+) kg soil) soil		
	0–15	15–30	Mean	0–15	15–30	Mean	0–15	15–30	Mean	0–15	15–30	Mean
Site-1	7.9	7.2	7.6	0.372	0.281	0.326	1.60	1.20	1.40	31.51	28.42	29.96
Site-2	8.2	8.2	8.2	0.392	0.545	0.468	1.30	1.30	1.30	21.80	23.76	22.78
Site-3	8.5	8.2	8.4	0.648	0.482	0.565	1.00	1.00	1.00	17.50	18.40	17.95
Site-4	8.2	8.1	8.2	2.500	1.200	1.850	1.90	1.80	1.85	33.0	30.11	31.55
Site-5	8.4	8.5	8.5	0.865	1.440	1.152	1.60	1.80	1.70	18.90	16.81	17.86
Site-6	8.5	8.5	8.5	1.300	0.178	0.739	0.60	0.67	0.63	10.30	7.50	8.90
Site-7	8.1	8.2	8.2	0.090	0.080	0.085	1.60	1.20	1.40	16.40	14.48	15.44
Control	6.0	6.1	6.1	0.040	0.050	0.045	0.54	0.42	0.48	6.85	11.12	8.98

(*Source:* Jeevan Rao and Shantaram, 1995g).

It is apparent from the above studies that the widely divergent compositions of urban wastes have significantly different effects on soil properties. Excessive metals can lead to phytotoxicity. This could also inhibit seed germination and the establishment of plants (Hortenstine and Rothwell, 1969; Wong, 1985; Jeevan Rao and Shantaram, 1995b). For this reason, there is a need to examine the composition of both wastes and the properties of soils before USW application to land.

Beneficial Effects of Urban Solid Wastes on Crop Yield

Almost all investigations using USW have reported a significant increase in crop yield. However, the extent of increase varied with the native soil nutrient content, type of plants grown, and toxic metal contents. These observations reflected in research conducted in the UK (Garner, 1966), US (Hortenstine and Rothwell, 1969; Terman et al., 1973), Chile (Sotomayor, 1979), China (Wong, 1985) and Australia (Naidu, 2001).

Similar studies in India by Jeevan Rao and Shantaram (1996b) in a pot culture experiment found an increase in dry matter yield of maize over control with application of USW. However, a decrease in dry matter production at the 44 t/ha. level of waste treatment compared with other treatments. This effect was attributed to the strong complexation reaction between the humic acid fraction of organic matter and nutrients and biological immobilization of native soil nitrogen and other macro- and micro-nutrients.

Elemental Uptake by Plants

Potential benefits of municipal solid wastes in providing plant nutrients such as N, P, Ca and organic matter have long been recognized (Chang et al., 1978). However, potentially harmful effects and introduction of heavy metals such as Cd, Cr or Hg into the food chain are of concern (Logan

Table 7.8: Total micronutrients content in soils treated with urban solid wastes

Location	Depth (cm)											
	Iron (Fe)			Zinc (Zn)			Copper (Cu)			Manganese (Mn)		
	0–15	*15–30*	*Mean*	*0–15*	*15–30*	*Mean*	*0–15*	*15–30*	*Mean*	*0–15*	*15–30*	*Mean*
	ppm											
Site-1	7677.00	7452.00	7564.50	160.00	154.00	157.00	60.20	54.00	57.10	380.00	385.00	385.50
Site-2	7616.40	4677.50	6146.95	87.40	64.20	75.80	26.70	24.00	25.35	470.00	230.00	350.00
Site-3	3550.00	3140.00	3345.00	69.40	72.60	71.00	20.40	22.50	21.45	200.00	130.00	165.00
Site-4	7859.30	7677.00	7768.39	178.00	175.40	176.70	70.40	73.30	71.85	360.00	370.00	365.00
Site-5	5556.30	4980.60	5268.45	167.90	151.90	159.90	64.30	62.20	63.25	310.00	360.00	335.00
Site-6	2870.00	2720.00	2795.00	40.90	36.90	38.90	4.10	4.30	4.20	90.00	91.00	90.50
Site-7	3430.00	3250.00	3340.00	35.10	31.20	33.15	6.20	5.40	5.80	170.00	164.00	167.00
Control	14600.00	1480.00	14700.00	15.00	12.10	13.80	8.50	6.50	7.50	180.00	210.00	195.00

(*Source:* Jeevan Rao and Shantaram, 1994a)

Table 7.9: Total heavy metal content in soils treated with urban solid wastes

Location	Depth (cm)														
	Lead (Pb)			Nickel (Ni)			Cobalt (Co)			Cadminum (Cd)			Chromium (Cr)		
	0–15	15–30	Mean	0–15	15–30	Mean	0–15	15–30	Mean	0–15	15–30	Mean	0–15	15–30	Mean
	ppm														
Site-1	172.00	168.20	170.10	11.00	9.50	10.25	7.00	6.50	6.75	0.80	0.70	0.75	36.90	31.50	34.20
Site-2	98.00	65.70	81.85	6.40	1.70	4.05	8.00	2.00	5.00	0.80	0.90	0.85	13.50	24.00	18.75
Site-3	64.20	78.90	71.55	8.70	3.30	6.00	3.00	3.00	3.00	0.80	0.80	0.80	36.90	35.00	35.95
Site-4	149.50	126.00	137.75	1.00	12.60	6.80	5.00	9.00	7.00	0.80	0.80	0.80	50.00	20.50	35.25
Site-5	124.50	104.00	114.25	8.40	8.90	8.65	2.00	5.00	3.50	0.70	0.90	0.80	46.00	49.70	47.85
Site-6	71.60	84.80	78.20	1.10	5.80	3.45	3.00	1.00	2.00	0.80	0.70	0.75	47.50	21.80	34.65
Site-7	12.00	10.60	11.30	1.00	0.90	0.95	7.00	6.50	6.75	0.50	0.30	0.40	6.50	5.40	5.95
Control	6.00	4.30	5.15	2.50	2.00	2.25	5.50	4.40	4.95	0.50	0.40	0.45	7.00	6.20	6.60

(*Source:* Jeevan Rao and Shantaram, 1994a)

Table 7.10: DTPA-extractable Micronutrients in Soils Treated with Urban Solid Wastes

Location	*Depth (cm)*											
	Iron (Fe)			*Zinc (Zn)*			*Copper (Cu)*			*Manganese (Mn)*		
	0-15	*15-30*	*Mean*	*0-15*	*15-30*	*Mean*	*0-15*	*15-30*	*Mean*	*0-15*	*15-30*	*Mean*
Site-1	57.00	48.00	52.50	15.20	14.20	14.75	28.60	26.40	27.50	19.80	20.10	19.95
Site-2	6.60	12.40	9.50	4.00	3.70	3.85	4.30	5.30	4.80	17.20	13.00	15.10
Site-3	16.00	12.00	14.00	6.20	5.20	5.70	6.60	6.20	6.40	12.20	4.00	8.10
Site-4	13.00	24.40	18.70	9.10	12.00	10.55	17.80	26.70	22.25	25.80	12.40	19.10
Site-5	7.20	11.60	9.40	7.30	7.00	7.15	16.80	18.30	17.55	11.00	8.80	9.90
Site-6	9.40	11.20	10.30	3.30	2.60	2.95	1.10	1.00	1.05	22.40	10.80	16.60
Site-7	4.50	3.50	4.00	2.50	1.40	1.95	0.92	0.81	0.86	3.60	2.80	3.20
Control	6.50	4.80	5.65	1.00	0.62	0.81	0.90	0.72	0.81	9.00	11.00	10.00

(*Source:* Jeevan Rao and Shantaram, 1994a)

and Chaney, 1983). The validity of such concern and, ultimately, the success or failure of USW utilization in agriculture depends on the properties of the municipal USW loading rates and on the chemical and biological characteristics of the soils to which organic solid wastes are applied.

The availability of metals and uptake by plants are related to their total concentrations, to their forms and associations in the soil, and to the various physical and chemical factors operating at the soil-root interface. The influence of plant species on the soil-plant relationship may be considerable (Fleming, 1965; Archer, 1971). Different species, and indeed different cultivars, regulate metal uptake at both the soil-root and root-shoot interfaces to varying degrees (Jeevan Rao and Shantaram, 1995a). In general, the trace element content of the whole plant tends to increase with increasing maturity, and seasonal influences are also important (Fleming, 1965). The effect of land disposal of wastes on levels of micro-elements in leaves and edible plant parts is important because this is the first step in the entry of toxicants into the food chain. Moreover, the presence of toxic substances in food crops also impairs the local and international marketability of produce. This is well regulated through national guidelines on toxic substances concentration in food, and it is duly quoted in US EPA guidelines.

The effect of USW on the elemental composition of crops has been investigated by many researchers throughout the world. In India, D'Souza and Ramachandran (1984) studied the uptake of Fe, Mn, Cu, Zn, Co, Pb, Ni, Cd and Cr by maize (*Zea mays* L.) grown on a laterite soil (Oxisol) and medium black soil (Vertisol) that were treated with different doses (0, 56, 112, 224 and 448 t/ha.) of city compost from Mumbai. The results revealed that the compost increased plant Zn concentration with higher doses in both the soil types. Significantly higher concentration of Cu, Co, Pb, Ni and Cd in plants were obtained in compost-treated Oxisol but not

Table 7.11: DTPA-extractable heavy metals in soils treated with urban solid wastes

Location	*Depth (cm)*														
	Lead (Pb)			*Nickel (Ni)*			*Cobalt (Co)*			*Cadminum (Cd)*			*Chromium (Cr)*		
	0–15	*15–30*	*Mean*	*0–15*	*15–30*	*Mean*	*0–15*	*15–30*	*Mean*	*0–15*	*15–30*	*Mean*	*0–15*	*15–30*	*Mean*
	ppm														
Site-1	26.00	24.10	25.05	1.00	0.80	0.90	0.34	0.30	0.32	0.23	0.20	0.21	0.27	0.22	0.24
Site-2	7.20	8.00	7.60	0.46	0.35	0.40	0.36	0.30	0.33	0.15	0.16	0.15	0.34	0.54	0.44
Site-3	7.00	8.00	7.50	0.40	0.31	0.35	0.28	0.22	0.25	0.15	0.15	0.15	0.29	0.15	0.22
Site-4	17.70	27.20	22.45	0.67	0.77	0.72	0.34	0.28	0.31	0.17	0.26	0.21	0.27	0.24	0.25
Site-5	13.40	12.70	13.05	0.37	0.40	0.38	0.22	0.22	0.22	0.18	0.17	0.17	0.50	0.36	0.43
Site-6	2.10	3.00	2.55	0.19	0.28	0.23	0.28	0.26	0.27	0.10	0.12	0.11	0.25	0.35	0.30
Site-7	0.10	0.10	0.10	0.12	0.10	0.11	1.00	0.80	0.90	ND	ND	ND	0.50	0.30	0.40
Control	0.04	0.02	0.03	0.05	0.02	0.03	0.04	0.02	0.03	ND	ND	ND	ND	ND	ND

ND—Not detected (*Source:* Jeevan Rao and Shantaram, 1994a)

Table 7.12: Mean relative availability (%) of heavy metals in soils treated with USW

Site-1	Cu (48.19) > Cd (28.66) > Pb (14.71) > Zn (9.39) > Ni (8.71) > Mn (5.21) > Co (4.73) > Cr (0.72) > Fe (0.69)
Site-2	Cu (19.09) > Cd (18.26) > Ni (13.88) > Pb (9.75) > Co (9.75) > Zn (5.16) > Mn (4.65) > Cr (2.38) > Fe (0.17)
Site-3	Cu (29.95) > Cd (18.75) > Pb (10.51) > Co (8.33) > Zn (8.04) > Ni (6.99) > Mn (4.58) > Cr (0.60) > Fe (0.42)
Site-4	Ni (36.55) > Cu (30.85) > Cd (26.87) > Pb (16.70) > Zn (5.97) > Mn (5.25) > Co (4.95) > Cr (0.85) > Fe (0.24)
Site-5	Cu (27.77) > Cd (22.29) > Pb (11.48) > Co (7.70) > Zn (4.47) > Ni (4.44) > Mn (2.99) > Cr (0.90) > Fe (0.18)
Site-6	Cu (25.03) > Mn (18.37) > Co (17.66) > Cd (14.82) > Ni (11.04) > Zn (7.55) > Pb (3.23) > Cr (1.06) > Fe (0.37)
Site-7	Cu (14.91) > Co (13.29) > Ni (11.55) > Cr (6.59) > Zn (5.80) > Mn (1.90) > Pb (0.88) > Fe (0.12) > CR (ND)
Control	Cu (10.64) > Zn (5.78) > Mn (5.11) > Ni (1.50) > Co (0.58) > Pb (0.56) > Fe (0.04) > Cd (ND) > Cr (ND)

Values in parentheses indicate mean relative availability (%) of elements; ND—Not detected (*Source:* Jeevan Rao and Shantaram, 1999)

in Vertisol. These investigations did not explain the difference in element uptake between the two soil types. Uptake of Cr was unaffected, and Mn and Fe concentrations in maize shoots were significantly reduced in the Oxisol with addition of compost but not in the Vertisol.

Jeevan Rao and Shantaram (1995c, 1996b) conducted a pot culture experiment by using different levels of USW (11, 22, 33 and 44 t/ha.) with maize as a test crop. The uptake of Fe by maize decreased at all the levels of waste application. Similarly, Cu uptake was also affected by organic wastes application. The Mn uptake increased up to 33 t/ha. waste addition, then decreased at 44 t/ha. wastes application. Plant Zn uptake by maize increased significantly at all levels compared with control. The data on heavy metal uptake by maize plant revealed that the plant lead and nickel uptake increased significantly over control. The cobalt uptake was also increased significantly. There was no significant differences in uptake of Cd and Cr; these two heavy metals occurred only in traces in maize plants. Although the contents of heavy metals in maize did not attain or exceed toxic dietary limits, the result should be viewed with caution because of the potential health hazard once they reach the food chain.

Jeevan Rao and Shantaram (1995a, 1995h) evaluated the status of heavy metals and micronutrients in crop plants grown on farmers' fields where fresh USW had been used regularly for the previous ten years or more. Iron content of plants varied, depending on plant species and plant parts studied. Not surprisingly, the Fe content of roots was greater than the plant tops. The Mn content of most of the plant tops was in the range 15–100 mg/kg. The Zn content of all the plant tops was within the range 15–200 mg/kg. The concentrations of Cu in the plant shoots was in the 4–15 mg/kg range. The Cu levels in plants increased more than 5 mg/kg, a fact which has to be viewed seriously as this might cause phytotoxicity due to continuous and uncontrolled application of USW on land. Generally, the contents of lead in plant tops were in the range of 0.1–10 mg/kg. Significant differences in Pb uptake and distribution in plant parts was observed. The nickel contents of most of the shoots were more than 1.0 mg/kg and differed markedly between species. The cobalt concentration of plant shoots studied were below 1.0 mg/kg and roots accumulated more cobalt than shoots and edible parts. The Cr and Cd concentration in the shoots of plant species studied were below 1.0 mg/kg. The edible portions of the plant species studied contained less Cr and Cd and, therefore, did not pose any serious problem when consumed either by animals or man. Roots accumulated more Cr and Cd than the above-ground parts, suggesting the need to carefully monitor the stage when root crops are cultivated on USW-treated soils. It was concluded that Fe, Mn, Cr, Cd and Co did not pose any pollution problem in the food chain, whereas elevated levels of Zn, Cu, Pb and Ni were observed in some of the plant species studied.

Chu and Wong (1987) studied the heavy metal contents of vegetable crops grown on loamy sand soil treated with at 25, 50, 75, 100, 125 and 150 t/ha. of refuse compost in pots and sown with *Brassica chinensis* (mustard), *Daucus carota* (carrot) and *Lycopersicon esculentum* (tomato) in a glasshouse. The heavy metals were found to be higher in the roots than in the aerial parts of mustard and tomato; but the reverse was found in carrot. It was concluded that in the edible tissue of the three crops, tomato accumulated metals to a lesser extent than the other two.

Petruzzelli et al. (1989) reported results of a four-year field experiment on the influence of repeated refuse compost additions at the maximum rate allowed by Italian legislation (30 t/ha/yr) on the transfer of heavy metals from the soil to a corn crop. Cadmium in corn plants increased particularly in the roots from 0.22 to 1.31 mg/kg. Concentration of Zn and Cu increased in grains from 26.8 to 35.8 and from 2.4 to 4.2 mg/kg, respectively. Relevant increase in the roots was detected for Zn from 34.6 to 146.8 mg/kg only in the fourth year. Ni concentrations increased in the root portion while the content of Pb and Cr in corn were generally unaffected by compost application.

Other Potential Problems

Most liquid and solid wastes contain significant contents of salt that can cause salinity with long-term application. To prevent salinity and alkalinity problems from organic waste application, the following precautions should be taken (Jeevan Rao and Shantaram, 1995g).

Soluble salt and sodium content of wastes should be determined before applying wastes on agricultural lands.

Soil characteristics need to be determined before and after each application of wastes.

The amounts of leaching water required to maintain a suitable EC level in the soil are to be determined.

ENVIRONMENT-FRIENDLY MANAGEMENT OF USW

The environment-friendly practices to dispose of USW are as follows (Jeevan and Shantaram, 1998b).

Landfills

At present, most of the municipal waste is dumped in low lying areas without proper precautions. This results in breeding of rats, flies and other vermin and pollution of surface water as well as groundwater. It is necessary to adopt sanitary landfilling techniques so that the sites can be put to beneficial use much earlier.

Energy Source

The calorific value of municipal refuse can be utilized by burning it in excess of air as in incineration, in the absence of air as in parolysis, or by biochemical decomposition to produce biogas. It is, therefore, necessary to standardize the method and popularize it so that at least the domestic energy requirement is taken care of (Bhide and Sundarasen, 1983).

Waste Recycling

Recycling is only the first step in a development process. Cities that recover their materials avoid paying disposal costs. But the real benefit to the local economy comes from converting scrap into useful products. Recycling has become a necessity, as many countries are unable to locate sites suitable for USW disposal since land is scarce. The US and Japan are among the leaders in the recycling trade. In Japan especially, recycling is very methodical due to a high level of civic awareness. Some 35 % of Japan's paper comes from the wastepaper basket.

Composting as a Source of Plant Nutrients

Compost from city refuse can be used with chemical plant nutrients as it improves the soil fertility. The solid waste could be recycled and reused for many beneficial purposes.

Manure by composting

Composting is a microbiological, non-polluting and safe method of disposal and recycling by converting wastes to manures. Composting is practised in Delhi, Mumbai, Kolkata and other Indian cities. In USW farms of East Kolkata, 25 varieties of vegetable are grown in a year without addition of chemical fertilizers (Christine, 1989). Manures from compost yards are sold to farmers in many of the towns and cities of Tamil Nadu.

Mechanical composting plants

Rural methods of composting are not proving suitable for handling wastes from towns and big cities. The mechanization of composting has several advantages such as environment sanitation to minimize pollution, recovery of discarded materials and production of compost in less time. It is considered to be better suited for bigger cities, particularly where USW is rich in organic material. The large metroplitan cities like Mumbai, Kolkata, Delhi and Chennai generate some 2000–4000 tonnes of urban wastes per day, causing disposal problems.

The ideal size of compost plants may be those with a daily intake capacity of about 100–150 t of USW. The extent of mechanization should be tailored so as to suit local conditions due to variations in the characteristics of urban wastes from city to city. Experience in India shows that in spite of being heavily subsidized by the government, mechanical compost plants have not been very successful for various reasons. A status report by the Ministry of Agriculture in 1985 showed the capacity utilization of plants created in several cities to be 8–43%, with most of them not operational above one-third capacity.

Quality of compost from city wastes

Composts produced at mechanical compost plants are poor in plant nutrients and generally contain about 0.5–0.8% N, 0.5–0.7% P and 0.5–0.8% K. To improve the quality of compost produced in city plants, better quality USW containing a high percentage of organic matter should be used. This will reduce the transportation of inert materials, help in effective capacity utilization and produce good quality compost. Blending of sewage and sludge during composting as a source of nutrients and moisture may be beneficial. Frequent analysis of the compost samples in order to monitor the quality should be carried out and enrichment of compost with cheaper sources of P (rock phosphate) to reduce the cost of production per unit of plant nutrient should be examined.

Standards for the composition and characteristics of compost in most cases are not available though this aspect is receiving attention. For example, Anthonis (1994) has suggested the following:

Minimum nutrient content: 1–3% N, 1.5–3% P_2O_5, 1–1.5% K_2O

Moisture content: not exceeding 15–25%

Organic matter: at least 20% carbon

C:N ratio: between 10:1–15:1

pH: around neutral (6.5-7.5)

Vermi-composting of USW

Vermi composting of wet waste by using earthworms is an ideal method of compost-making on a small scale. One can use two different types of earthworms; one is a surface feeder and the other a burrower. The worm castings (vermin-compost) is an ideal pot culture medium and it converts the wet wastes into a useful value-added product. Using earthworms to convert USW to manure has been tried by the Mumbai Green Society. The native species *Pheretima elongata* has been used. The Corporation of Mumbai now uses this method to produce manure from 35 tonnes of USW a day.

Possible hazards and toxicities

The presence of pathogens, undesirable heavy metals, toxic concentration of micronutrients and nitrate hazards are some of the major problems of recycling of solid and liquid municipal wastes. Although the methodologies for treatment of excreta, sewage and sludges have been partly established, little has been done to establish and maintain the treatment plants in many megacities of India. Suitable spaces are not available for setting up treatment plants close to the various discharge points and pumping stations.

Both liquid and solid wastes contain a variety of hazardous heavy metals, including Pb, Cd, Cr, and Ni, added to high concentration of micro-elements such as Fe, Mn, Cu, and Zn which may accumulate in soil or in plant bodies or even leach into the groundwater. Nitrate that is highly mobile may pollute both surface waters and groundwaters.

While the installations of waste treatment plants and their operation and maintenance are costly propositions, the amount spent on the establishment of treatment plants would be positively balanced by the removal of pathogens, reduction of diseases in the community, improvement of human health, and maintenance of an eco-friendly environment.

Use in Urban Forestry

USW can be used either in urban forestry or in energy plantation programmes. Energy plantations may play a significant role in the solution of municipal waste disposal problems in metropolitan cities. Most products from these plantations are not consumed by humans although some animals eat forage. There is no reason to suspect that products like timber, fuel and paper prepared from trees grown under municipal wastes applications would represent a health hazard. Another important feature of these plantations is the relative long cropping periods. Moreover, these wastes are potentially useful in nursery production of planting stock to regenerate trees in green belts.

Chemical Products

Cellulose in USW can be converted into glucose enzymatically or through chemical hydrolysis. The glucose can be fermented to produce ethanol. Other useful chemicals like acetone, organic acids and glycerol can also be extracted in this manner.

Lightweight Aggregate

An Australian research and development company has evolved a process of converting municipal waste into lightweight aggregate. Here, the municipal solid and liquid waste are mixed with clay to produce a ceramic rock called Neutralite, graded for use in construction and building.

Refuse-derived Fuel (RDF)

The combustible refuse-derived pellets which have a 10% ash content (against 35% in coal) and do not contain sulphur but only carbon dioxide could well be a suitable alternative to coal for textile mills and chemical industries. They can also be used as a domestic fuel. There is an urgent need to have joint-action forums to educate the people for safe disposal of USW. Concerted efforts by government and non-government agencies along with the people's participation can make any city clean.

THE PATH AHEAD

There are some problems of environmental pollution due to the disposal of urban solid wastes on land. This calls for further research in the following areas.

- No proper established criteria are available for recommending the optimum loading rates of soils to receive urban solid wastes as a source of organic matter and nutrients.
- The build-up of soluble salts and heavy metal contents and ways to mitigate this situation need to be worked out.
- Reactions of organic constituents of USW and their microbial breakdown products with metals need further investigation.
- Development of guidelines for USW disposal in relation to toxic heavy metals on agricultural land is essential before promoting its utilisation in agriculture.
- Data needs to be more widely available on relationships between the concentration of metals in soil and uptake by plants, and the availability of metals in polluted soils over a period of time.
- Data needs to be more available on critical concentrations of heavy metals in food plants in relation to suggested dietary limits.
- Studies on waste, soil and crop management practices and their role in minimizing the pollution are required.
- Biological pollution of soils in relation to land disposal of USW needs to be investigated.
- Similarly, physical changes of the soil associated with use of USW need to be studied.
- There is a need to standardize soil-testing methods for polluted soils and organic wastes.
- The information on retention mechanisms and factors influencing the form and long-term behaviour of metals in soils is essential to develop practicable recommendations and management techniques for the application of wastes to soils.
- Maximum loading rates need to be determined, based on the capability of the soil to absorb these wastes without creating a pollution hazard or adversely affecting the soil's future productive capacity.

- Rate of soil organic matter (SOM) decomposition and the dynamics of soil organic matter in different soils under different soil, water and crop management situations need to be studied.
- Characterization of organic matter of different soils and identification of specific organic constituents responsible for their specific role on nutrient supply, growth promotion, yield and quality improvement is necessary.
- Microbial biomass (or soil biomass) forms an important component of soil organic matter. It is perhaps the seat for Mineralization–Immobilization–Turnover (MIT) and forms the 'organic pool' for the so-called 'isotopic exchange' of C and N, which could be an apparent cause for the relatively low recovery of ^{15}N from applied fertilizer and/or an inflated uptake of soil N. (The other cause for more apparent recovery of fertilizer N without use of tracers is 'priming action'.)
- How do we as soil scientists help maintain agricultural sustainability by appropriate soil management that would mean a productive and sustainable agriculture?
- How could the organic residues be managed so that the four major attributed benefits of organic matter are obtained—nutrient retention and supply, water availability, detoxification (through chelation of toxic metals and degradation of pesticides) and soil conservation through reduced erosion losses by regulating processes involved such as decomposition, ion exchange, detoxification, chelation and by manipulating the choice of residues in appropriate combinations?
- Research is needed to quantify the decline in organic matter vis-à-vis tillage operations.
- There is evidence that net mineralization of N from added organics depends on lignin, polyphenol and N content of the material. Polyphenol/N and (lignin + polyphenol)/N ratios have an inverse relation with net N mineralization. There is a need, therefore, to analyse the organic materials for these components.
- Research is needed to evaluate the effect of organic materials on increasing the efficiency of fertilizers and also to evaluate crop yield response to various combinations of organic and chemical fertilizers for a sustainable production system.
- There is also a need to improve the efficient and effective use of organic materials in cropping systems (best mode/method of application—surface mulched, ploughed down, dished in, side-dressed or placed), proper rate and time of application, best sequence of crops to be grown and rate of release of nutrients.

CONCLUSIONS

It is evident that using USW as fertilizer is an attractive proposition, but this process requires a careful evaluation of the possible effects on soil and plant quality. There is a large potential to recycle the nutrients contained in USW provided that care is exercised to closely monitor the hazards like accumulation of salts, sodium, toxic heavy metals and biological pollutants. Most of the potential problems associated with USW application on land can be avoided by judicious management through proper selection of wastes, soils and crops and proper reliance on soil and soil-plant testing.

REFERENCES

Anthonis, G., 1994, Standards for organic fertilizers. *Agro-chemical News-in-Brief* 17(2): 12.

Archer, F.C., 1971, Factors affecting the trace element content of pastures. In: Ministry of Agriculture and Fisheries Federation. Trace Elements in Soils and Crops. *Technical Bulletin* 21. HMSO, London: 150–157.

Bhide, A., 1984, Urban Solid Waste Management—An Assessment. In: National Environmental Engineering Research Institute, Silver Jubilee Comemoration Volume. Nehru Marg, Nagpur: 219–225.

Bhide, A.D., Alone, B.Z., Titus, S.K., Trivedi, R.C. and Dave, J.M., 1972, Quality and quantity of refuse from Kolkata City. *Indian Journal of Environmental Health* 14: 80–87.

Bhide, A.D. and Muley, V.U., 1973, Studies on pollution of ground water by solid wastes. In: Saraf, R.K. (Ed), *Proceedings of Symposium on Environmental Pollution*. Nagpur: 236–243.

Bhide, A.D. and Sundaresan, B.B., 1983, Solid Waste Management in Developing Countries. INSDOC State of the Art Report Series-2, New Delhi.

Biswas, T.D., Das. B., and Varma, H.K.G., 1984, Effect of organic matter on some physical properties of soils in the permanent manurial experiments. *Bulletin, National Institute of Science,* India, 26: 142–147.

Chang, A.C., Page, A.L., Lund, L.J., Pratt, P.R. and Bradford, G.R., 1978, Land application of sewage sludge—A field demonstration. Final Report for Regional Wastewater Solids Programme. Department of Soil and Environmental Science. University of California, Riverside.

Chaney, R.L., 1973, Crop and food chain effects of toxic elements in sludges and effluents. In: Proceedings of Joint Conference on Recycling Municipal Sludges and Effluents on Land. National Association of State Universities and Land Grant Colleges, Washington DC: 129–141.

Christine, F., 1989, Appropriate technology for urban wastes in Asia. *Biocycle,* July: 56–57.

Chu, L.H and Wong, M.H., 1987, Heavy metal contents of vegetable crops treated with refuse compost and sewage sludge. Plant and soil: 103 : 191-197.

Cottrell, N.M., 1975, Disposal of municipal wastes on sandy soils: Effects on plant nutrient uptake. MS Thesis, Oregan State University, Corvallis, Oregan, US.

D'Souza, T.J. and Ramachandran, 1984, Plant uptake of micronutrients and toxic heavy metals from two contrasting soils amended with municipal sewage sludge and city compost. In: Seminar on Soil Resources and Productivity Management. Abstract of Papers. IARI, New Delhi: 104–105.

Fleming, G.A., 1965, Trace elements in plants with particular reference to pasture species. *Outlook Agriculture* 4: 270–285.

Gaur, A.C., Neelakantan, S. and Darsan, S., 1984, *Organic Manures*. I.C.A.R., New Delhi: 107 pp.

Gallardo-Lara, F. and Nogales, R., 1987, Effect of the application of town refuse compost on the soil-plant system: A review of biological wastes 19: 35–62. *Soils and Fertilizer Abstracts* 508: 1987.

Garner, H.V., 1966, Experiments on the direct, cumulative and residual effects of town refuse, manures and sewage sludge at Rothamsted and other centres, 1940-47. *Journal of Agricultural Science* 67: 223–233.

Giusquiani, P.L., Maracchini, C.B. and Usinelli, M., 1988, Chemical properties of soils amended with compost of urban waste. *Plant and Soil* 109: 73–78.

Glaub, J.C. and Golueke, C.G., 1989, Municipal organic wastes and composts for arid areas. *Arid Soil Research and Rehabilitation* 3: 171–184.

Golueke, C.G., 1977, *Biological Reclamation of Solid Wastes*. Rodale Press Emmaus PA: 1–3.

Gopal, D, Singh, R.P. and Kapoor, R.C., 1991, Ground water pollution by solid wastes. A case study. Pollution Research, *Environ. Media* 10: 111–116.

Gupta, M.D., Chatterjee, N. and Gupta, S.K., 1984, Utilization of Kolkata City Wastes as Manures. In: *Seminar on Soil Resources and Productivity Management*. Indian Agricultural Research Institute, New Delhi, 32 pp.

Gupta, M.D., Chattopadhyay, N., Gupta, S.K. and Banerjee, S.K., 1986, Characterization of Kolkata city waste compost with particular reference to organic matter. *Journal of Indian Society of Soil Science*. New Delhi 34: 736–742.

Haan, S. De, 1981, Results of municipal waste compost research over more than 50 years at the Institute for Soil Fertility at Haren/Groningon. *The Netherlands Journal of Agricultural Science* 29: 49–61.

Hernando, S.L. and Polo, A., 1989, Effect of the application of a municipal refuse compost on the physical and chemical properties of a soil. *Science of the Total Environment* 81 & 82: 589–596.

Hoeks, J., 1977, Mobility of pollutants in soil and ground water near waste disposal sites. *IAHS Publication* 123: 380–388.

Hoertling, J.W., 1989, Trace metal pollution from a municipal waste disposal site at Pangnirtung, North-West Territories. Arctic 42: 57–61. *Soils and Fertilizer Abstracts* 146, 1990.

Hortenstine, C.C. and Rothwell, D.F., 1969, Evaluation of composed municipal refuse as a plant nutrient source and soil amendment on Leon Fine Sand. Proceedings of Soils and Crop Science Society of Florida 29: 312–319. *Soils and Fertilizers Abstracts* 422, 1971.

ISI, 1983, *Specification for Drinking Water*. IS. 10500. Indian Standards Institution, New Delhi.

Jeevan Rao, K. and Shantaram, M.V., 1993a, Physical characteristics of Urban Solid Wastes of Hyderabad. *Indian Journal of Environmental Protection* 13(10): 721–725.

Jeevan Rao, K. and Shantaram, M.V., 1993b, Chemical composition of Urban Solid Wastes of Hyderabad and their use in agriculture. *Indian Journal of Environmental Protection* 13(11): 813–816.

Jeevan Rao, K. and Shantaram, M.V., 1994a, Heavy metal pollution of agricultural soils due to application of USW. *Indian Journal of Environmental Health* 36(1): 31–39.

Jeevan Rao, K. and Shantaram, M.V., 1994b, Soil properties at landfill sites used for USW disposal at Hyderabad, India. *Proceedings 15th World Congress of Soil Science*, Mexico, Symposium 1(b) vol. 2b: 255–256.

Jeevan Rao, K. and Shantaram, M.V., 1994c, Chemical characteristics of landfill runoff water. *Indian Journal of Environmental Health* 36(4): 282–283.

Jeevan Rao, K. and Shantaram, M.V.1994d, Effect of long term application of USW on relative availability of heavy metals in soils. Proceedings of Diamond Jubilee National Seminar of ISSS, IARI, New Delhi: 636-638.

Jeevan Rao, K. and Shantaram, M.V., 1994e, USW disposal problem in Hyderabad. Presented at UGC sponsored National Symp. and Workshop on Environmental Education in University Curricula, JNTU, Hyderabad.

Jeevan Rao, K. and Shantaram, M.V., 1994f, Environmental pollution from urban solid wastes—challenges and strategies. *National Symp. on Environmental Pollution*, AV College, Hyderabad 100 pp.

Jeevan Rao, K. and Shantaram, M.V., 1995a, Contents of heavy metals in crops treated with urban solid wastes. *Journal of Environmental Biology* 16(3): 225–232.

Jeevan Rao, K. and Shantaram, M.V., 1995b, Bioassay for predicting pollution of soil profiles due to disposal of USW at landfill site. *Indian Journal of Environmental Protection* 15(2): 154–155.

Jeevan Rao, K. and Shantaram, M.V., 1995c, Effect of urban solid wastes application on concentration of micronutrients and heavy metals in soil and maize plant parts. *Indian Journal of Environmental Protection* 15(3): 201–209.

Jeevan Rao, K. and Shantaram, M.V., 1995d, Ground water pollution from open refuse dumps at Hyderabad. *Indian Journal of Environmental Health,* 37(3): 197–204.

Jeevan Rao, K. and Shantaram, M.V., 1995e, Concentrations and relative availability of heavy metals in urban solid wastes of Hyderabad, India. Bioresource Technology, *Elsevier Applied, Sc.* England 53(1): 53–55.

Jeevan Rao, K. and Shantaram, M.V., 1995f, Distribution of total and extractable heavy metals in soils profiles under USW dumps—A correlation study. *Proceedings of International Conference on Organic Mineral Interactions in Sediments and Soils,* University of Newcastle, UK.

Jeevan Rao, K. and Shantaram, M.V., 1995g, Nutrient changes in agricultural soils due to application of USW. *Indian Journal of Environmental Health* 37(4): 265–271.

Jeevan Rao, K. and Shantaram, M.V., 1995h, Effect of long-term application of USW on heavy metal contents of soils and plants—A correlation study. *XII International Symposium on Environmental Biogeochemistry,* Rio de Janeio, Brazil.

Jeevan Rao, K. and Shantaram, M.V., 1995j, Water pollution due to open refuse dumps—A review. *Journal of the Institution of Public Health Engineers India* 1: 7–11.

Jeevan Rao, K. and Shantaram, M.V., 1995k, Effect of application of USW on soil plant system—A review. *Agricultural Reviews,* ARCC Karnal India 16(3): 105–116.

Jeevan Rao, K. and Shantaram, M.V., 1996b, Effect of urban solid wastes application on dry matter yield/ uptake of micronutrients and heavy metals by maize plant. *Journal of Environmental Biology* 17(1): 25–32.

Jeevan Rao, K. and Shantaram, M.V., 1996c, Soil pollution due to disposal of urban solid wastes at landfill sites, Hyderabad. *Journal of Environmental Protection* 5: 373.

Jeevan Rao, K. and Shantaram, M.V., 1996d, Heavy metals and micronutrient content and their relative availabilities in stabilized urban solid waste profiles of Hyderabad. *Indian Journal of Environmental Protection* 16(9): 692–699.

Jeevan Rao, K. and Shantaram, M.V., 1997a, Relationship between heavy metals and soil properties in municipal solid waste treated soils. *4th International Conference on the Biogeochemistry of Trachea Elements* (June 23–26). University of California, Berkeley, US.

Jeevan Rao, K. and Shantaram, M.V., 1998a, Correlation matrix between heavy metals in soils treated with municipal solid waste. *Proceedings 16th World Congress of Soil Science,* Symposium, 28, Montpellier, France, 529 pp.

Jeevan Rao, K. and Shantaram, M.V., 1998b, Urban solid waste management with reference to Hyderabad. In: Hosetii, B.B. and Kumar, A. (Eds.), *Environmental Impact Assessment and Management.* Daya Publications, New Delhi: 148–171.

Jeevan Rao, K. and Shantaram, M.V., 1999, Relative availability of heavy meatls in Municipal solid waste treated soils. Agricultural Science Digest: 19(2): 129–133.

King, L.D., Rudgers, L.A. and Webber, L.R., 1974, Applications of municipal refuse and liquid sewage sludge to agricultural land I. Field Study. *Journal of Environmental Quality* 3: 361–366.

Kirkham, M.B., 2001, Personal Communication.

Kowald, R., Bahtiyar, M. and Ditter, P., 1982, Effect of refuse compost on the properties and the crop yield of an arable field 23: 178–189. *Soils and Fertilizer Abstracts* 671, 1982.

Liptak, E.G. (Ed.), 1974, Land pollution. In: *Environmental Engineers Handbook,* volume 3, Chilton Book Company, Rednor Pennsylvania, USA.

Logan, T.J. and Chaney, R.L., 1983, Utilization of municipal wastewater and sludge on land—Metals. In: Page, A.L. (Ed.), *Proceedings of the Workshop on Utilization of Municipal Wastewater and Sludge on Land.* University of California, Riverside: 235–326.

Mays, D.A., Terman, G.L. and Duggan, J.L., 1973, Municipal compost, effects on crop yields and soil properties. *Journal of Environmental Quality* 2: 89–92.

Murray, J.P., Rouse, J.V. and Carpenter, A.B., 1981, Ground water contamination by sanitary landfill leachate and domestic wastewater in carbonate terrain. Principal source diagnosis. *Chemical Transport Characteristics and Design Implications* 15: 745–757.

Murillo, J.M., Hernandez, J.M., Barroso, M. and Lopez, R., 1989, Production versus contamination in urban compost utilization. Anales de Edafologia Agrobilogia, Spain 48: 143–160. *Soils and Fertilizer Abstracts* 591, 1990.

Musmeci, L., Zavattiero, E. and Castagnoli, O., 1985, Behaviour of heavy metals in municipal landfill leachate and ground water. In: *Heavy Metals in the Environment.* International Conference, Athens, Volume II: 622–624.

Mutatkar, V.K, 1985, Prospects for organic–inorganic fertilizers. *Fertilizer News* 12: 60–67.

Naidu, R., 2001, Personal Communication.

Nandkishore, T.G., 1980, Studies on urban solid waste composing. MSc (Ag) Thesis submitted to University of Agricultural Sciences, Hebbar, Bangalore. 150 pp.

NEERI, 1970, Feasibility studies on alternate methods of USW disposal for Kolkata City. Central Public Health Engineering Research Institute Report. National Environmental Engineering Research Institute, Nagpur, 120pp.

NEERI, 1983, *Solid Waste Management—A Course Manual* (Bhide, A.D. Ed.). Nehru Marg. National Environmental Engineering Research Institute, Nagpur. 100 pp.

Nicholson, R.V., Cherry, J.A. and Reardon, E.J., 1983, Migration of contaminants in ground water at a landfill—A case study 6. Hydro Geochemistry. *Journal of Hydrology,* Netherlands 63: 131–176.

Nogales, R., Gomez, M. and Gallardo-Cara, F., 1985, Town refuse compost as a potential source of Zn for plants. In: *Heavy Metals in Environment.* International Conference, Athens, Vol. I: 487–489.

Olaniya, M.S. and Saxena, K.L., 1977, Groundwater pollution by open refuse dumps at Jaipur. *Indian Journal of Environmental Health* 19: 176–188.

Patel, M.K. and Tiwari, T.N., 1990, Physico-chemical characterisation of solid wastes from Industrial Area of Rourkela. *Indian Journal of Environmental Protection* 10: 257–266.

Pelletin, J., Letard, M., Barbo, P., Quillecs, L.E., Amiard, J.C., Berthet, B. and Melayer, C., 1986, Use of town refuse composts and sewage sludge composts in vegetable production. Cahiers du CTIEL 24: 81. *Soils and Fertilizer Abstracts* 1253, 1987.

Petruzzelli, G., Lubrano, L. and Guidi, G., 1989, Uptake by corn and chemical extractability of heavy metals from a four-year compost treated soil. *Plant and Soil* 116: 23–27.

Radia, K., Naseem, F.U. and Shahid, S.H., 1990, Evaluation of recyclable materials in municipal waste. *Karachi, Biological Wastes* 31: 113122.

Ramachandran, V. and D'Souza, T.J., 1990, Transformation of added chromium compounds in two soils. *Journal of Indian Society of Soil Science,* New Delhi 38: 419-425.

Sawhney, B.L. and Raube, J.A., 1986, Groundwater contamination: Movement of organic pollutants in the Grauby landfill. In: *Bulletin*, Connecticut Agricultural Experiment Station, New Harem CT, US.

Schossing, C., 1983, Refuse compost application on illuvial sands and silty marsh soils of Eastern Friesland. Mitteilungen der Deutschen Bodeu Kunlichen Gesell Schaff 38: 697–702. *Soils and Fertilizers Abstract* 641, 1987.

Shekdar, A.V., 1999, Municipal solid waste management—The Indian perspective. *Journal of Indian Association for Environmental Management* 26: 100–108.

Sotomayor, R.I., 1979, Refuse compost as a source of organic fertilizer compared with chemical fertilizer. Part I. Agricultural Technica, Chile, 39: 152–157. *Soils and Fertilizer Abstracts* 271, 1981.

Srinivasan, V.S., 1977, The influence of USW dumps on ground water. *Scientifica*: 111–117.

Srivastava, O.P., Mann, G.S. and Bhatia, I.S., 1971, The evaluation of rural and urban compost. *Journal of Research*, Punjab Agricultural University, Ludhiana 8: 451–455.

Talashilkar, S.C. and Vimal, O.P., 1985, Effect of city compost alone and mixed with sewage sludge (1:1) in combination with urea/superphosphate on the dry matter yield and nutrient uptake by wheat. *Indian Journal of Environmental Health* 27: 110–117.

Terman, G.L., Soilean, J.M. and Allen, S.E., 1973, Municipal waste compost: Effects on crop yields and nutrient content in greenhouse pot experiments. *Journal of Environmental Quality* 2: 84–89.

Tester, D.T. and Harker, R.J., 1982, Groundwater pollution investigations in the Gretouse Basin II solid waste disposal. *Water Pollution Control* 81: 38–328.

Tietjen, C. and Hart, S.A., 1968, Compost for agricultural land. Journal of Sanitary Engineers. *Proceedings of American Society of Civil Engineering* 95: 269–287.

Trivedi, P.R. and Gurdeep, R., 1992a, Disposal of solid waste. In: *Solid Waste Pollution*. Akashdeep Publishing House, New Delhi: 116–186.

Trivedi, P.R. and Gurdeep, R., 1992b, Management of solid wastes. In: *Management of Pollution Control*. Akashdeep Publishing House, New Delhi.

Varadarajan, S. and Elangovan, S., 1992, Problems and impacts. In: *Environmental Economics*. Speed Publication, Madurai: 125–130.

WHO, 1984, *Guidelines for Drinking Water Quality*, Vol. 1. Recommendations. World Health Organization, Geneva.

Winant, W.H., Menser, H.A. and Bennett, O.L., 1981, Effects of Sanitary Landfill Leachate on some soil chemical properties. *Journal of Environmental Quality* 10: 318–322.

Wong, M.H., 1985, Phytotoxicity of refuse compost during the process of maturation. *Environmental Pollution* A 37: 159–174.

8

Benefits and Risks Associated with Biosolids Application to Agricultural Production Systems—Experiences from New South Wales, Australia

D.L. Michalk[1], *M. S. Whatmuff*[3], *G.J. Osborne*[2] and *T.S. Gibson*[3]

INTRODUCTION

Management of residuals from municipal and industrial wastewater treatment is a major health and environmental problem. Countries around the world are committing increasing resources to find acceptable long-term solutions. However, finding methods that are environmentally compatible, socially acceptable and economical viable is a continuing challenge, particularly as air, water and food quality standards become more stringent in response to community demands. Faced with annual increases in the amount of wastewater solids for disposal due to population growth and to improved extraction of solids from effluent, authorities have progressively turned from incineration, landfill and ocean dumping to alternative methods that emphasize beneficial use of wastes (Clapp et al., 1994).

Wastewater solids may be either in the form of a slurry or as a dewatered product that may be further stabilized by lime treatment or through composting. Dewatered wastewater solids are known as biosolids, the accepted term for sewage sludge processed to meet environmental standards for beneficial use. The resource value of these products is shown by their application to land, where they supply plant nutrients and organic matter to the soil for crop and pasture production (Smith, 1996). However, biosolids also contain heavy metals, persistent organic contaminants and pathogens that have the potential to pollute surface water as also groundwater, inhibit plant growth, and accumulate in the food chain

[1] NSW Agriculture, Orange Agricultural Institute, Orange, NSW 2800
[2] Australian Water Technologies Pty Ltd, Sydney Water Corporation, PO Box 365, Guildford NSW 2161
[3] NSW Agriculture, Organic Waste Recycling Unit, Richmond, NSW 2753

with negative consequences for the environment and human health (Page and Chang, 1994).

The potential for adverse effects through land application of biosolids is determined by the:

- concentration of contaminants in the waste source;
- amount and frequency of application;
- nutrient requirements of the target crops and pastures; and
- properties of the soils to which they are applied.

Extensive study of these aspects over the past 25 years has produced sound regulations in the United States (US EPA, 1993) and Europe (Council of the European Communities, 1986; Department of the Environment, 1989) that foster the safe use of biosolids in commercial production on non-acidic soils. These regulations have dispelled much of the public uncertainty surrounding the transfer of potentially-harmful contaminants to consumers through water and food (Machno, 1996) and led to widespread land application of biosolids products in North America and Europe.

Land application in Australia has been a more recent development. Until the early 1990s, the large metropolitan centres (all coastal) either discharged primary treated waste into the ocean or stockpiled it on dedicated land. In response to increasing community concerns about beach and point source pollution in the late 1980s (Australian Water Resources Council, 1992; Gough and Fraser, 1995), State Governments have enacted legislations similar to that of the Council of the European Communities (CEC) Directive on urban wastewater treatment (Council of European Communities, 1991). This legislation prohibits ocean discharge of solids requires improved treatment of effluent discharged to rivers and through ocean outfalls and recommends guidelines for better containment of stockpiles and landfill sites. Due to the scarcity and cost of landfill containment, and air pollution and subsequent disposal of ash from incinerated wastewater solids, land application has emerged as an attractive cost-effective option for biosolids management in Australia.

While biosolids may boost agricultural production by improving the chemical fertility and physical structure of many of our uniquely impoverished Australian soils, land application is not without its problems because of the pathogens, heavy metals and other contaminants that are associated with biosolids products. Due to the lack of locally-generated data, early guidelines developed to regulate the use of biosolids in Australia were based on overseas research (for example, Awad et al., 1989; Ross et al., 1991). However, due to major differences in soil chemistry, it was not appropriate to extrapolate regulations on allowable contaminant concentrations based on North American and European research to Australian conditions (Whatmuff and Osborne, 1992). Thus, it was necessary to develop regulations which would be specific for Australia in order to promote the sustainable use of biosolids in Australian agriculture and to maintain a 'clean, green' status for both local consumption and export markets.

New South Wales (NSW) was the first Australian state to release guidelines for biosolids usage (Awad et al., 1989). These guidelines have subsequently been revised several times (Ross et al., 1991) and released as the current Environmental Guidelines: Use and Disposal of Biosolids Products (NSW EPA, 1997). These guidelines are supported by the findings of a major multi-disciplinary research programme commenced in 1992 to determine the potential benefits and risks associated with the use of biosolids in a wide range of agricultural production systems.

NSW Agriculture undertook the $11 million programme with funds provided by the Sydney Water Corporation (Sydney Water). The guidelines also underpin the turnaround in biosolids management by Sydney Water, in which 99% of biosolids are now beneficially used on land. This chapter reviews the outcomes from this programme and discusses impacts of the research on the development of Sydney Water's extensive beneficial use programme in NSW.

BIOSOLIDS IN NEW SOUTH WALES

Wastewater from domestic and industrial sources is about 99.9% water (Water Board, 1991). The remaining 0.1% solids, separated from influent at the Wastewater Plant (WWP) by screening, sedimentation and microbial action, is composed of grit, grease and solids such as plastic, paper and human faecal matter (Water Board, 1991). The majority of the solid content is cellulose and protein, and is rich in nitrogen, phosphorus and sulphur (Court, 1989). The bulk of the metals and other contaminants entering the wastewater processing plant is also retained in the solids. These solids are processed further by anaerobic and aerobic digestion so as to produce biosolids, a stabilized product suitable for beneficial use in land application.

Biosolids Production

It is estimated that 300,000 product tonnes of biosolids are produced annually in NSW, two-thirds of which is produced by the four million people in the Sydney Basin. Sydney Water is the authority responsible for providing water, sewage and drainage services for the 13,000 km^2 area defined as the Sydney Region. Wastewater is collected at WWPs strategically located throughout the region that extends from the Hawkesbury River in the north to Gerroa in the south, and from the Pacific Ocean westward to Mount Victoria in the Blue Mountains. Sydney

Water currently treats about 500,000 ML of wastewater that produces 200,000 product tonnes annually. This is expected to increase to >300,000 product tonnes over the next ten years, mainly as a result of improved equipment at the major ocean plants which increase the extraction of solids from the effluent.

The degree of treatment ranges from basic primary treatment at the four major ocean WWPs to tertiary treatment with nutrient removal processes at most of the inland WWPs that discharge into the Hawkesbury-Nepean river systems (Hope and McDougall, 1996). To be suitable for beneficial use and to avoid human health risks, biosolids require stabilization in order to reduce pathogens, odours and attraction of vectors. At the major WWPs, sludge is subjected to primary mesophilic anaerobic digestion for at least a 12-day period at 25–35°C followed by either mechanical dewatering to approximately 20% solids or storage in lagoons for varying periods (7 days to 12 months) before use. Lagoon biosolids can be directly injected into the soil (10% solids). Following the addition of polymers, centrifugation or belt presses are used to produce dewatered biosolids at some of the smaller plants.

In regional NSW, there are about 270 WWPs operated by 128 local Councils that collectively service a population of 1.7 million people (Huxedurp et al., 1996). About half of these plants serve populations of 3000 people, while fewer than ten plants have a capacity to serve 50,000 (NSW DLWC, 1997). It is difficult to estimate the amount of biosolids produced by these country plants because of the range in the treatment processes used. However, by using the expected dry solid output/person for different types of WWPs (for example, trickling filter plants and extended aeration plants), Huxedurp et al. (1996) estimated that about 100,000 product tonnes of wastewater solids is produced annually from country NSW.

Biosolids Products, Quality and Regulatory Constraints

Sydney Water produces several products that are stabilized in order to meet biosolids environmental guidelines for beneficial use (NSW EPA, 1997). Liquid biosolids (LB) from lagoon storages and dewatered biosolids (DWB) produced from digestion processes and dewatering are the main outputs, but some additional processing may be undertaken by commercial companies to manufacture specialized products. These include composted biosolids (CB) formulated by mixing DWB and green waste, and alkali-stabilized or lime-amended biosolids (LAB) produced by combining DWB or raw primary-treated solids with cement kiln dust or calcium oxide.

Clearly, concentrations of both nutrients and pollutants must be specified for individual batches of biosolids to allow assessment of their suitability for beneficial use. To meet the guidelines for land application, quality grading of these products has been established by comparing the contaminant concentrations of the product with the 'Contaminant Acceptance Concentration Threshold' for Grades A to D in NSW EPA Guidelines (Table 8.1). Each contaminant is graded, and the lowest grade for any one contaminant becomes the grade for the biosolids product. Tables 8.2 and 8.3 give the quality profile of Sydney Water's biosolids (based on DWB) and show how this has changed over a period of time. All WWPs with the exception of Port Kembla WWP (serving a heavy industrial area) generally meet the contaminant thresholds for beneficial use (NSW EPA, 1997) provided that they also meet the stability requirements (Hope and McDougall, 1996).

Table 8.1: NSW EPA Contaminant Acceptance Concentration Thresholds (mg/kg) for biosolids (NSW EPA, 1997)

Contaminant	*Acceptance Concentration Threshold*			
	Grade A	*Grade B*	*Grade C*	*Grade D*[a]
Arsenic	20[b]	20	20	30
Cadmium	3	11	20	32
Chromium	100	500	500	600
Copper	100	750	2000	2000
Lead	150	150	420	500
Mercury	1	9	15	19
Nickel	60	145	270	270
Selenium	5	14	50	90
Zinc	200	1400	2500	3300
DDT/DDD/DDE	0.5	0.5	1.0	1.0
Aldrin	0.02	0.2	0.5	1.0
Dieldrin	0.02	0.2	0.5	1.0
Chlordane	0.02	0.2	0.5	1.0
Heptachlor	0.02	0.2	0.5	1.0
HCB	0.02	0.2	0.5	1.0
Lindane	0.02	0.2	0.5	1.0
BHC	0.02	0.2	0.5	1.0
PCBs	0.30	0.3	1.0	1.0

[a]Contaminant concentrations which exceed Contaminant Acceptance Concentration for Grade D shall be graded Contaminant Grade E;

[b]Tabulated concentrations are not means. The mean concentrations of biosolids samples must be below the tabulated values by at least two standard deviations based on the historical data from the STP.

Biosolids are not homogenous products and their composition varies widely with time, the amount and type of industry in the wastewater catchment area, and the exact nature of treatment process (Table 8.3). This has important implication for land application because differences in the composition can alter the metal uptake by plants (Chaney, 1983). In a survey of 20 Australian biosolids, De Vries (1983) found that the

Table 8.2: Percentage of Sydney Water's biosolids by contaminant grade

Contaminant grade	Limiting use	% of Sydney Water's biosolids	
		1995[a]	1999
A	Home lawns and gardens	0	0
B	Public contact sites and urban landscaping	2	20
C	Agriculture	62	78
D	Forestry	28	0
E	Landfill disposal	8	2

[a]Gough and Fraser (1995)

Table 8.3: Change in metal concentrations in biosolids from selected STPs, 1983–1999 (mg/kg)

STP	Cu	Zn	Cd	Ni	Pb
Malabar 1983[1]	872	2190	32	318	552
Malabar 1993[2]	1381	2606	12	159	315
Malabar 1999[3]	716	1313	7	84	202
Glenfield 1983[3]	1015	1242	10	41	263
Glenfield 1999[3]	431	579	1	21	42
Port Kembla 1983[1]	987	5510	285	20	173
Port Kembla 1999[3]	692	1441	20	24	81

[1]de Vries (1983); [2]Michalk et al. (1996); [3]Sydney Water

highest concentrations of Cd, Cu, Ni and Zn were from industrial areas. However, residential sewage can also be high in some metals. Copland (1990), for example, reported higher levels of Se, Zn and Cu in Sydney's residential areas than from some of the industrial areas. Significant progress has been made in reducing the Cu and Zn derived from plumbing by pH buffering of drinking water (Table 8.3) and lowering the P levels by promoting the use of environmentally-friendly detergents. Sydney Water also manages a rigorous policy aimed at increasing the recycling of metals and organics by industry. This has reduced the release of contaminants into the waste stream and has increased the potential range of uses of the biosolids produced (Table 8.3).

Raw wastewater contains a range of pathogenic organisms, including bacteria, viruses, parasites (protozoa and helminthes) and fungi (WHO, 1981). These potentially disease-causing agents that are discharged in faeces from infected humans and animals reflect the health of the community and the standards of hygiene that prevail (Smith, 1996). of the range of potential health hazards, salmonellosis (*Salmonella* spp.) and bovine cysticercosis are the infectious diseases most likely to affect humans and livestock as a result of sludge contamination in NSW (Ross et al., 1992), while roundworms (*Ascaris lumbricoides*) and beef tapeworm (*Taenia saginata*) are the two helminthic parasites most likely to be found in biosolids treated by Sydney Water (Ross, 1992).

A prime function of wastewater treatment is to eliminate or reduce the pathogen load in biosolids and wastewater (Angle, 1994). However, the reduction achieved depends upon the type of treatment applied to the biosolids and the range of organisms present. Due to the expense and danger associated with work on pathogenic organisms, monitoring programmes usually assay only for indicator organisms in biosolids (Angle, 1994). Eamens et al. (1996) used *Escherichia coli*, faecal streptococci, *Clostridium perfringens* spores and *Salmonella* spp. as indicator bacteria and the most probable number technique (MPN) to assess the reduction in bacteria populations entering and leaving mesophilic anaerobic digesters operated by Sydney Water. Digestion effectively reduced *E. coli*, faecal streptococci and *Salmonella* spp. populations enough to meet Grade B stabilization standards, but did not affect the level of *Cl. perfringens* spores (Figure 8.1). In contrast, biosolids produced from WWPs, using extended aeration processes (for example, biological nutrient removal or intermittently decanted aerated lagoon, IDAL), have poor product stability resulting in unacceptable levels of odour, poor dewaterability, and failure to meet the NSW Guidelines for SOUR (Specific Oxygen Uptake Rate) standards (Gough and Fraser, 1995; Hope and McDougall, 1996).

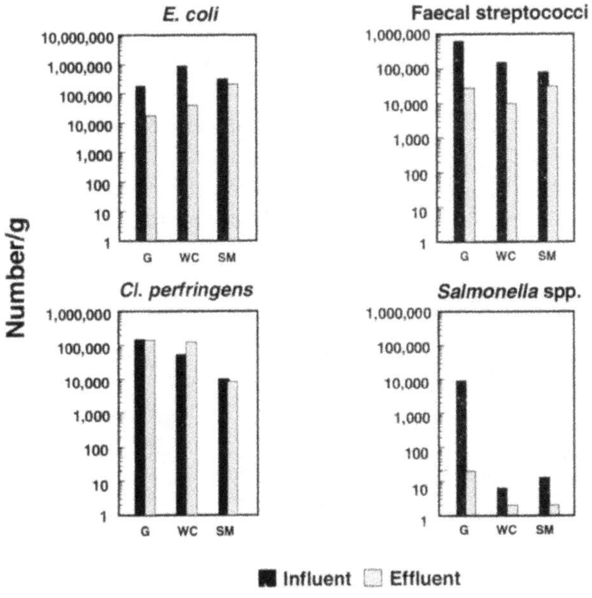

Fig. 8.1: Bacterial numbers (Most Probable Number method) in STP digester influent and effluent at Glenfield (G), West Camden (WC) and St Mary's (SM) (Modified from Eamens et al., 1996).

Further sanitation using short-term storage (about 8 weeks) was not an effective means of stabilizing DWB, irrespective of the treatment process used, but holding biosolids in lagoons for several months lowered bacterial populations relative to DWB (Eamens et al., 1996). Since no Sydney Water biosolids meets microbiological standards for unrestricted use (Table 8.4), beneficial use of DWB and LB biosolids is restricted to agriculture, forestry, and land rehabilitation activities. However, composting (Fahy et al., 1996) and lime amendment (Kempton, 1996) offer opportunities for Grade A stablization standard to be met. An RDP Envessel pasteurization process was installed at North Head WWP in 1995 to stabilize approximately 25% of Sydney's primary digested wastewater solids (Hope and McDougall, 1996). It combines heat with CaO to reduce the amount of quicklime to 30% solids content, compared to 100% for other alkali-stablization processes (Kempton, 1996).

Table 8.4: Stabilization Grade A microbiological standards for unrestricted use biosolids (NSW EPA, 1997)

Parameter	*Standard*
E. coli	<100 MPN per gram (dry weight)
Faecal coliforms	<1000 MPN per gram (dry weight)
Salmonella spp.	Not detected per 50 grams of final product (dry weight)

Note: MPN = most probable number

For helminthic parasites, Johnson et al., (1996a) also reported species differences in die-off with large numbers of roundworm eggs surviving anaerobic mesophilic treatment (Figure 8.2), whereas tapeworm eggs were killed when subjected to anaerobic digestion for more than one week.

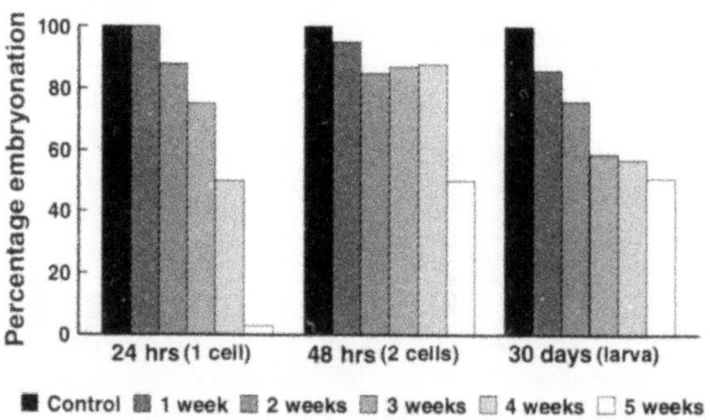

Control samples stored at 4°C before incubation. Not immersed in digester tank.

Fig. 8.2: Viability of roundworm (*Ascaris suum*) eggs determined by in vitro culture and microscopic detection of embryos and larvae recovered between 1 and 5 weeks after immersion in a mesophilic anaerobic digester (Johnson et al., 1996b).

IDAL was less effective in reducing populations of both parasites. After six weeks, 95% of roundworm eggs in the IDAL process were still viable, and it took six weeks to kill off tapeworm eggs (Johnson et al., 1996a). However, further processing of biosolids by composting (10–16 weeks) or alkali-stabilization that generated temperatures of 40–55°C proved to be lethal to both parasites, irrespective of the wastewater treatment process used (Johnson et al., 1996b).

Since the human ascarid species (*A. lumbricoides*) occur in negligible levels in Sydney's urban population, further treatment specifically for roundworm is not warranted.

Change in Methods of Wastewater Solids Disposal

Prior to 1989, all wastewater solids collected in the Sydney region were either incinerated (8%) or discharged to the ocean (92%). However, in response to community concerns about pollution of Sydney beaches and impacts on the marine environment, the NSW Government passed a legislation in March 1989 that prohibited all ocean disposal of solids after october 1993. This resulted in a major shift in practices in the Sydney region. The policy was implemented in July 1989 when the discharge from the inland and minor ocean plants was discontinued and the solids diverted to compost manufacturers. In March 1990, solids from the first major ocean WWTP at Bondi were also progressively diverted to composting, and in January 1992, the incinerators at North Head STP were turned off and replaced with a lime stabilization plant. In December 1992, Malabar was the last STP to cease ocean disposal when the first 6400 tonnes of solids collected as DWB was used in a large grazing experiment established at Goulburn as part of the NSW Agriculture-Sydney Water Biosolids Research Programme (Michalk et al., 1996a).

Subsurface LB injection was initially viewed as the preferred method for land application because it reduced odours and the perceived risk of run-off pollution (Awad et al., 1989). However, injected LB was only cost-effective when suitable land was located within 20 km from the WWP (Gough and Fraser, 1995). Since finding suitable land in the urbanised Sydney region proved to be a major problem, Sydney Water's beneficial use programme was changed to emphasize the transportation of digested DWB to inland cropping areas up to 300 km from the WWPs (Hope and McDougall, 1996). While this strategy provides little opportunity for profit at this stage, some cost offsetting is possible (Gough and Fraser. 1995), which makes land application of DWB competitive with alternatives such as landfill (Hope and McDougall, 1996).

Prior to the release of the current Biosolids Guidelines (NSW EPA, 1997), no regulations governed disposal of biosolids on WWP sites. Consequently, most local authorities in regional NSW disposed of biosolids by stockpiling on dedicated WWP land (58%) or at landfills (26%), where there were no constraints on quantity or quality of biosolids (Huxedurp et al., 1996). There was little incentive for councils to develop programme for beneficial use in agriculture or compost manufacture, which require compliance with NSW land application guidelines (Ross et al., 1991). Currently, only about ten per cent of biosolids produced in regional NSW are put to beneficial use (NSW DLWC, 1997).

Regional Councils have not changed their biosolids disposal practices since 1996 for two reasons. First, the Biosolids Guidelines still allow disposal of all classes of biosolids without constraint within the bounds of a WWP or lands owned by the Council and zoned for disposal. Second, even the largest country cities produce limited quantities of biosolids (typically <1500 dry tonnes/year from the largest regional centres); this does not justify the expense of undertaking the characterization to determine whether their biosolids meet the reuse criteria and to monitor environmental impacts following land application. However, regional Councils have made significant progress in improving the quality of effluent discharged from WWPs to meet legislative requirements of the NSW Protection of the Environment Operations Act (POEO, 1997) by installing nutrient removal processes. This is considered to be a more important priority for inland WWPs because of the increasing frequency of toxic blue-green algae outbreaks linked to N and P pollution in the rivers of the Murray-Darling Basin (Harvey, 1993).

DEFINING RESEARCH NEEDS AND DEVELOPING A RESEARCH PROGRAMME

When Sydney Water seriously considered applying biosolids to land in the early 1990s, there was little Australian information available to specify the fertilizer value and other benefits of biosolids to the agricultural systems or to determine the likely environmental impacts associated with their use in NSW (Hope and McDougall, 1996). To answer these important questions, Sydney Water committed $8 m to support an extensive, applied research programme in NSW that commenced in 1992. A further $3 m was approved to complete the research. An important first step was to undertake a review of the international literature on land application of biosolids. This review provided a unique opportunity to learn from 30 years of research in North America and Europe on the beneficial use of biosolids and to establish priorities for research to ensure safe and sustainable application of biosolids to agricultural land in NSW (Osborne and Stevens, 1992).

A major finding from the literature was that most of the international research that commenced in the 1970s targeted soils with various combinations of high pH, CEC and organic matter that effectively reduce metal bioavailability and nitrate leaching (Osborne and Stevens, 1992). However, soils within a reasonable distance from Sydney are acidic (pH <5.5) with low CEC (<15 meq%) and low organic matter (<2%). These chemical features are often associated with thin, seasonally hard-setting topsoils overlying the clayey subsoil of low porosity that reduce the soil's capacity to retain biosolids and increase risks of pollution of surface water by run-off. These differences meant that it was not possible to apply the outcomes of overseas research directly to agricultural systems in NSW. However, many of approaches taken and techniques used in the overseas research were useful for the design and implementation of experiments that aimed to: quantify, and if possible improve, the value of wastewater solids as soil conditioners and fertilizers for agricultural production on soils with $pH_{Ca} < 5.5$; and investigate and where necessary, strengthen guidelines governing the use of biosolids to ensure protection of human health (heavy metals and pathogens) and the environment (heavy metals and excess nutrients).

The assessment of benefits and definition of impacts associated with land application of biosolids requires a thorough knowledge of the soil ecosystem (Dowdy et al., 1991) because biosolids applications will influence the way the physical, chemical and biological processes interact (Oberle and Keeney, 1994). Soil pH and organic matter have been identified as the most important factors controlling the availability of contaminants in biosolids-treated soils (Logan and Chaney, 1983; Kiekens, 1984; Alloway, 1990; McBride, 1995).

The risk pathway model developed by the US EPA (1989) provided an excellent scientific framework to evaluate the potential of transferring biosolids-derived contaminants to the most exposed individual through a number of environmental pathways relevant to biosolids application to agricultural land in NSW. Table 8.5 shows the US EPA environmental pathways targeted by the 20 projects undertaken by NSW Agriculture with Sydney Water funding.

Table 8.5: Environmental pathways of concern identified for application of biosolids to agricultural land

	Pathway	*Limiting metal*	*Relevant project*[1]
1F	Biosolids Soil Plant Human	Cd, Pb	3, 4, 5, 7, 8, 9, 10, 18
3	Biosolids Soil Plant Animal Human	Cd	18
4	Biosolids Soil Animal Human	Cd	
5	Biosolids Soil Plant Animal	Se, Mo, Cd, Zn	18
6	Biosolids Soil Animal	Cd	18
7	Biosolids Soil Plant	Cu, Ni, Zn	3, 4, 5, 7, 8, 9, 10,18
8	Biosolids Soil Soil biota	Cu (As, Cr, Zn)	13, 19
9	Biosolids Soil biota Predator	Pb, Zn	
10	Biosolids Soil Airborne dust Human	None	
11	Biosolids Soil Surface water Human	None	18
12	Biosolids Soil Air Human	None	
12W	Biosolids Soil Groundwater Human	Cd, Zn, NO_3	18

[1]Numbers refer to projects described in column 1 of Table 8.7.

Since it was not possible to achieve both aims of the NSW biosolids programme within the same experiments, some projects were concerned with demonstrating the fertilizer value of biosolids in a range of agricultural production systems, while others focused on setting guideline parameters (Table 8.6). Different research approaches to biosolids application were adopted for these two types of project. Projects aimed at assessing the fertilizer value of biosolids used those application rates that only provided sufficient nutrients for the crops grown. Bioavailability and uptake of metals was assessed to provide data on what to expect in commercial practice. By contrast, to provide data to set guideline parameters, soil metal levels were raised above those that would occur in commercial applications by using high biosolids application rates. The relationship between metal loading and the amount accumulated in agricultural products was used to set maximum metal concentrations in soil that would protect both the environment and the consumer.

Table 8.6: Areas of research by the NSW Agriculture-Sydney Water Biosolids Research Programme 1993–1999

Fertilizer value of biosolids use in agriculture:
Coastal (dairy) and extensive (sheep) pastures
Deciduous horticulture
Vegetable production (biosolids-based composts)
Cereals (summer and winter) and sugarcane
Chemically-treated biosolids—effects on P availability
Determination of risks with biosolids use in agriculture:
Heavy metal movement and uptake
Nutrient movement and environmental pollution
Survival of pathogens and parasites during treatment processes and after land application
Effects on N fixation and soil microbial processes
Green waste/biosolids-based compost use in agricultural systems:
Development of quality standards
Identification and development of sustainable markets
Soil and plant sampling procedures
Economics of biosolids use in agriculture

Accordingly, the 20 research projects undertaken in NSW between 1992 and 1999 represented a wide range of soil types with different pH and organic carbon levels (Table 8.7). As the same protocols were used for data collection at all these research sites, the results provide a valuable resource for testing and developing models that describe the biosolids-soil-plant system (Hird and Bamforth, 1996).

Table 8.7: Rainfall, landform and soil characteristics of sites in NSW where biosolids research was conducted between 1993 and 1999 (adapted from Hird and Bamforth, 1996)

Location (Project number)	*Annual rainfall (mm)*	*Landform*[1]	*Soil type*				
			Type[2]	*Texture class*	*pH*[3]	*C%*	*CEC*[4]
Berry (2)		Black plain	Vertisols	Silty loam	4.3	4.5	14.4
Bathurst (3)		Foothill slope	Yellow podsolic	Sandy clay loam	5.1	11.1	5.0
Orange (4)		Upper hill slope	Krasnozem	Loam	5.2	1.6	7.7
Grafton (5)		Black plain	Vertisols	Silty clay loam	4.2	2.3	10.9
Mt Tomah (6)		Mid hill slope	Yellow earth	Loam	5.9	8.9	33.0
Rydalmere (7)		Upper hill slope	Red podsolic	Clay loam	4.6	3.9	8.9
Somersby (7)		Bench	Earthy sand	Loamy sand	4.6	1.2	1.9
Somersby (7)		Hill slope	Podsol	Sand	4.5	1.1	1.4
Gilgandra (8)		Hill slope	Solodic	Sandy loam	4.7	1.0	8.5
Narromine (8)		Foothill slope	Solodic	Sandy loam	4.7	0.9	4.4
Glenfield (9,10)		Lower hill slope	Red podsolic	Clay loam	4.4	1.9	8.3
Camden (12, 20)		Hill slope	Yellow podsolic	Sandy loam	4.7	1.7	9.8
Tomingley (17)		Foothill slope	Yellow solodic	Sandy clay loam	4.4	2.0	4.2
Condobolin (17)		plain	Red brown earth	Loam	?	?	?
Goulburn (18)	669	Undulating hill	Soloth/Solodic	Sandy loam	4.5	2.6	5.0

Notes:
[1]Landform using Speight (1991)
[2]Soil taxonomy based on Great Soil Group system (Stace et al., 1968)
[3]pH measured in $CaCl_2$
[4]CEC measured in $cmol^{(+)}/kg$

BENEFICIAL EFFECTS OF BIOSOLIDS ON AGRICULTURAL PRODUCTION

Biosolids has considerable potential as a fertilizer because of the significant amounts of nutrients that it contains (Table 8.8). The guiding principle for effective land application of biosolids is to supply the nutrients in the required quantities when the crops and pastures need them.

Agricultural systems in NSW are heavily dependent on P and N fertilizers (NFF, 1995) and the slow-release nature of biosolids makes them an attractive alternative to inorganic fertilizers. Several factors that

Table 8.8: Nutrient contents of dewatered biosolids from industrial and residential WWPs in the Sydney region

Biosolids source	*Total nutrients (%)*						*Reference*
	N	*P*	*Ca*	*Mg*	*K*	*S*	
Port Kembla 1983	3.90	0.76	2.33	3.76	0.35	na	de Vries (1983)
Port Kembla 1993	1.6	2.1					Whatmuff (1996)
Malabar 1983	2.60	0.86	1.86	1.90	0.08	na	de Vries (1983)
Malabar 1993	2.75	1.41	2.81	0.30	0.10	1.23	Michalk *et al.* (1996a)
Glenfield 1983	3.87	3.62	2.10	0.44	0.03	na	Awad *et al.* (1989)

influence the potential plant responses to biosolids include the availability of the nutrient contained in biosolids, chemical and physical properties of soils, and rate of biosolids application. The impacts of these factors on yield responses of field crops, pastures and horticultural crops are described in the following sections primarily from the research programme outlined in Table 8.5.

Availability of N and P in Biosolids

The N and P concentrations in biosolids vary considerably with source, treatment process and the type of biosolids produced (Tables 8.8 and 8.9). The N content of Australian biosolids varies between one and six per cent (de Vries, 1983), with an average of about 3.5% (Tables 8.8 and 8.9). Higher concentrations of more readily available N are produced from WWPs using biological removal (BR) processes than is possible with anaerobic digestion (Table 8.9, Corbin et al., 1994). The wastewater treatment process also affects the form of N with aerobic digestion mineralizing organic N to produce both NH_4^+ and NO_3^- whereas only NH_4^+ is produced by anaerobic digestion (Sommers and Giordano, 1984). Research suggests that although these soluble N forms account for 15–60% of the total N content in digested biosolids, up to 80% can be lost during the dewatering process (CIWEM, 1995b) or ammonia volatilization in the first two weeks after biosolids are surfaced-applied, as was the case in forestry reported by Robertson and Polglase (1996). Biosolids produced by BNR (aerobic) treatment are usually unstable and difficult to dewater.

Table 8.9: Nitrogen and phosphorus content in biosolids from four Sydney WWPs (Corbin et al., 1994 cited by NSW DLWC, 1997)

Nutrient	*Biological removal*	*STPs with Anaerobic Digestion*		
(%)	*(Penrith STP[1])*	*Liverpool STP[2]*	*Riverstone STP[3]*	*St Marys STP[4]*
Nitrogen	4.6 (32)	2.9 (22)	3.2 (11)	3.3 (52)
Phosphorus	3.4 (32)	2.9 (22)	1.3 (11)	3.9 (53)

[1]No chemical dosing
[2]$FeCl_2$ dosing at aeration tanks
[3]$FeCl_2$ dosing at inlet of sedimentation tank, alum dosing post-clarifer
[4]$FeCl_2$ dosing at end of sedimentation. Figures in brackets denote number of samples tested.

Once applied to the soil, the exact temporal patterns of N release from biosolids and subsequent behaviour in the soil are important for determining application rates that match the N requirement of the agricultural activity. Havilah et al. (1996) showed, for example, that over a 12-month period, only 16% of N applied in DWB was available to maize and ryegrass crops grown in sequence at Berry in coastal NSW, and had a relative efficiency of only 26% compared to N applied as urea. Robertson and Polglase (1996) showed that for aerobic-digested DWB half the N release occurred within a three-week period after application, whereas it took 25 weeks for anaerobically-digested DWB to mineralize the same amount of N. These large differences in N mineralization of DWB derived from different wastewater treatment processes could result in an under-supply or over-supply of N to the soil system if an average mineralization of 25% is assumed for the first year after application. However, only stabilized biosolids can be used under the NSW EPA (1997) Guidelines and mineralisation of N from these products appears to be close to the 16% reported by Havilah et al. (1996). N behaviour is further affected by temperature and rainfall, by microbial activity, and by soil type (NSW DLWC, 1997); therefore, these factors must also be taken into account in determining the appropriate biosolids application rate. The environmental implications of variable mineraliztion rates are discussed later.

P concentrations in Sydney biosolids are also variable, ranging from <0.5 to 7% (de Vries, 1983; Finney and Rawlinson, 1990). Like N, the amount and form of P in biosolids is determined by the wastewater treatment process, especially where chemical dosing with iron or aluminium salts is undertaken in order to reduce P concentrations to meet NSW POEO Act standards for effluent discharged into receiving waters (POEO, 1997). While dosing significantly increases the total P captured from influent wastewater (Tables 8.8 and 8.9), it may reduce P availability to plants due to strong adsorption of P by Al and Fe ions. This is illustrated in Table 8.10, where application of chemically-dosed-DWB from St Mary's STP had a non-significant effect on available soil P as determined by Bray test, even though it contained twice the quantity

Table 8.10: Available P concentration in four different soil types following application of three rates of dewatered biosolids from St Mary's (ferric chloride dosed) and Malabar (no dosing) WWPs (modified from Osborne, 1996)

Soil information		*Biosolids*			*Bray P concentration (mg/kg) [Rate of DWB application (dry t/ha.)]*			
Type[1]	*pH*	*Source*	*%Fe*	*%P*	*Nil*	*Low*	*Medium*	*High*
Red brown earth	4.9	St Mary's STP	15	3.8	13	13 [4]	14 [11]	15 [14]
Yellow solodic [Dy2.83]	4.4	St Mary's STP	15	3.8	10	9 [2]	13 [7]	19 [14]
Solodic [Db1.53]	4.7	Malabar STP	<2	1.4	16	33 [6]	53 [12]	57 [24]
Solodic [Dy2.52]	4.7	Malabar STP	<2	1.4	7	21 [6]	25 [12]	37 [24]

[1]Soil types described using Great Soil Group system (Stace et al., 1968) [Principal Profile Form (Northcote, 1979)]

of P. In contrast, application of biosolids from Malabar WWP, where chemical nutrient removal was not practised, increased (Probability<0.05) soil P available to plants. Current studies indicate that in chemically-dosed biosolids, the Fe concentration reduces the recovery of biosolids-P in plant tops (J. Davis and G. Blair, pers. comm.). Low P availability may compromise the value of Fe-dosed biosolids as a fertilizer in NSW on two counts. First, most of the soils targeted for land application in NSW are P-deficient and require readily-available P to sustain plant production. In a future commercial market, slow P release from biosolids will make biosolids less attractive to producers because the costs of handling, storage and application are increased relative to traditional fertilizers for the equivalent amount of P. At present, these concerns do not arise because biosolids are currently provided at negligible direct costs to producers in NSW. Second, the acid nature of the soils targeted for land application will maintain P in unavailable or slowly-available forms unless limed to pH_{Ca} 5.5 (Brady, 1990). However, to counter these effects, Sydney Water has modified Fe-pickling processes at WWPs and recommends the use of starter fertilizers in crops grown on biosolids-treated soils. Starter fertilizers are particularly important where biosolids have been applied immediately before sowing the crop as the slow-release nature of the biosolids would not have produced sufficient plant-available N and P for crop establishment and early plant growth. Even so, each case should be assessed relative to paddock history and climatic conditions before sowing the crop.

Impacts of Biosolids on Soil Properties

The physical properties of biosolids (water-holding capacity, drainage characteristics and structural stability) may be just as important as the nutrients supplied, particularly when used on poorly-structured soils. Most of the agricultural and pastoral soils in NSW have thin, seasonally hard-setting surface soils with low organic matter overlying subsoils of low porosity. An increase in organic matter content usually improves soil structure, decreases bulk density, and increases air porosity, soil moisture retention and hydraulic conductivity of such soils. While the magnitude of these changes has been well researched (Khaleel et al., 1981; Metzger and Yaron, 1987), little is known about the underlying mechanisms. Are these changes simply a direct physical additive effect of biosolids or do they result from actual soil structural improvement through the development of stable aggregates formed when soil mineral particles are bound together by organic 'glues' released by microbial breakdown of organic matter? The answer to this question has implications for biosolids application policy in NSW. If these desirable changes were mainly due to direct physical effects, regular application of biosolids (every 3–5 years) may be necessary to maintain the improvement to soil.

Studies on several soil types in NSW (Joshua et al., 1996) showed that biosolids modified the soil physical properties by decreasing bulk density and increasing infiltration and moisture retention (Table 8.11). The magnitude of changes in bulk density was directly related to the amount of undecomposed biosolids present in sampled cores. This implies that in the early stages after land application, any decrease in bulk density was due to a dilution effect caused by physical mixing of lower density biosolids with mineral soil of higher bulk density rather than to improved soil aggregation. Hydraulic conductivity also increased (Table 8.11) leading to greater infiltration (Figure 8.3), although the results were variable due to the random inclusion of intact biosolids pieces in the samples used for physical analyses. In turn, while these changes increased moisture retention, the effect on available water content varied with soil type (Table 8.11).

Table 8.11: Physical characteristics of soil one year or more after application of dewatered biosolids (Joshua et al., 1996)

Location/Soil type[1]	Application rate (dry t/ha.)	Bulk density (g/cm^3)	Moisture retention (cm^3/cm^3) 10 kPa	1500 kPa	AWC[3]	Ksat (mm/hr)
Glenfield/Red podsolic	0	1.38	0.32	0.24	0.08	?
	400[2]	1.24	0.31	0.20	0.11	?
	1200[2]	1.14	0.38	0.24	0.14	?
Bathurst/Yellow podsolic	0	1.56	0.24	0.11	0.13	?
	60	1.41	0.27	0.14	0.13	?
Goulburn/Soloth	0	1.46	0.26	0.13	0.13	32
	60	1.25	0.30	0.13	0.17	53
Goulburn/Solodic	0	1.35	0.29	0.09	0.20	17
	60	1.26	0.28	0.07	0.21	38
Goulburn/Red earth	0	1.21	0.25	0.11	0.14	10
	60	1.15	0.33	0.11	0.22	35

[1] All three soils have clay content >40% in B-horizon
[2] Biosolids applied as liquid (LB)
[3] Available water content

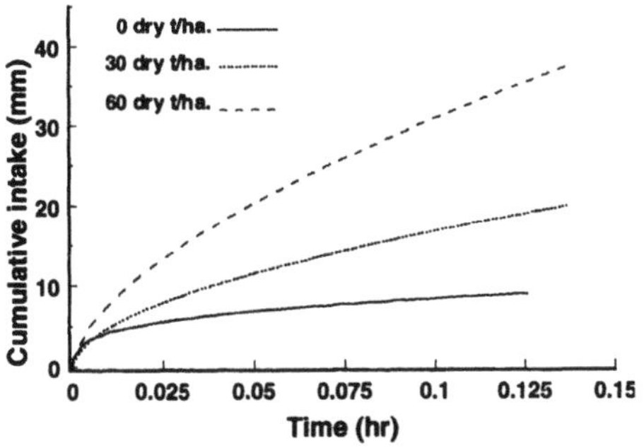

Fig. 8.3: Initial water intake rates for 0, 30 and 60 dry t/ha. of dewatered biosolids applied at Bathurst (Joshua et al., 1996)

However, despite the difficulties encountered in measuring these physical parameters due to the variability in biosolids distribution within the soil surface, the combined impacts of physical and chemical changes were evident at Goulburn where the growing season of pastures was extended by application of >60 t/ha. DWB (Michalk et al., 1996a). Long-term experiments at Glenfield suggest that it may take several years before the improved soil structure due to biosolids is evident in podsolic soils (Joshua et al., 1996).

Plant Production Response to Biosolids Application

From the producer's viewpoint, the real benefit of biosolids is the additional agricultural production obtained from land application. The demand for biosolids by the agricultural sector in NSW has been overwhelming because of the production increases demonstrated with biosolids used in a wide range of food and livestock activities.

Field crops

Cereal production is a major agricultural industry in NSW. Research in central NSW has demonstrated that biosolids applied at rates as low as 6 t /ha. DWB can increase wheat yield by 40% relative to standard fertilizer practice at some sites, but had little effect at others (Figure 8.4). These variable responses reflect the differences in the availability of P in the biosolids products used. At the low-response sites, little available P was supplied by the Fe-treated (15% Fe) biosolids applied whereas at the responsive sites the DWB with less than 2% Fe significantly increased available soil P (Table 8.10). This highlights the importance of the quality of biosolids towards increasing the crop yield.

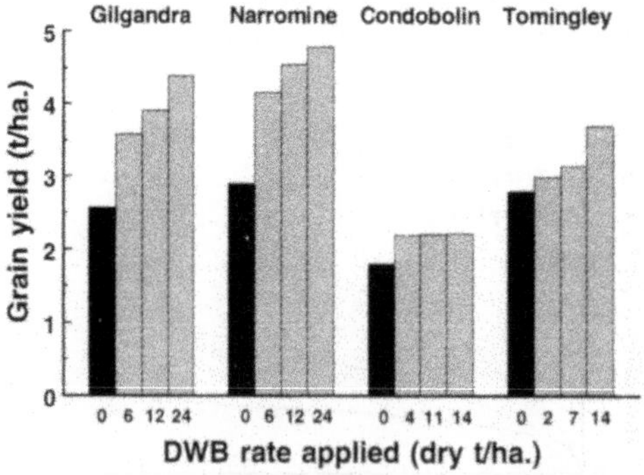

Fig. 8.4: Response of grain yield wheat (*Triticum aestivum*) treated with different rates of dewatered biosolids (redrawn from Cooper (1996) and Bamforth (1996b))

Application of alkali-stabilized biosolids was less successful, producing no short-term response in a range of summer crops (cotton, maize, sorghum, soybeans and forage millet) or wheat (Bamforth 1996a, b), although these products were as effective as lime when applied at equivalent neutralising rates in ameliorating soil pH.

Biosolids are also of interest to supply nutrients and restore structure for the sugarcane industry on soils that have been excessively cultivated and compacted by the passage of heavy machinery. Hughes and Hirst (1996) showed that 60 t/ha. DWB produced the same fresh stalk-weight (186 t/ha.) and sugar yield (24 t/ha.) as 180 kg N/ha. applied as inorganic fertilizer. These results have created considerable interest in the use of biosolids in the cane industry although their demand would depend on cost, availability and environmental constraints.

Vegetables

Composts produced from biosolids generally produce a positive response when applied to horticultural crops through nutrient supply, improved soil condition and disease suppression (Hoitink and Fahy, 1986). In NSW, Cresswell et al. (1996) produced a range of composted biosolids (CB) to determine the production responses of tomato (*Lycopersicon esculentum* var. *commune*), beetroot (*Beta vulgaris*), broccoli (*Brassica ruvo*) and eggplant (*Solanum melongena*) in relation to commercial fertilizers. Strong responses were recorded for all these vegetable crops with yield increases of 10–158% obtained with CB relative to commercial rates of inorganic fertilizers.

The higher growth rate of vegetables grown on CB-amended soil was attributed to improved supplies of nutrients and soil water. However, the level of response varied with the bulking agent used in CB formulation, because the latter determined the availability of N. For example, the lowest response measured with compost made with sawdust where the large surface area per unit volume and a C:N ratio of 30:1 effectively limited plant response by immobilizing N. In contrast, CB with readily-available N depressed yield or product quality by promoting effective vegetative growth. Tighter process control, particularly bulking agent particle size distribution and quality will reduce N immobilization problems (Fahy et al., 1996) while excessive N supply to N-sensitive crops can be controlled through the application rate (Cresswell et al., 1996).

Disease suppression may increase the future value of CB in vegetable and flower production as chemical fumigants are phased out in order to meet environmental standards (Tesoriero et al., 1996). However, the data obtained gave inconsistent disease suppression and based on the data obtained, it is unlikely that the reported ability of CB to suppress diseases could be used as a factor to market these products.

Fruit

Almost all of the 7600 ha. of apple, pear and stone-fruit orchards in NSW are planted on soils with low pH or fertility problems, or both. These problems are compounded by the development of high-density orchards where heavy use of N fertilizers increase soil acidification (Koffmann et al., 1996). Experiments at Bathurst evaluated the effects of biosolids products on the production of deciduous orchards (apples, pears and peaches) and their impact on soil water quality.

Compared to the use of inorganic fertilizer, DWB applied at 10 dry t/ha. or greater amounts either maintained or improved tree growth rates (as measured by tree butt area) and produced equivalent or larger crops of apples, pears and peaches as compared to a fertilized control. Peach was the most responsive crop with 30 dry t/ha. DWB doubling fruit production per tree and increasing the average fruit weight by 23g compared with that produced using standard inorganic fertilizers.

In a subsequent experiment on the same trial site, biosolids and green waste-based composts were applied as mulches to the five-year-old peach and apple trees. Growth and yield measured over three seasons confirmed the fertilizer value of organic amendments in these production systems, particularly for peaches. other measured positive effects with potential to influence production included improved soil water characteristics and reduced evaporative loss.

Pasture and livestock

Grassland is an important outlet for biosolids land application because of its year-round availability (Smith, 1996). Due to its proximity to Sydney, degraded pastoral land in southern NSW, used primarily for extensive wool and sheep production, was a potential target area for land application. one aim of a large grazing experiment established at Goulburn in the Southern Tablelands of NSW was to assess the potential of biosolids products to increase output from a sheep-breeding enterprise (Michalk et al., 1996a).

Pasture production responded significantly to nutrients supplied by DWB at rates of 30, 60 and 120 dry t/ha. During winter, when feed supplies are most critical for sheep production, DWB-treated pastures consistently had higher total and green biomass than pastures treated with superphosphate and lime (Figure 8.5) because of the combined effect of the high N input (744–3528 kg/ha. for the 30 and 120 t/ha. treatments, respectively) and the improved moisture status of DWB-treated soils. Measurements undertaken in the dry autumn of 1995 showed that the 30, 60 and 120 t/ha. DWB-treated plots contained 11%, 24% and 46% more water than the control in the topsoil (0–20 cm). The combination of

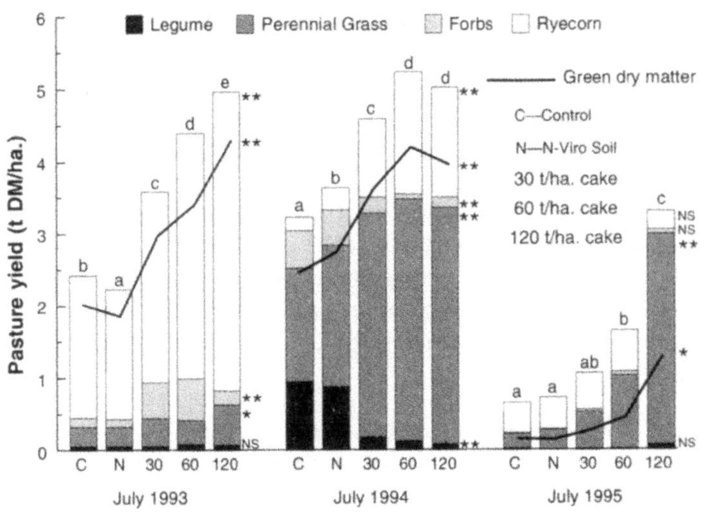

Fig. 8.5: Effect of dewatered biosolids and fertilizers on total yield, green yield and botanical composition of pastures grown on duplex soils measured in mid-winter at Goulburn (Michalk et al., 1996a). Histograms with different letters are significantly different

increased soil water and available nutrient resulted in the 120 t/ha. DWB plots producing five times the total yield and 15 times the yield of green pasture during the dry winter compared with the control (Figure 8.5).

Botanical composition has changed significantly over time (Figure 8.5). Initially, the pasture was dominated by ryecorn (*Secale cereale*), but this was replaced after 12 months by perennial grasses composed mainly of cocksfoot (*Dactylis glomerata*). However, phalaris (*Phalaris aquatica*) responded to N supplied by biosolids and by 1995 made up about 20% of the perennial grass component in pasture treated with 120 t/ha DWB. White clover (*Trifolium repens*) and subterranean clover (*T. subterraneum*) accounted for about one-third of winter production in the control in 1994, but little legume established in the DWB treatments (Figure 8.5). Application of DWB also increased the concentrations of N, P, S, Ca and Mg in the top growth of all pasture species compared to inorganic fertilizers. Yield and composition of pastures treated with alkali-stabilized biosolids (N-Viro Soil®) were similar to pastures treated with agricultural lime and superphosphate.

Higher pasture production and quality resulted in large increases in wool and meat production. Ewes and lambs grazing in pastures treated with >60 t/ha. DWB were heavier than those grazing in pastures with other treatments even when differential higher stocking rates were introduced in 1996 to place additional grazing pressure on the more productive DWB-treated pastures. The combined effect of higher stocking rate (30% in 120 t/ha. DWB pastures) and the maintenance per head of wool production increased the wool production per hectare by 4 kg/ha. (Table 8.12). DWB applications also produced upto 66 kg/ha. more lamb. Taken together, these production increases are significant economically. More importantly, Michalk et al. (1996c) report that while pastures treated with inorganic fertilizer had reached the upper limit for the stocking rate (at about 6 ewes/ha.), DWB-treated pastures have the potential to be grazed at stocking rates well above 10 ewes/ha. No adverse effects of biosolids on sheep live weight gain or reproductive performance were observed. Livestock performance on pastures treated with N-Viro Soil® was similar to that measured with conventionally-fertilized pastures (Michalk et al., 1996c).

Table 8.12: Total annual ewe wool production (kg/ha.) and lamb production (kg LW/ha.) for sheep grazing on pastures with biosolids applied at Goulburn (Michalk et al., 1996b)

Treatment	*Wool production (kg/ha./year)*			*Lamb production (kg LW/ha./year)*[3]		
	1994	*1995*	*1996*	*1994*	*1995*	*1996*
Control[1]	12	22	22	75	82	88
30 dry t/ha. DWB[2]	11	21	22	55	91	125
60 dry t/ha. DWB	11	23	26	53	128	135
120 dry t/ha. DWB	11	23	26	51	128	154

[1]Fertilized initially with 2.5 t/ha. agricultural lime and 400 kg/ha. 50:50 lime:superphosphate and topdressed with 250 kg/ha. as an annual dressing
[2]Biosolids applied as a one-off application
[3]LW denotes live weight

RISKS OF CONTAMINATION ASSOCIATED WITH USE OF BIOSOLIDS FOR CROP AND PASTURE PRODUCTION

The use of biosolids in agriculture has measurable benefits for improving the soil condition and production. However, biosolids also contain heavy metals, persistent organic compounds and pathogens that have the potential to adversely affect long-term productivity, cause spoilage of agricultural products and pose health problems in humans and livestock. A major problem encountered by agencies in guideline development for land application programmes is the behaviour of biosolids-derived heavy metals in soil and their uptake by plants and animals (Lue-Hing et al., 1994). Metals may also inhibit the functions of soil microbes involved in nutrient cycling. Potential health risk associated with pathogens that survive both wastewater treatment and a period after land application is another problem that must be addressed in the guidelines for biosolids use.

In NSW, 14 million ha. of land has a surface soil pH (measured in 0.01 *M* $CaCl_2$, pH_{Ca}) of below 5.0 (half with a soil pH_{Ca} <4.5) with an additional 6 million ha. with a pH_{Ca} between 5.0 and 5.5. These soils are common in agricultural areas near the Sydney Basin, targeted by Sydney Water's biosolids beneficial use programme. Biosolids-derived metal contaminants, added to these low pH soils, have the potential to be mobile and bioavailable. Further constraints are placed on the use of biosolids in Australia because national health authorities have set Maximum Permissible Concentrations (MPCs) for contaminants in foodstuffs (for import and domestic consumption) which for Cd, are lower than those allowed in many other countries (ANZFA, 1996). The MPC for Cd in root crops, leaf crops and cereals is 0.1 mg/kg fresh weight.

Whatmuff (2002) has demonstrated that the adoption in Australia of guidelines for biosolids use based on guidelines which have been developed in other parts of the world are not appropriate. This was particularly the case for Part 503 regulations developed by the US EPA (1993). Rates of metal uptake in vegetable crops were 10-fold greater for Cd and 20-fold greater for Zn than rates predicted from metal uptake dose response curves used in the US EPA biosolids guidelines. Cd and Zn uptake in glasshouse-grown plants was also more than twice that in field-grown plants. of more concern was the excessive uptake measured when the plants were grown on soils with Cd and Zn concentrations well below soil contaminant ceiling concentrations specified in the US EPA equal to the maximum soil limits set in the United Kingdom biosolids guidelines. A significant implication of the results was that the adoption of the US EPA dose-response relationships to set metal limits for acid soils in NSW would potentially reduce production through metal phytotoxicity and compromise markets through metal contamination in foodstuff above acceptable standards.

Clearly, these results highlight the danger of adopting overseas regulations or relying on glasshouse experiments as a basis for NSW biosolids guidelines. Thus, a field research programme on soils common in NSW was necessary in order to define contaminant uptake dose-response curves for the Australian unique soils. The flaws of using glasshouse trials for risk assessment are well recognised (Page et al., 1989) and the US EPA now adopt a field-based approach to refine their biosolids guidelines (US EPA, 1993; 1995).

Metal Uptake by Edible Crops and Pasture Species

Metals of major concern

Cu and Zn are the most common metal contaminants found in Sydney biosolids, but Cd is the metal that poses the greatest potential hazard to health when biosolids are applied to agricultural land (Ryan et al., 1982). This is particularly so in Australia, where a long history of applying superphosphate contaminated with Cd has already elevated Cd concentration in highly fertilized soils (Williams and David, 1976) to levels where contamination of livestock products may occur (Langlands et al., 1988). Whatmuff (1996) reported that where the soil Cd level was high (1.4 mg Cd/kg), Cd accumulated in the edible portion of some plants to levels exceeding 0.1 mg Cd/kg tissue (fresh weight), the Australian Maximum Permissible Concentration (MPC) for human foodstuffs (ANZFA, 1996).

Soil chemical factors affecting metal bioavailability

Our research has confirmed the importance of soil pH, total metal loading, soil Zn:Cd ratio and the non-additive effect of repeated biosolids applications on metal uptake as factors that influence metal uptake and accumulation in the edible portion of crops. Of these, soil pH is the most important because with decrease in soil pH, there is a corresponding increase in metal solubility (Lindsay, 1979; Kiekens, 1984) and metals become more available for plant uptake (Smith, 1994; Sanders et al., 1986; Davis and Carlton-Smith, 1984).

In general, Cd uptake by plants was negatively correlated with soil pH and metal uptake increased with metal load (Figures 8.6 and 8.7). However, this uptake pattern was influenced by soil Zn concentration, which had an inhibiting affect on Cd uptake when the soil Zn:Cd ratio exceeded 100 (Figure 8.8).

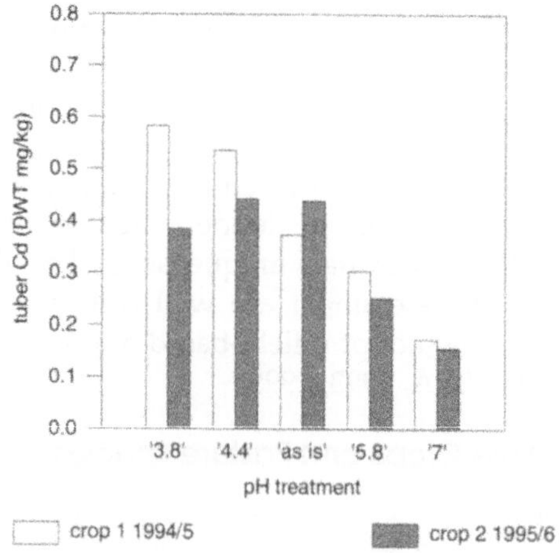

Fig. 8.6: Effect of soil pH on the Cd concentration in potato (*Solanum tuberosum*) tubers grown on biosolids-amended soils

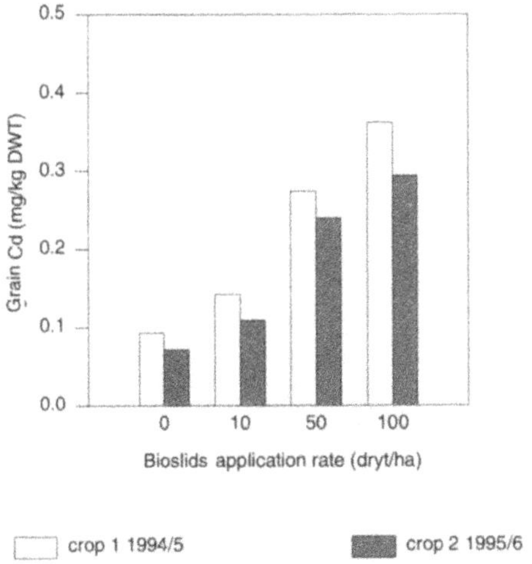

Fig. 8.7: Effect of application rate on the Cd concentration in wheat (*Triticum aestivum*) grain in plants grown on biosolids-amended soils

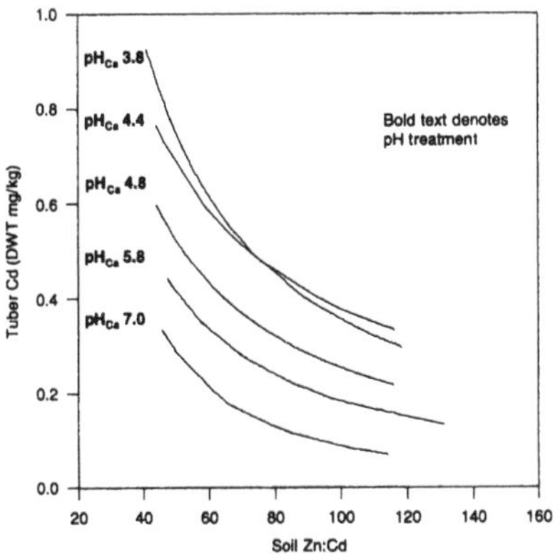

Fig. 8.8: Effect of soil pH on the relationship between soil Zn:Cd and Cd accumulated by potato (*Solanum tuberosum*) tubers grown on biosolids-amended soils

Repeated biosolids application did not result in additive metal uptake by plants, especially at soil pH_{Ca}>4.8 (Figure 8.9). The metal uptake response to biosolids-derived Cd by plants grown on soils previously amended with biosolids was compared to plants grown on soils that received recent, single applications only (Figure 8.10). Previous biosolids applications reduced the rate of metal uptake from more recent application by upto 50%. However, Cd and Zn applied with the previous applications remained bioavailable some seven years after the applications ceased. These studies provided no clear answer to claims (Chaney and Ryan, 1993) of the persistence of the effect of biosolids organic matter on metal sorption.

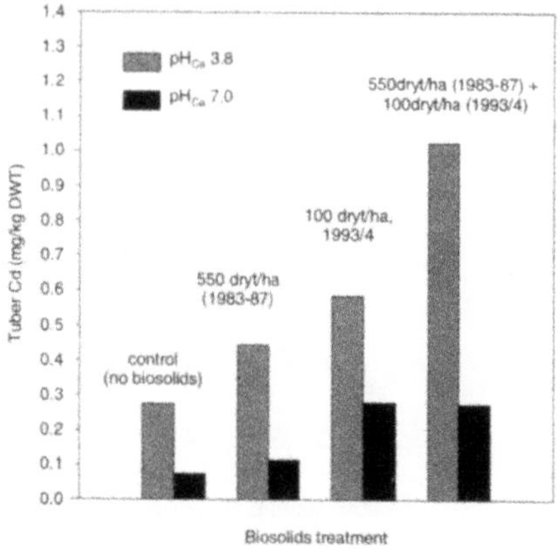

Fig. 8.9. Non-additive accumulation of biosolids-Cd by potato (*Solanum tuberosum*) tubers grown on repeat application and newly-amended soils

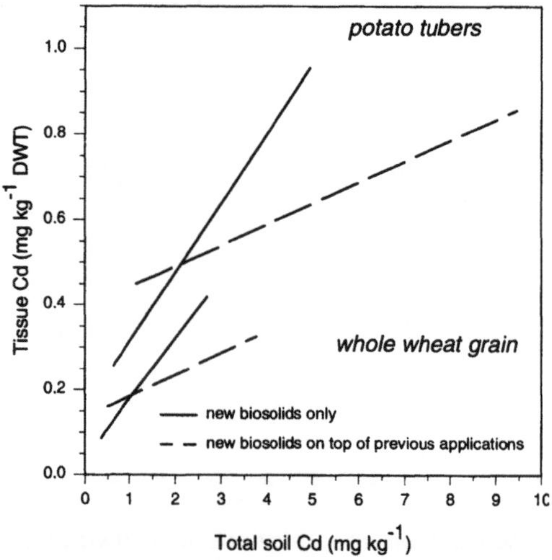

Fig. 8.10. Effect of repeat application of biosolids on the uptake response to biosolids-Cd by potato (*Solanum tuberosum*) tubers and wheat (*Triticum aestivum*) grain at pH_{CaCl_2}4.4

Plant factors affecting metal uptake and accumulation

Whatmuff (1996) studied the plant uptake of metals by applying different rates of biosolids to produce a wide range of Cd (0.4–9.7 mg/kg) and Zn (51–577 mg/kg) concentrations in the surface soil (0–15 cm) in the field that exceeded the Maximum Allowable Soil Contaminant Concentrations in the NSW guidelines (NSW EPA, 1997). Metal uptake and accumulation differed between species and plant part. Overall, leafy vegetables accumulated the highest levels of Cd and Zn and pasture species (especially grasses) the least (Table 8.13).

Table 8.13: The uptake of Cd and Zn by various plants grown on soils amended with biosolids and composted biosolids at pH_{Ca} 4.2–4.8

Plant	*Part*	*Treatment*	*Cd (mg kg⁻¹)*		*Zn (mg kg⁻¹)*	
			oven dry wt Soil [1] [3]	*dry wt plant*	*oven dry wt soil [200]* [3]	*dry wt plant*
lettuce	leaves	biosolids (0–650 dry t $ha.^{-1}$)	0.5-7.0	1.4-11.4	48-407	7-6540
	leaves (inner)	biosolids compost[1] (0, 62.5, 125 dry t $ha.^{-1}$)	0.12-6.5	0.65-4.6	185-437	113-267
	leaves (outer)	biosolids compost[1] (0, 62.5, 125 dry t $ha.^{-1}$)	0.12-6.5	1.2-9.3	185-437	91-410
beetroot	'root'	biosolids compost[1] (0, 62.5, 125 dry t $ha.^{-1}$)	0.12-6.5	0.12-0.65	185-437	60-180
eggplant	fruit	biosolids compost[1] (0, 62.5, 125 dry t $ha.^{-1}$)	0.12-6.5	0.4	185-437	28-34
potato	leaves	biosolids (0–650 dry t $ha.^{-1}$)	0.4-0.97	3.1-22	51-577	49-1210
	peel	biosolids (0–650 dry t $ha.^{-1}$)	0.4-0.97	0.6-2.5	51-577	29-221
	tuber	biosolids (0–650 dry t $ha.^{-1}$)	0.4-0.97	0.3-1.2	51-577	25-86
radish	leaves	Biosolids (0–650 dry t $ha.^{-1}$)	0.5-7.0		48-407	43-880
	root	biosolids (0–650 dry t $ha.^{-1}$)	0.5-7.0	0.22-3.6	48-407	
silverbeet	leaves	biosolids (0–650 dry t $ha.^{-1}$)	0.5-7.0	0.63-17	48-407	84-1470
wheat	leaves	biosolids (0–650 dry t $ha.^{-1}$)	0.45-3.18	0.06-0.69	51-230	21.5-89
	grain	biosolids (0–650 dry t $ha.^{-1}$)	0.45-3.18	0.045-0.28	51-230	36.5-115
sugarcane	leaves	biosolids (0–60 dry t $ha.^{-1}$)	<0.6-1.02	0.31-0.5	54-76	15-23
	stem	Biosolids (0–60 dry t $ha.^{-1}$)	<0.6-1.02	0.02-0.14	54-76	9-17
maize	silage	Biosolids (0–650 dry t $ha.^{-1}$)	0.4-0.97	0.045-0.52	51-577	0.5-39
	kernels	biosolids (0–650 dry t $ha.^{-1}$)	0.4-0.97	BD[2]	51-577	38-145
canola	leaves	biosolids (0–650 dry t $ha.^{-1}$)	0.4-0.97	0.615-1.85	51-577	196-410
	oil	biosolids (0–650 dry t $ha.^{-1}$)	0.4-0.97	BD[2]	51-577	0.06-0.11
	meal	biosolids (0–650 dry t $ha.^{-1}$)	0.4-0.97	0.04-0.2	51-577	79-115
ryegrass	leaves	biosolids (0–650 dry t $ha.^{-1}$)	0.45-3.18	0.05-1.1	51-230	42-212
other grasses	leaves	biosolids (0–120 dry t $ha.^{-1}$)	<0.5-2.2	0.05-0.74	10-120	20-234
ryecorn	leaves	biosolids (0–120 dry t $ha.^{-1}$)	<0.5-2.2	<0.01-0.15	10-120	25-165
weeds	leaves	biosolids (0–120 dry t $ha.^{-1}$)	<0.5-2.2	0.11-4.5	10-120	24-415
clover	leaves	biosolids (0–120 dry t $ha.^{-1}$)	<0.5-2.2	0.07-0.8	10-120	18-470

[1]biosolids compost comprising 2 parts biosolids, 1 part sawdust and 1 part green waste
[2]BD = below detection
[3]Numbers in [bold] indicate maximum allowable soil concentration

Leafy vegetables, such as lettuce and silverbeet, accumulated very high levels of Cd (>9 mg/kg, Table 8.13) at soil concentrations well below 1 mg/kg. Silverbeet (*Beta vulgaris*) was most efficient at extracting soil Cd with leaf concentrations exceeding the MPC in control plots where past fertilizer use had increased the total soil Cd to 0.4 mg/kg. Addition of bulking agents (sawdust and green waste) with the composted biosolids did not reduce the response to biosolids-Cd compared to biosolids alone (Figure 8.11). This may be a consequence of the coarse nature of these bulking agents and their generally inert chemical nature.

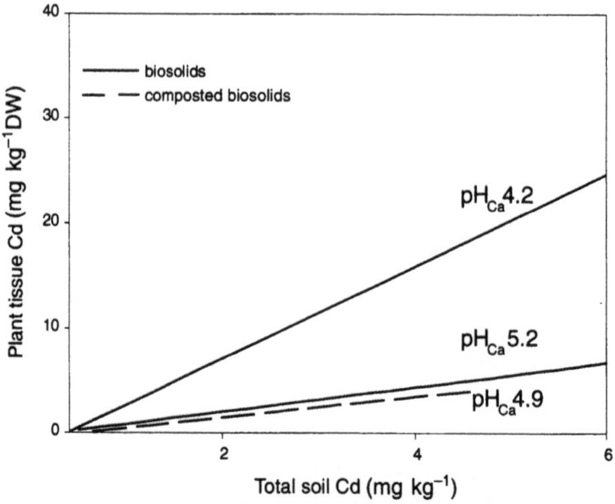

Fig. 8.11: A comparison of Cd-uptake response by lettuce grown on soils amended with either biosolids or biosolids compost

Metals also tended to accumulate in the foliage of all plants (Figure 8.12). The MPC for Cd was exceeded in potato (*Solanum tuberosum*) tubers where the total soil Cd concentration was >5 mg Cd/kg and soil pH_{Ca} < 5.8, whereas radish (*Raphanus sativus*) 'roots' exceeded the Cd MPC at total soil Cd concentration >1.5 mg Cd/kg at a similar soil pH. Cd accumulated in wheat (*Triticum aestivum*) grain above the MPC where total soil Cd exceeded 1 mg Cd/kg soil and/or at soil pH_{Ca} <5.8.

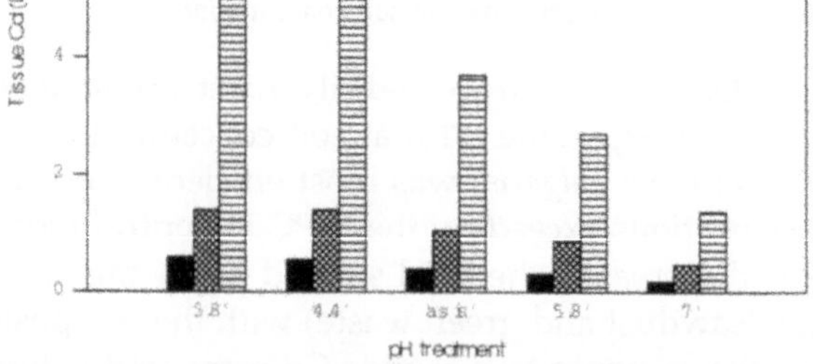

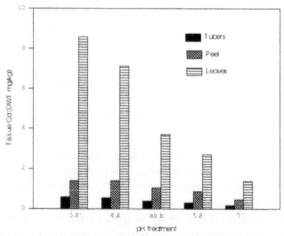

Fig. 8.12: Effect of soil pH on the distribution of Cd in various parts of potatoes (*Solanum tuberosum*) grown in biosolids-amended soils.

Analysis of pasture biomass after biosolids application showed that only broadleaf species such as capeweed consistently accumulated Cd at concentrations that exceeded the limit of 0.5 mg/kg dry wt in stockfeed set by the US National Research Council (NRC, 1980). However, under dry season conditions, Cd levels in perennial grass and clover also exceeded this limit (Michalk et al., 1996c).

Based on these studies, Zn was the only metal that could potentially be a phytotoxicity hazard for plants grown in biosolids-amended soils. Phytotoxicity symptoms were observed in leafy vegetables and potatoes only on biosolids amended soils (<pH_{Ca} 5.8), where soil Zn concentration exceeded 200–250 mg/kg soil. Comparison of the experimental data with critical values (Table 8.14) indicates that the upper range Zn concentrations shown in Table 8.13 are potentially damaging to sensitive plants. The degree of phytotoxicity increased with decreasing soil pH, where soil metals were more available for plant uptake. However, appreciable accumulation of Cd or Zn was measured in apple, pear or peach fruit or foliage at rates of biosolids used in the horticultural experiment described earlier (Koffmann et al., 1996).

Table 8.14: Critical concentrations of heavy metals in soils and plants

Element (mg/kg)	*Critical soil concentration*[1]	*Normal range in plants*[1]	*Critical concentration in plants*[2]
As	20-50	0.02-7	5-20
Cd	3-8	0.1-2.4	5-30
Cr	75-100	0.03-14	5-30
Cu	60-125	5-20	20-100
Ni	100	0.02-5	10-100
Pb	100-400	0.2-20	30-300
Se	5-10	0.001-2	5-30
Zn	70-400	1-400	100-400

[1]Range of values above which toxicity is considered to be possible
[2]Levels above which toxicity effects are likely (adapted from Alloway, 1990)

Metal Uptake by Grazing Animals

Biosolids-derived contaminants can be transferred to livestock products either by direct ingestion of biosolids-treated soil or by grazing plants with elevated levels of metals and organic pollutants. Where biosolids are applied to the soil surface, ingestion of biosolids particles adhering to plant surfaces accounts for a larger intake of contaminants than that contained within herbage tissue (Smith, 1996). In NSW, however, consumption of contaminated forage is a more important pathway; direct ingestion is minimized because the guidelines require incorporation into the soil of all biosolids used in agricultural production, including grassland systems (NSW EPA, 1997).

The effect of long-term exposure on metal levels in livestock tissue was investigated in the Goulburn grazing experiment . After five years of continuously grazing on pastures treated with different rates of DWB, results confirmed overseas studies, such as Stark et al. (1995), with no adverse effects being measured on ewe liveweight gain or reproductive performance (Michalk et al., 1996c). Of crucial importance to consumers was the absence of change in metal concentrations in meat of the ewes and lambs grazing in biosolids-treated pastures, with levels of Cd, Ni, Pb, Cu and Hg well below standards for carcass meat entering the human food chain.

However, higher concentrations of Cd in kidney and Cu in liver tissues were measured in sheep grazing in to biosolids-treated pastures (Figure 8.13). Unlike many overseas studies where these metals accumulate in liver and kidney tissue in direct proportion to the amount ingested (Bray et al., 1985; Brams et al., 1989), a higher accumulation was measured at lower soil-metal loading (Figure 8.13). This result may be due to the antagonistic effects between elements found in diet mixtures that reduce metal adsorption by animals (Smith, 1996). In the Goulburn experiment, high Zn content in forage from biosolids-treated pastures (Michalk et al., 1996b) was several times the daily sheep requirement (NRC, 1980), and this may have interfered with accumulation, as shown in other studies by Bremner et al. (1976) for Cu and McKenna et al. (1992) for Cd.

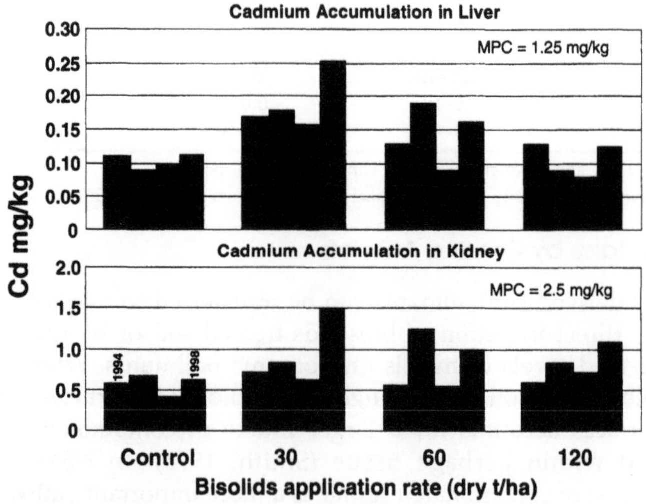

Fig. 8.13: Cadmium levels in ewe liver and kidney after 4 years of exposure to biosolids treated pastures (D. Michalk, unpublished data)

In practical terms, the Cd and Cu levels in livestock tissues (even in the most exposed tissues) do not constitute a health risk since the average life span for breeding sheep is <7 years and much shorter for meat-producing animals. No increases in metal or pesticide levels were measured in milk and wool samples collected from ewes continuously grazing in biosolids-treated pastures which contained metal concentrations above the accepted NSW EPA limits (Michalk et al., 1996c).

Effects of Heavy Metals on Soil Biota

Heavy metals in soils can have a major negative effect on populations, biodiversity and function of soil microorganisms (McGrath, 1994). However, it has been difficult to attribute either a positive or negative effect of biosolids-derived metals on soil microbial parameters from research undertaken in NSW.

Studies at a research site that had been first treated with biosolids in 1986 showed that microbial biomass and soil respiration did not respond to the additional organic carbon (OC) applied with biosolids treatments (Rogers, 1996). However, whether this is an effect of heavy metals or the form and availability of biosolids OC is unclear. At the same site, a decline in the activity of the two enzymes (phosphodiesterase and sulphatase) was measured on soils with the lowest soil pH, but the data was inconclusive (Rogers, 1996).

Since Australian agriculture depends on biological fixation to provide N to crops and pastures, it is especially important to ascertain the impact of biosolids application on *Rhizobium* bacteria and their ability to fix atmospheric N. The effect of biosolids-metals on the intrinsic symbiotic effectiveness (ISE) of *Rhizobium* spp. depends on soil type and application rate (Evans et al., 1996). The ISE of *R. trifolii* and *R. meliloti* on two biosolids-amended yellow podsolics, a sandy loam and a grey clay, were unaffected, whereas in a Kraznozem and a red-brown earth, the ISE was reduced (Evans et al., 1996). The simple hypothesis of a critical heavy metal concentration reducing ISE is not consistent with these observations. It was also found that the N mineralization potential was not reduced relative to unamended soils even with very high biosolids application rates (250 dry t/ha.), where soil moisture was not limiting (Munn et al., 1997). Further studies are needed to assess the effects of changes in microbial parameters on soil fertility and productivity.

Health Risk of Pathogen Survival after Application

The potential transmission of pathogens has been minimized through effective wastewater treatment and restrictions for landuse based on stability-grading specifications in the NSW EPA (1997) Biosolids Guidelines. However, some pathogens surviving in dewatered biosolids may pose a health risk following land application. Light, temperature and soil moisture content are the principal environmental factors influencing pathogen survival in biosolids-amended soils (Coker, 1983). In NSW, Eamens et al. (1996) sampled biosolids lumps in the soil as a worst-case scenario. They reported that after five months, it was only necessary for indicator bacteria (*Escherichia coli, Salmonella* spp., *Clostridium perfringens* and faecal streptococci) in DWB to die back to pre-application levels with summer application, but at least a month longer was needed with early spring applications. Liquid biosolids had 10–100 times less bacteria per gram than DWB and reached background levels in a shorter time after application. Similar trends were evident in surface soils and in soils from lower depths in the profile.

To pose a health risk, surviving pathogens must be present at dose levels sufficient to produce clinical disease symptoms. For example, Moore (1971) reported that the infective dose of *Salmonella* for humans is more than 10^4 organisms per gram. Eamens et al. (1996) showed that bacterial populations were generally less than 10^4 organisms per gram of biosolids six weeks after land application. Storage for upto eight weeks before application also significantly reduced bacteria survival in DWB to below this infective threshold. No ill health was observed in livestock at Goulburn where sheep grazed pastures treated with high biosolids rates (Michalk et al., 1996c), suggesting that there is minimal risk of infecting the grazing animals.

ENVIRONMENTAL PROBLEMS ARISING FROM OFF SITE MOVEMENT OF BIOSOLIDS

The movement of contaminants and nutrients in biosolids-amended soils from the application site by surface run-off or leaching to groundwater, or both, is a potential source of environment pollution from land application of biosolids (Smith, 1996). The extent of these environmental consequences will depend on the specific biophysical characteristics (climate, landform and soil) prevailing at the site of application. Management factors such as type of biosolids product, the method of incorporation, and the type of agricultural production can be used to minimize off-site impacts. Although considerable experience has been gained overseas on off-site impacts of land application, the information may not be relevant to NSW because of the physical and chemical nature of the soils. Unlike Australia, soils of North America and Europe, where most of the studies have been conducted, have higher soil pH and fertility, and rarely exhibit the strong textural contrast that is typical of agricultural soils in NSW (Hird and Bamforth, 1996). Low pH, low cation exchange capacity and strong textural contrasts within the profile have important influence on minimizing the potential off-site impacts of the low biosolids rates recommended for use on NSW soils.

Contamination of Surface Water

There is a public perception that land-applied biosolids are potentially susceptible to displacement by runoff from rainfall, causing significant off-site pollution, particularly when biosolids are left exposed at the soil surface. Surface-exposed biosolids can occur even when incorporation is practised; especially when commercially-available disc ploughs are used (Figure 8.14). However, measurements under natural (Kelling et al., 1977) and simulated (Bruggeman and Morstaghimi, 1993) rainfall show that biosolids application increases soil infiltration, and reduces runoff and losses of sediment and nutrients relative to untreated areas. In NSW, Joshua et al. (1998) also found that application of DWB incorporated into the surface soil reduced runoff and increased retention of rainfall with increasing rate of biosolids (Figure 8.15). These results were consistent over a period of time, except that runoff was further reduced at the highest biosolids rate (120 dry t/ha. DWB) three years after application due to better plant growth and soil cover.

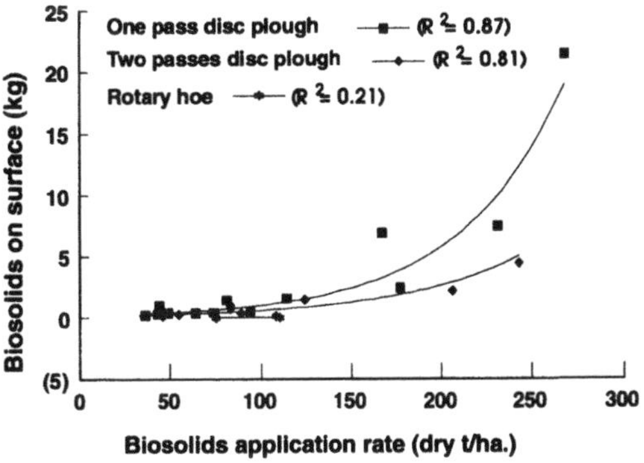

Fig. 8.14: Dewatered biosolids (dry t/ha.) remaining on the soil surface after incorporation with a disc plough (adapted from Salt et al., 1996)

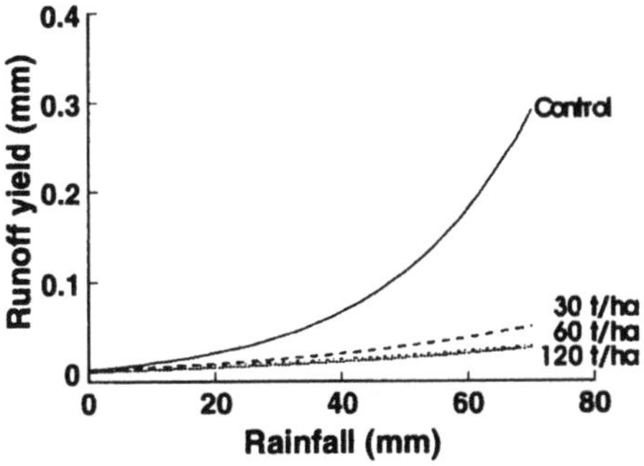

Fig. 8.15: Runoff yield (mm) under natural rainfall from plots with 0–120 dry t/ha. DWB applied to soloth at Goulburn (Joshua et al., 1998)

Analysis of water collected from 12 runoff events at the Goulburn grazing site by Joshua et al. (1998) showed that all metals except for Fe, Al and Mn were well below Guideline standards for stock water on most occasions. Analysis of 1:5 soil-solution extracts from untreated soil confirmed that biosolids were not the source of these metals in the runoff.

Where elevated metal levels were recorded (mainly Zn, Ni, Pb and Cu), they occurred as spikes shortly after biosolids application and the data were inconsistent. However, in terms of the total metal movement, the quantities of metals transported from biosolids-treated pastures in the surface runoff were either the same as or less than untreated areas, because of the significant reduction in the volume of runoff from biosolids-treated pastures with increasing application rate. Nitrate concentration in runoff was occasionally higher for biosolids-treated pastures rather than untreated pastures, but the results were variable (Joshua et al. 1998). Most pastures, including pastures not treated with biosolids, exceeded the ANZECC (1994) guidelines at least once in monitoring over three years, but were generally lower. Joshua et al. concluded that biosolids did not contribute significant amounts of nutrient or metal pollutants to runoff and posed little environmental risk to surface waters adjacent to biosolids applications.

This conclusion was tested by Michalk et al., (1996b) at the same site with an intensively bi-weekly sampling of 24 fixed locations in the two downstream catchments and one upstream catchment from the grazing site where no biosolids had been applied. Pre-treatment monitoring showed that all three catchments had relatively high concentrations of N and P prior to biosolids application (Figure 8.16), and there was no difference in the concentrations of heavy metals between sampling sites. Post-treatment monitoring showed no measurable effect of high rates of biosolids application on N and P concentrations. Zn was the only important biosolids contaminant present in higher concentrations in dams located on the Goulburn site relative to the surrounding sites.

However, lime:super with a measured Zn of 303 mg/kg (Michalk et al., 1996c) is the more likely source of this contamination because Cu levels would also have been high if movement of biosolids particles was the point-source of pollution. High levels of faecal coliforms (Figure 8.16) and *E. coli* (Michalk et al., 1996c) in all three catchments indicates that grazing livestock are the major source of faecal contamination in surface waters and that biosolids application had not increased the bacteria loading in dams and creeks surrounding the Goulburn grazing experiment.

Contamination of Soil and Groundwater by Downward Movement

Environmental hazards derived from biosolids-derived metals and nutrients are linked closely to their movement in soil profiles because

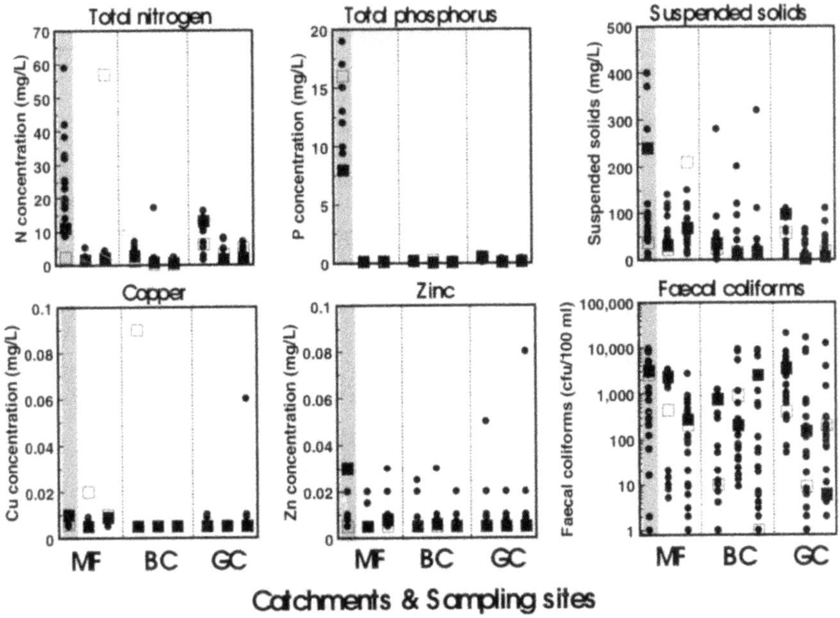

Fig. 8.16: Changes in quality of surface water measured in three catchments (Murray's Flat—MF; Boxer's Creek—BC; Gundarry Creek—GC) and Experimental site—ES surrounding the Goulburn study site over a 12-month period (Michalk et al., 1996c)

Open squares indicate starting values measured on 20 January 1993; closed squares denote the value at the last measurement take on 18 January 1994, and closed circles denote the data collected between these dates.

even slow transport through soil and subsoil may eventually pollute the groundwater (Li and Shuman, 1996). Coarse surface texture, and low pH, CEC, and organic matter are characteristics that pre-dispose soils in NSW to downward movement of N and metal contaminants.

However, due to the existence of a heavy clay B-horizon underlying the coarser-textured topsoil in many of the target soils for land application of biosolids in NSW, leached N and other biosolids-derived contaminants may be retained in the soil profile, thereby reducing the potential to pollute the groundwater. Medium-term experiments (4–10 years) were conducted to determine the pollution potential in these soils.

The extent of nitrate movement was measured after biosolids application to three soil types at Goulburn. Concentration of NO_3-N increased to a depth of 30 cm five months after treatment (Figure 8.17), and to 70 cm after one year. The actual nitrogen content of the soil due to this movement increased with application rate, but was low for all treatments (Joshua et al., 1998).

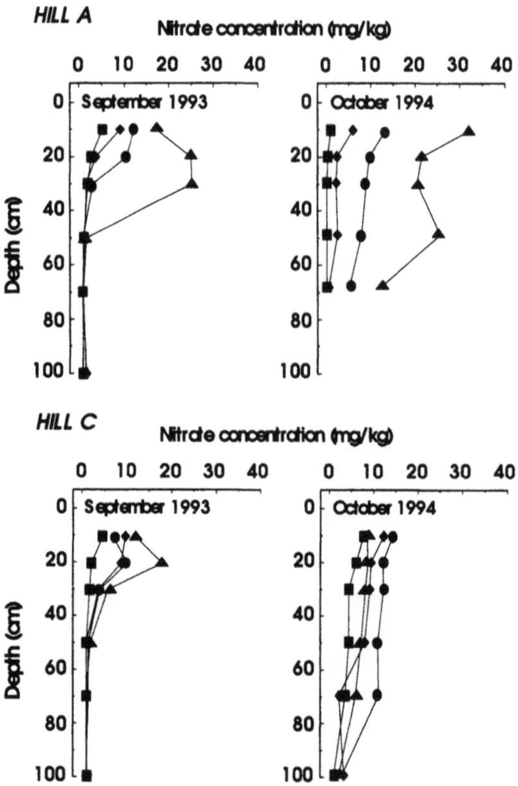

Fig. 8.17: Nitrate concentrations (mg/kg) in soil profiles at Goulburn in September 1993 (5 months after application) and october 1994 with applications of 0 dry t/ha. (solid dot), 30 dry t/ha. (solid square), 60 dry t/ha. (solid vertical triangle) and 120 dry t/ha. (solid inverted triangle) applied to pastures on Hill A (soloth) and Hill C (red earth) (modified from Joshua et al., 1998)

The presence of a clay horizon of low permeability in two duplex soils, a soloth and a solodic, restricted the nitrate movement in this horizon (>20 cm) as shown by the higher NO_3-N content in comparison to that in an adjacent red earth for the upper depths (Figure 8.17). High infiltration of the surface horizon and the very low hydraulic conductivity of the subsurface clay horizon of duplex soils induced a small amount of subsurface lateral nitrate movement.

At the same Goulburn site, Cd, Zn and Cu did not move beyond the plough layer even at high DWB rate of biosolids application (>60 t/ha.) five years after application. No lateral movement of Zn or Cu was detected by comparison between metal concentrations measured at incremental depth in treated and untreated enclosures (Joshua et al., 1998). The retention of metals in the topsoil might be attributable to complexing with soil organic matter. This implies that management practices that maintain high organic matter levels in NSW agricultural systems will play a key role in the retention of all metals in the plough layer of biosolids-treated soils. There were no heavy metals in leachate water collected from horticultural crops treated with dewatered biosolids applications upto 60 dry t/ha. (Koffmann et al., 1996).

IMPACT FOR BIOSOLIDS USE IN NSW

One of the major factors contributing to the success of the beneficial use programme in NSW over the past decade has been the applied research programme that identified the fertilizer value of various biosolids products (Osborne, 1996) and assessed the environmental hazards of biosolids when applied using the standard practice. This information has been used by Sydney Water to establish good operational management practices along with a comprehensive communication strategy that has allayed many of the concerns about the impacts of pathogens, heavy metals and organic pollutants on the environment and human health. With this growing public acceptance, the biosolids industry is now recognized as a responsible and important contributor towards waste recycling in NSW (Walsh and Rawlinson, 1995).

This experiment in NSW has created a growing interest in the commercial use of biosolids and Sydney Water now beneficially recycles 99% of biosolids from the Sydney region to agriculture, forestry, landscaping and land rehabilitation markets (Figure 8.18). In the agricultural sector, biosolids account for >60% of land application, and the demand is exceeding the supply under current arrangements wherein Sydney Water supplies, transports and spreads the product free of charge. In 1996, for example, over 50 000 t of alkali-stabilized biosolids was

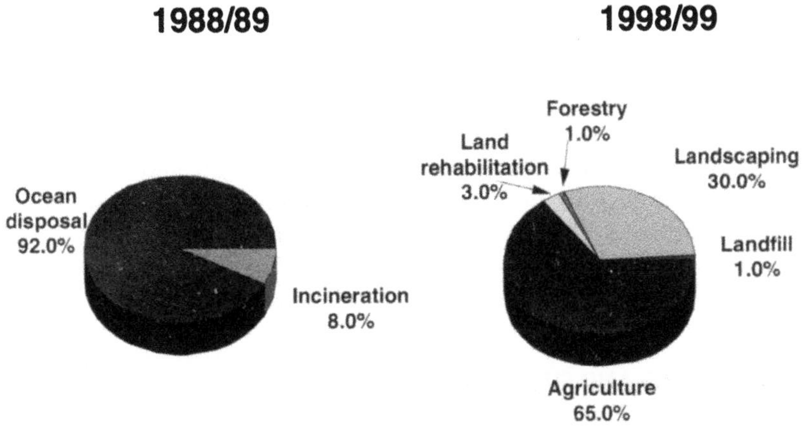

Fig. 8.18: Disposal and recycling of biosolids in NSW, 1989–1999

beneficially used as a liming material at rates of 3–15 t/ha. to cropland (Kempton, 1996) on about 80 farms located up to 300 km from Sydney. This usage still represented a significant economic saving because the costs to transport and spread biosolids on properly licensed and managed sites was significantly less than the cost of disposal to a lined landfill in Sydney.

Considerable pressure is being generated by commercial interests to quickly deregulate the biosolids industry in order to allow private companies the opportunity to develop fully commercialized markets (Beckett, 1995). However, Kanak et al. (1995) doubt that the benefits will totally cover the cost of transportation from the Sydney region to suitable cropping areas. While some cost recovery of the fertilizer value of biosolids products is sought to offset operational costs associated with the current on-farm applications, it is too soon for Sydney Water to provide guaranteed and consistent supplies for biosolids (Hope and McDougall, 1996). Since there are sufficient outlets for Sydney's long-term production of biosolids, Sydney Water is placing an immediate emphasis on improving the quality of biosolids and application efficiencies of the current programme with partial cost recovery rather than on the development of a fully-commercial industry.

Results from NSW research, together with experience gained over the past seven years in commercial land application, have also assisted the NSW EPA with the release of their Environmental Management Guidelines for the Use and Disposal of Biosolids Products (NSW EPA, 1997). However, there is still a need to further refine these Guidelines based on the current findings to facilitate on-farm application. For example, the results reported by Michalk et al. (1996c) and Joshua et al. (1998) on the environmental impact of surface runoff and sub-surface flow indicate that the current requirement for a 5m fence and other buffers in treated paddocks places an unwarranted restriction on the use of biosolids in commercial cropping systems, especially where the paddock size is small, as is the case on the NSW Tablelands.

The requirement for incorporation of all biosolids products could also be reviewed in certain circumstance such as the location where biosolids are applied to permanent pastures. Research by Michalk and Curtis (1998) showed that surface application poses no greater risk to the environment than incorporated DWB. Endorsement of surface application in the Guidelines would extend the beneficial use of biosolids to rehabilitating degraded pastures rather than restricting their application to newly-sown pastures where incorporation forms part of the establishment process.

CONCLUSIONS AND FUTURE DIRECTIONS

Beneficial use of biosolids is now well established in NSW and Sydney Water enjoys widespread community support for its application programme that is both beneficial to agricultural production and environmentally sustainable at current application rates. NSW Government authorities responsible for regulating biosolids application to agricultural land now have a large body of NSW data that can be used to support and refine the current guidelines. Where biosolids have been applied at 3–30 t/ha. DWB, no detrimental effects on plant and livestock products or contamination of waterways or groundwater have been recorded over a range of soil types and agricultural activities. This should encourage local councils in country NSW, where less than ten per cent of biosolids are beneficially used to develop land application, as a viable alternative to current stockpiling and landfilling practices (NSW DLWC, 1997).

Where higher rates (>100 dry t/ha.) are applied either as a single application or accumulated through sequential applications over a period of time, the downward movement of nitrate and heavy metals do occur. Under these conditions, accumulation of Cd in edible portions of leafy vegetables poses the greatest hazard for Cd entry to the human food chain. At the same time, however, other than in vegetable production areas, there is minimal risk of these metals or nutrients contaminating either surface water or groundwater on the duplex soils that are the dominant soil group in the agricultural areas targeted for application. Nevertheless, careful long-term monitoring of metal bioavailability is still required to fully explain the reactions of biosolids in the unique soil compositions in NSW.

ACKNOWLEDGMENTS

The NSW Agriculture Biosolids Research Programme would not have been possible without the sustained financial support of Sydney Water and the vision of the late Eric Corbin, former General Manager of OWRU. The outcomes are also a testimony to the collaboration of many research officers within NSW Agriculture who were involved in the programme from the mid-1980s to 1999.

REFERENCES

Alloway, B.J., 1990, Cadmium. In: Alloway, B.J. (Ed.). *Heavy Metals in Soil*. Blackie, Glasgow: 100–124.

Angle, J. Scott, 1994, Sewage sludge: Pathogenic considerations. In: Clapp, C.E., Larson, W.E. and Dowdy R.H. (Eds). *Sewage Sludge: Land Utilization and the Environment*. Soil Science Society of America Miscellaneous Publication, Madison: 35–39.

ANZECC [Australian and New Zealand Environment and Conservation Council], 1994, National Water Quality Management Strategy: Water Quality Management — An outline of the Policies.

ANZFA [Australia and New Zealand Food Authority], 1996, Proposal P144, Review of the Maximum Permitted Concentrations of Cadmium in Food (July).

AWRC [Australian Water Resources Council], 1992, Guidelines for Sewerage Systems: Sludge Management (draft). 34 pp.

Awad, A.S., Ross, A.D. and Lawrie, R.A., 1989, *Guidelines for the Use of Sewage Sludge on Agricultural Land* (1st Edition). NSW Agriculture, Orange, NSW, Australia.

Bamforth, I., 1996a, The growth and yield response of a range of summer crops to biosolids products. In: Osborne, G.J., Parkin, R.L., Michalk, D.L. and Grieve, A.M. (eds). *Biosolids Research in NSW*. NSW Agriculture, Organic Waste Recycling Unit, Richmond: 95–102.

Bamforth, I., 1996b, The growth and yield response of wheat to biosolids products. In: Osborne, G.J., Parkin, R.L., Michalk, D.L. and Grieve, A.M. (Eds), *Biosolids Research in NSW*. NSW Agriculture, Organic Waste Recycling Unit, Richmond: 103–111.

Beckett, J., 1995, Closing the gap on resource depletion: commercial incentives within the biosolids industry. *Proceedings of the Sixteenth Australian Water and Wastewater Association Federal Convention*: 65–70.

Brady, N.C., 1990, *The Nature and Properties of Soils* (10th edition). Macmillan Publishing Company, New York.

Brams, E., Anthony, W. and Weatherspoon, L., 1989, Biological monitoring of an agricultural food chain: Soil cadmium and lead in ruminant tissues. *Journal of Environmental Quality* 18: 317–323.

Bray, B.J., Dowdy, R.H., Goodrich, R.D. and Pamp, D.E., 1985, Trace metal accumulations in tissues of goats fed silage produced on sewage amended soil. *Journal of Environmental Quality* 14: 114–118.

Bremner, I., Young, B.W. and Mills, C.F., 1976, Protective effect of Zn supplementation against copper toxicosis in sheep. *British Journal of Nutrition*, 36: 551–561.

Bruggeman, A.C. and Mostaghimi, S., 1993, Sludge application effects on runoff, infiltration and water quality. *Water Resources Bulletin*, American Water Resources Association 29: 15–25.

CES [Consultants in Environmental Service Limited], 1992, UK Sewage Sludge Survey. Final report. CES, Gateshead. (Cited by Smith, S.R. 1996. *Agricultural Recycling of Sewage Sludge and the Environment*. CAB International, Oxford, p. 382.)

Chaney, R.L., 1983, Potential effects of waste constituents on the food chain. In: Parr, J.F. Marsh, P.B. and Kla, J.M. (Eds), *Land Treatment of Hazardous Wastes*. Noyes Data Corp, Park Ridge, New Jersey: 152–240.

Chaney, R.L., 1990, Twenty years of land application research. Part 1. *Biocycle* 31: 54–59.

Chaney, R.L. and Ryan, J.A., 1993, Heavy metals and toxic organic pollutants in MSW-composts: Research results on phytoavailability. In: Hoitink, H.A.J. and Keener, H.M. (Eds). *Science and Engineering of Composting: Design, Environmental, Microbiological and Utilization Aspects*. Renaissance Publications, Worthington, Ohio: 451–506.

CIWEM [Chartered Institution of Water and Environment Management], 1995a, *Sewage Sludge—Utilisation and Disposal*. CIWEM, London.

CIWEM [Chartered Institution of Water and Environmental Management], 1995b, *Sewage Sludge: Introducing Treatment and Management*. CIWEM, London, 118 pp.

Clapp, C.E., Dowdy, R.H., Linden, D.R., Larson, W.E., Hormann, C.M., Smith, K.E., Halbach, T.R., Cheng, H.H. and Polta R.C., 1994, Crop yields, nutrient uptake, soil and water quality during 20 years on the Rosemount sewage sludge watershed. In: C.E. Clapp, Larson, W.E. and Dowdy, R.H. (Eds). *Sewage Sludge: Land Utilization and the Environment*. Soil Science Society of America Miscellaneous Publication, Madison: 137–148.

Coker, E.G., 1983, Biological aspects of the disposal-utilisation of sewage sludge on land. *Advances in Applied Biology* 9: 257–322.

Cooper, J.L., 1996, The agronomic value of biosolids in cropping systems on acid soils in central NSW. In: Osborne, G.J., Parkin, R.L., Michalk, D.L. and Grieve, A.M. (Eds), *Biosolids Research in NSW*. NSW Agriculture, Organic Waste Recycling Unit, Richmond: 87–94.

Copland, B., 1990, Domestic Catchment Samples Program Report. Trade Waste Unit, Sydney Water Board, NSW.

Corbin, E.J., Bamforth, I., Cooper, J., and Osborne, G.J., 1994, Beneficial use of nutrients in sewage sludge. Proceedings of the 2nd Australian Conference on Biological Nutrient Removal from Wastewater. *Australian Waste and Wastewater Association:* 249–255.

Council of the European Communities, 1986, Council Directive of 12 June 1986 on the protection of the environment, and in particular of soil, when sewage sludge is used in agriculture (86/278/EEC). *Official Journal of the European Communities* No. L 181/6–12.

Council of the European Communities, 1991, Council Directive of 21 May 1991 concerning urban wastewater treatment (91/271/EEC). *Official Journal of the European Communities* No. L 135/40–52.

Court, J., 1989, Sludge management overview. *Proceedings of the Australian Water and Wastewater Association.* NSW Regional Conference, Bowral, NSW, Australia.

Cresswell, G., Fahy, P.C. and Tesoriero, L., 1996, Growth yield and heavy metal uptake of vegetables in soil amended with composted biosolids. In: Osborne, G.J., Parkin, R.L., Michalk, D.L. and Grieve, A.M. (Eds), *Biosolids Research in NSW.* NSW Agriculture, Organic Waste Recycling Unit, Richmond: 201–208.

Davis, R.D., and Carlton-Smith, C.H., 1984, An investigation into the phytotoxicity of zinc, copper and nickel using sewage sludge of controlled metal content. *Environmental Pollution* (Series B) 8: 163–185.

Department of the Environment, 1989, *Code of Practice for Agricultural Use of Sewage Sludge.* HMSo, London.

De Vries, M.P.C., 1983, Investigations on twenty Australian sewage sludges—their evaluation by means of chemical analysis. *Fertilizer Research* 4: 75–87.

Dowdy, R.H., Page, A.L. and Chang, A.C., 1991, Management of agricultural land receiving wastewater sludges. In: *Soil Management for Sustainability.* Soil and Water Conservation Society, Edmonton: 85–101.

Eamens, G.J., Lavis, A.M. and Ross, A.D., 1996, Survival of pathogenic and indicator bacteria in biosolids applied to agricultural land. In: Osborne, G.J., Parkin, R.L., Michalk, D.L. and Grieve, A.M. (Eds), *Biosolids Research in NSW.* NSW Agriculture, Organic Waste Recycling Unit, Richmond: 129–138.

Evans, J., Munn, K.J. and Chalk, P.M., 1996, Effects of metal-contaminated biosolids on biological N fixation and mineralisation of legume nitrogen. In: Osborne, G.J., Parkin, R.L., Michalk, D.L. and Grieve, A.M. (Eds), *Biosolids Research in NSW.* NSW Agriculture, Organic Waste Recycling Unit, Richmond: 152–180.

Fahy, P.C., Noble, D., Greswell, G. and Leake, S., 1996, Development of quality standards for the production and use of composted biosolids. In: Osborne, G.J., Parkin, R.L., Michalk, D.L. and Grieve, A.M. (Eds), *Biosolids Research in NSW.* NSW Agriculture, Organic Waste Recycling Unit, Richmond: 189–195.

Finney, I. and Rawlinson, L., 1990, Sludge and wastewater re-use. In: Elmes, D. and Mallen Cooper, J. (Eds), *The Agricultural and Horticultural Use of Sewage Products.* The Australian Institute of Agricultural Science, New South Wales Branch, AIAS Occasional Paper No. 55.

Gough, D. and Fraser, R., 1995, Long term planning for biosolids management in Sydney. In: Kolarik L.O. and Priestley, A.J. (Eds), *Modern Techniques in Water and Wastewater Treatment.* CSIRO Publishing, East Melbourne: 177–181.

Harvey, P., 1993, Control of nutrient inputs. In: Algal Management Strategy, Background Papers. Algal Management Working Group, Murray Darling Basin Commission: 9–20.

Havilah, E.J., Rawlinson, L.V. and Osborne, G.J., 1996, Biosolids as nitrogen fertilizers in coastal NSW. In: Osborne, G.J., Parkin, R.L., Michalk, D.L. and Grieve, A.M. (Eds), *Biosolids Research in NSW.* NSW Agriculture, Organic Waste Recycling Unit, Richmond: 55–61.

Hird, C. and Bamforth, I., 1996, Selection of sites suitable for application of biosolids products. In: Osborne, G.J., Parkin, R.L., Michalk, D.L. and Grieve, A.M. (Eds), *Biosolids Research in NSW*. NSW Agriculture, Organic Waste Recycling Unit, Richmond: 9–17.

Hoitink, H.A.J., 1994, Beneficial effects induced by composted biosolids in horticultural crops. In: Clapp, C.E., Larson, W.E. and Dowdy R.H. (Eds), *Sewage Sludge: Land Utilization and the Environment*. Soil Science Society of America Miscellaneous Publication, Madison: 95–100.

Hoitink, H.A.J. and Fahy, P.C., 1986, Basis for the control of soil borne plant pathogens with composts. *Annual Review of Phytopathology* 24: 93–114.

Hope, P.S. and McDougall, C.R., 1996, The biosolids revolution in Sydney Water: the land application program. In: Polglase, P.J. and Tunningley, W.M. (Eds). *Land Application of Wastes in Australia and New Zealand: Research and Practice*. CSIRO Forestry & Forest Products, Canberra: 67–72.

Hughes, R.M. and Hirst, P., 1996, Effectiveness of N-Viro Soil®, dewatered biosolids, cane mill mud, lime and inorganic fertilizers as nutrients and ameliorants for sugarcane soils. In: Osborne, G.J., Parkin, R.L., Michalk, D.L. and Grieve, A.M. (Eds), *Biosolids Research in NSW*. NSW Agriculture, Organic Waste Recycling Unit, Richmond: 112–115.

Huxedurp, L., Mikulandra, M. and Rawlinson, L., 1996, Biosolids management: Implications of new EPA Guidelines for NSW Country Councils. WaterTECH. *Australian Water & Wastewater Association:* 136–143.

Johnson, P.W., Dixon, R. and Ross, A.D., 1996a, Parasitic worms in processed biosolids – Do they survive? WaterTECH. *Australian Water & Wastewater Association:* 516–522.

Johnson, P.W., Dixon, R. and Ross, A.D., 1996b, Survival of tapeworm (*Taenia hydatigena*) and roundworm (*Ascaris suum*) eggs in processed biosolids. In: Osborne, G.J., Parkin, R.L., Michalk, D.L. and Grieve, A.M. (Eds), *Biosolids Research in NSW*. NSW Agriculture, Organic Waste Recycling Unit, Richmond: 139–151.

Joshua, W.D., Salt, M. and osborne, G.J., 1996, Changes in soil physical properties due to biosolids application to agricultural lands. In: Osborne, G.J., Parkin, R.L., Michalk, D.L. and Grieve, A.M. (Eds), *Biosolids Research in NSW*. NSW Agriculture, Organic Waste Recycling Unit, Richmond: 28–33.

Joshua, W.D., Michalk, D.L., Curtis, I.H., Salt, M. and Osborne, G.J., 1998, The potential for soil contamination of soil and surface waters from sewage sludge (biosolids) in a sheep grazing study. *Geoderma* 84: 135–156.

Kanak, A., Osborne, G. and Swinton, E.A., 1995, Beneficial use of biosolids in the Sydney region. *Water* 22(3): 9–12.

Kelling, K.A., Peterson, A.E. and Walsh, L.M., 1977, Effect of wastewater sludge on soil moisture relationship and surface runoff. *Journal of Water Pollution and Federation* 51: 1698–1703.

Kempton, T.J., 1996, Lime stablisation—Market driven selection of a biosolids stabilisation process. In: Polglase, P.J. and Tunningley, W.M. (Eds). *Land Application of Wastes in Australia and New Zealand: Research and Practice*. CSIRO Forestry & Forest Products, Canberra: 50–55.

Khaleel, R., Redy, K.R. and Overcash, M.R., 1981, Changes in soil physical properties due to organic matter waste application: A review. *Journal of Environmental Quality* 10: 133–141.

Kiekens, L., 1984, Behaviour of heavy metals in soils. In: Berglund, S., Davis, R. D., and L'Hermite, P. (Eds). *Utilisation of Sewage Sludge on Land: Rates of Application and Long-term Effects of Metals*. D. Reidel Publishing, Dordrecht: 126–134.

Koffmann, W., Nicol, H. and Menzies, R., 1996, Biosolids as soil ameliorants and fertilizers in deciduous fruit orchards. In: Osborne, G.J., Parkin, R.L., Michalk, D.L. and Grieve, A.M. (Eds), *Biosolids Research in NSW*. NSW Agriculture, Organic Waste Recycling Unit, Richmond: 116–125.

Langlands, J.P., Donald, G.E. and Bowles, J.E., 1988, Cadmium concentrations in liver, kidney and muscle in Australian sheep and cattle. *Australian Journal of Experimental Agriculture* 28: 291–297.

Li, Zhenbin and Shuman, L.M., 1996, Heavy metal movement in metal-contaminated soil profiles. *Soil Science* 161: 656–666.

Lindsay, W.L., 1979, *Chemical Equilibria in Soils.* John Wiley & Sons Inc., New York.

Logan, T.L. and Chaney, R.L., 1983, Utilization of Municipal wastewater and sludges on land—Metals. In: Page A.L., Gleeson, T.L., Smith, J.E., Iskander, I.K., and Sommers, L.E. (Eds). *Proceedings 1983 Workshop on Utilization of Municipal Wastewater and Sludge on Land.* University of California, Riverside: 235–323.

Lue-Hing, C., Pietz, R.I., Granato, T.C., Gschwind, J. and Zenz, D.R., 1994, Overview of the past 25 years: operator's perspective. In: Clapp, C.E., Larson, W.E. and Dowdy, R.H. (Eds). *Sewage Sludge: Land Utilization and the Environment.* Soil Science Society of America Miscellaneous Publication, Madison: 7–14.

McBride, M.B., 1995, Toxic metal accumulation from agricultural use of sludge: Are USEPA regulations protective? *Journal of Environmental Quality* 24: 5–18.

McGrath, S.P., 1994, Effects of heavy metals from sewage sludge on soil microbes in agricultural ecosystem. In: Ross, S.M. (Ed.). *Toxic Metals in Soil-Plant Systems.* John Wiley Sons Ltd., Chichester: 247–274.

Mckenna, I.M., Chaney, R.L., Tao, S., Leach, R.M., and Williams, F.M., 1992, Interactions of plant Zn and plant species on bioavailability of plant Cd to Japanese quail fed lettuce and spinach. *Environmental Research* 57: 73–87.

Machno, P.S., 1996, Biosolids 2000: Public acceptance of biosolids recycling. WaterTECH. *Australian Water & Wastewater Association:* 54–60.

Metzger, L. and Yaron B., 1987, Influence of sludge organic matter on soil physical properties. *Advances in Soil Science* 7: 142–163.

Michalk, D.L. and Curtis, I.H., 1998, Surface application of biosolids to pastures—preliminary results of environmental degradation. In: Michalk, D.L. and Prately, J.E. (Eds.). Agronomy—Growing a Greener Future. *Proceedings of the Ninth Australian Agronomy Conference:* 757–760.

Michalk, D.L., Curtis, I.H., Langford, C.M., Simpson, P.C. and Seaman, J.T., 1996a, Effects of sewage sludge on pasture production and sheep performance. In: Mohammad Asghar (Ed.), Agronomy—Science with its sleeves rolled up. *Proceedings of the 8th Australian Agronomy Conference:* 429–432.

Michalk, D.L., Curtis, I.H., Langford, C.M., Simpson, P.C. and Seaman, J.T., 1996b, Evaluation of biosolids for use in a pasture ecosystem grazed by breeding sheep. WaterTECH. *Australian Water & Wastewater Association:* 9–98.

Michalk, D.L., Curtis, I.H., Langford, C.M., Simpson, P.C. and Seaman, J.T., Joshua, W. and Hird, C., 1996c, Evaluation of sewage sludge products for use in extensive livestock production. Final Report to Sydney Water of Phase 1. NSW Agriculture, Orange, NSW, Australia, 64 pp.

Moore, B., 1971, The health hazards of pollution. In: Sykes, G. and Skinner, F.A. (Eds). *Microbial Aspects of Pollution.* Academic Press, New York: 11–32.

Munn, K.J., Evans, J., Chalk, P.M., Morris, S.G. and Whatmuff, M., 1997, Symbiotic effectiveness of *Rhizobium trifolii* and mineralisation of legume nitrogen in response to past amendment of a soil with sewage sludge. *Journal of Sustainable Agriculture* 11: 23–37.

NFF [National Farmers Federation], 1995, *Australian Agriculture. The Complete Reference on Rural Industry.* (5th Edition, 1995/1996). Morescope, Melbourne.

Northcote, K.H., 1979, *A Factual Key for the Recognition of Australian Soils.* Rellim Technical Publications, Glenside.

NRC [National Research Council], 1980, *Mineral Tolerances of Domestic Animals.* National Academy of Sciences, Washington, DC.

NSW DLWC [Department of Land and Water Conservation], 1997, *Biosolids Management for Country NSW.* Department of Land and Water Conservation, Sydney, 164 pp.

NSW EPA [Environmental Protection Authority], 1997, *Environmental Management Guidelines for the Use and Disposal of Biosolids Products (October).* Environmental Protection Authority, Sydney. ISBN 0-7310-3792-8.

Oberle, S.L and Kenney, D.R., 1994, Interactions of sewage sludge with soil-crop-water systems. In: Clapp, C.E., Larson, W.E. and Dowdy R.H. (Eds). *Sewage Sludge: Land Utilization and the Environment.* Soil Science Society of America Miscellaneous Publication, Madison: 17–20.

Osborne, G.J., 1996, Fertilizer value of biosolids—A review. In: Osborne, G.J., Parkin, R.L., Michalk, D.L. and Grieve, A.M. (eds). *Biosolids Research in NSW.* NSW Agriculture, Organic Waste Recycling Unit, Richmond: 227–236.

Osborne, G. and Stevens, M. (Eds), 1992, *A Review of Published Literature on the Land Application of Sewage Sludge.* NSW Agriculture, Organic Waste Recycling Unit, Richmond, NSW. ISBN 0-7305-6680-3.

Page, A.L. and Chang, A.C., 1994, Overview of the past 25 years: Technical perspective. In: Clapp, C.E., Larson, W.E. and Dowdy R.H. (Eds), *Sewage Sludge: Land Utilization and the Environment.* Soil Science Society of America Miscellaneous Publication, Madison: 3–14.

Page, A.L., Logan, T.J., and Ryan, J.A. (Eds), 1989, W-170 Peer Review: Standards for the Disposal of Sewage Sludge. US EPA Proposed Rule 40 CFR, Parts 257 and 503 (February 6, 1989 Federal Register). Cooperative State Research Service, Technical Committee W-170. University of California, Riverside.

POEO [*Protection of the Environment Operations Act 1997*], NSW Environmental Protection Authority. http://www.epa.nsw.gov.au/legal/aboutpoeo.htm

Robertson, M.B. and Polglase, P.J., 1996. Release and leaching of nitrogen from biosolids applied to a pine plantation. In: Polglase, P.J. and Tunningley, W.M. (Eds), *Land Application of Wastes in Australia and New Zealand: Research and Practice.* CSIRO Forestry & Forest Products, Canberra: 26–31.

Rogers, S.L., 1996, Effects of biosolids on nutrient cycling. Microbial activity and population diversity in soils from the Glenfield long-term biosolids application trial. In: Osborne, G.J., Parkin, R.L., Michalk, D.L. and Grieve, A.M. (eds). *Biosolids Research in NSW.* NSW Agriculture, Organic Waste Recycling Unit, Richmond: 181–186.

Ross, A.D., 1992, Human health and microbiological hazards associated with sewage sludge applied to land. In: Osborne, G.J. and Stevens, M.S. (Eds), *A Review of Published Literature on the Land Application of Sewage Sludge.* NSW Agriculture, Organic Waste Recycling Unit, Richmond. ISBN 0-7305-6680-3.

Ross, A.D., Lawrie, R.A. Keneally, J.P. and Whatmuff, M.S., 1992, Risk characterisation and management of sewage sludge on agricultural land—implications for the environment and the food-chain. *Australian Veterinary Journal* 69: 177–1181.

Ross, A.D., Lawrie, R.A., Whatmuff, M.S., Keneally, J.P. and Awad, A.S., 1991, *Guidelines for the use of Sewage Sludge on Agricultural Land.* NSW Agriculture, Orange, Australia, 17 pp.

Ryan, J.A., Pahren, H.R. and Lucus, J.B., 1982, Controlling cadmium in the human food chain: A review and rationale based on health effects. *Environmental Research* 28: 251–302.

Salt, M., Hird, C. and Bamforth, I., 1996, Assessment of biosolids application rates, degree of incorporation and movement of mineral nitrogen in biosolids treated plots. In: Osborne, G.J., Parkin, R.L., Michalk, D.L. and Grieve, A.M. (Eds), *Biosolids Research in NSW.* NSW Agriculture, Organic Waste Recycling Unit, Richmond: 18–27.

Sanders, J.R., Adams, T.McM. and Christensen, B.T., 1986, Extractability and bioavailability of zinc, nickel, cadmium, and copper in three Danish soils sampled 5 years after application of sewage sludge. *Journal of Science, Food and Agriculture* 37: 1155–1164.

Smith, S.R., 1994, Effect of soil pH on availability to crops of metals in sewage-sludge treated soils. II. Cadmium uptake by crops and implications for human dietary intake. *Environmental Pollution* 86: 5–13.

Smith, S.R., 1996, Agricultural Recycling of Sewage Sludge and the Environment. CAB International. ISBN 0-85198-980-2.

Stace, H.C.T., Hubble, G.D., Brewer, R., Northcote, K.H., Sleeman, J.R., Mulcahy, M.J. and Hallworth, E.G., 1968, *A Handbook of Australian Soils*. Rellin Technical Publications, Glenside.

Stark, B., Suttle, N. Sweet, N. and Brebner, J., 1995, Accumulation of PTEs in Animals Fed Dried Grass Containing Sewage Sludge. Final Report to the Department of the Environment. WRc Report No. DoE 3753/1. WRc, Medmenham, Marlow.

Sommers, L.E. and Giordano, P.M., 1984, Use of nitrogen from agricultural, industrial and municipal wastes. In: Hauck R.D. et al. (Eds). *Nitrogen in Crop Production*. American Society of Agronomy, Madison: 207–220.

Tesoriero, L., Fahy, P.C., Cresswell, G. and Brown, F., 1996, Suppression of plant diseases in soils amended with composted biosolids. In: Osborne, G.J., Parkin, R.L., Michalk, D.L. and Grieve, A.M. (Eds), *Biosolids Research in NSW*. NSW Agriculture, Organic Waste Recycling Unit, Richmond: 196–200.

US EPA [United States Environmental Protection Agency], 1989, Standards for the disposal of sewage sludge: proposed rules. 40 CFR Parts 257 and 503. Federal Register 54: 5746–5901. US Governmental Printer office, Washington, DC.

US EPA [United States Environmental Protection Agency], 1993, Standards for the use or disposal of sewage sludge. Federal Register, 58(32): 9248–9412. US Governmental Printer Office, Washington, DC.

US EPA [United States Environmental Protection Agency], 1995, A Guide to the Biosolids Risk Assessment for the EPA Part 503 Rule. EPA/832-B-93-005. USEPA, Washington DC.

Water Board, 1991, Sludge management for the Sydney region. A public discussion document. Water Board, Sydney.

Walsh, J. and Rawlinson, L.V., 1995, Best practice in biosolids land application. *Water*, 22(3): 13–15.

Whatmuff, M.S., 1996, Biosolids application to agricultural land: Considerations of contamination by heavy metals. In: Osborne, G.J., Parkin, R.L., Michalk, D.L. and Grieve, A.M. (Eds), *Biosolids Research in NSW*. NSW Agriculture, Organic Waste Recycling Unit, Richmond: 237–247.

Whatmuff, M.S., 2002, Applying biosolids to acid soils in NSW: Are guideline soil metal limits from other countries appropriate? Biosolid soil metal limits for NSW acid soils. *Australian Journal of Soil Research* 40: 1941–1956.

Whatmuff, M.S. and Osborne, G.J., 1992, The behaviour of heavy metals in soils. Chapter 7. In: Osborne, G.J. and Stevens, M.S. (Eds), *A Review of Published Literature on the Land Application of Sewage Sludge*. NSW Agriculture, Organic Waste Recycling Unit, Richmond, NSW. ISBN 0-7305-6680-3.

WHO [World Health Organization], 1981, The Risk to Health of Microbes in Sewage Sludge Applied to Land. Euro Reports and Studies 54, World Health Organization, Copenhagen.

Williams, S.H. and David, D.J., 1976, The accumulation in soil of cadmium residues from phosphate fertilizer and their effects on the cadmium content of plants. *Soil Science* 121: 86–93.

9

Land Application of Municipal Wastewater and its Influence on Soil Physical Properties and Nitrate Leaching

G.N. Magesan

INTRODUCTION

Worldwide, large quantities of municipal, industrial and agricultural wastewater are generated every year. Most of these wastewaters are treated to remove solids, pathogens and other contaminants, and the resulting relatively clean water is recycled into the environment. However, these treated wastewaters may still retain large concentrations of organic matter, nutrients such as nitrogen and phosphorus, as well as other contaminants. In many countries, municipal wastewater is discharged into receiving waters such as rivers, lakes and the sea (UNEP, 1993). Such discharges can result in depletion of dissolved oxygen and, ultimately, eutrophication.

Land application is becoming popular across the world as an effective management option for the treatment of wastewater (Feigin et al., 1991). In New Zealand, legislation (for example, the *Resource Management Act 1991*) and Maori cultural preferences also contribute towards this practice. To the Maori, water is the essential ingredient of life—a priceless treasure left by ancestors for the life-sustaining use of their descendants—and must, therefore, remain pure and unadulterated to provide life for those that follow. In such a culture, the mixing of human wastes with natural water is a grievous wrongdoing (Taylor and Patrick, 1987).

The objective of land treatment of wastewater is to utilize the physical, chemical and biological properties of the soil-plant system so as to assimilate waste components without adversely affecting the soil quality or the wider environment. It is important to understand and predict the interactions between wastewater quality and soil properties if land treatment systems are to function sustainably in the long term. Research

Forest Research, Private Bag 3020, Rotorua, New Zealand
E-mail: gujja.magesan@forestresearch.co.nz

has shown that soil-plant systems can effectively treat wastewater. The degree of renovation of wastewater, however, depends on the nature of the waste components, soil characteristics, agronomic practices, and irrigation loading (Kardos and Sopper, 1973; Linden et al., 1984).

Land application can have a variety of beneficial and detrimental effects, depending on the characteristics of the wastewater and the soil. Some beneficial effects include improved wastewater quality, a means of wastewater disposal, irrigation to supply moisture and nutrients for plant growth, and the recharging of groundwater. Some detrimental effects include soil deterioration, damage to vegetation, the build-up of salts and heavy metals, undesirable odours, pollution of groundwater, and surface runoff of pollutants. Successful wastewater application, however, requires suitable soil, topography and hydrological conditions, suitable climate, efficient system design, and good management.

Irrigation of wastewater onto the soil could change its physical, chemical and biological properties. These soil properties play an important role in the transformation, retention and movement of nutrients present in the applied wastewater. In particular, physical properties of soil such as texture, structure, porosity, and hydraulic conductivity will influence soil moisture content and aeration and respiration. These factors control the type and rate of soil microbial activity and chemical reactions. Of the many publications about land disposal of wastes, only a few cover the effect of such disposals on the soil's physical properties (Pagliai and Antisari, 1993; Cameron et al., 1997) and on nitrate leaching (Magesan et al., 1998). Leaching of nitrate and other solutes from lands irrigated with wastewater can lead to the degradation of groundwater and surface waters (Bond, 1998; Magesan et al., 1999). Moreover, Cameron et al. (1997) reported that the processes that control the fate of wastewater in the soil, such as rate of release of nutrients and leaching through macro-pores, are complex and poorly understood. More research is required to understand the processes that affect the fate of wastewater in the soil. The objective of this chapter is to review the impact of municipal wastewater irrigation on some soil physical properties and nitrate leaching.

PHYSICAL PROPERTIES OF SOIL

Soil Texture

The different particles in soils are classified into groups of various sizes on the basis of their equivalent diameter. There are three main classes: sand (0.02–2 mm), silt (0.002–0.02 mm) and clay (<0.002 mm). The clay fraction has a significant influence on many physical and chemical processes that occur in the soil, mainly because the small particles have such a large and reactive surface area. In contrast, the sand and silt fractions typically do not have much influence on chemical processes, and their smaller surface areas do not adsorb or retain water to the same degree as clay fractions (Jury et al., 1991).

In general, irrigation of municipal wastewater onto land does not affect the soil texture; in other words, no significant difference in sand, silt and clay fractions between wastewater-irrigated and control plots is expected. For example, Hinrichs et al. (1974) reported that the application of wastewater from beef feedlots to cropland for more than two years did not change the size distribution of soil particles. However, some studies have reported changes in soil texture after wastewater irrigation. Abd Elnaim et al. (1987) reported textural changes in profiles due to prolonged irrigation with sewage water. This could be attributed to the redistribution of particle sizes due to colloidal or particulate movement through macropores during effluent irrigation.

Bulk and Particle Density

Bulk density is the dry mass per unit bulk volume of soil, including all voids. In general, increase in soil organic matter (due to land application of organic wastes or high suspended solids and BOD in wastewater) will decrease the bulk density of the soil. Various studies have shown that the addition of organic amendments such as sewage sludge and organic manure decreases the bulk density of the soil (Sopper and Kerr, 1979). Mbagwu (1992) and Obi and Ebo (1995) reported that land application of organic amendments or organic wastes on a sandy soil increased soil organic matter and significantly decreased the soil bulk density.

Only a few studies have examined the effect of wastewater application on bulk density. Mathan (1994) and Magesan (2001) reported that irrigation of primary-treated municipal wastewater significantly decreased the bulk density of sandy soils. In a long-term study, Mathan (1994) reported that the longer the application period, the less was the bulk density.

Some studies, however, have shown that irrigation of wastewater did not change the bulk density. For example, Sopper and Richenderfer (1979) reported that municipal wastewater spray irrigation had little effect on the bulk density of forested soils. Hinrichs et al. (1974) also reported that wastewater application did not change soil bulk density. In New Zealand, Magesan (2001) reported that the irrigation of tertiary-treated wastewater for four years on volcanic soils under forestry did not change its bulk density in the surface layers. At that site, however, there was a slight increase in bulk density at the lower depth (10–20 cm), which might be due to the increase in the clay fraction in that particular layer. Mathan (1994) suggested that clay migration from surface layers might lead to an increase in bulk density in the lower layers.

Particle density is the density of primary soil particles (mass per unit volume, excluding all voids). In most mineral soils, the mean particle density is 2.6–2.7 Mg m^{-3}, close to the density of quartz, which is often prevalent in sandy soils (Hillel, 1980). Magesan (2001) reported that after wastewater irrigation on a sand dune soil, the particle density significantly decreased, probably due to the presence of organic matter (Hillel, 1980).

Organic waste materials, which are frequently utilized as soil amendments for their fertilizer value, may enhance the aggregation of soil particles. In other words, land application of organic wastes can increase soil aggregate stability (Weil and Kroontie, 1979; Khaleel et al., 1981) but the effect depends on the texture of the soil. Sandy soils with low stability respond more than clay soils with inherently high stability. Epstein et al. (1976) noted that sludges and sludge compost increased the number of stable aggregates and the water retention of a silt loam soil. However, not all wastes will improve soil aggregate stability. Lieffering and McLay (1996) have shown that soil aggregate stability can be significantly reduced by the application of alkaline solutions typically present in high pH industrial wastes. This reduction in aggregate stability was attributed to the dissolution of organic matter by the high pH solution, which resulted in a loss of inter-particle bonding.

Total Porosity and Macro-porosity

Total porosity is the proportion of a unit volume of soil occupied by air or water and derived from bulk and particle densities. In general, addition of organic matter to soils lowers the bulk density and contributes to increased porosity (Khaleel et al., 1981). Mbagwu (1992) and Obi and Ebo (1995) reported that the addition of organic amendments or organic wastes to sandy soil significantly increased porosity, due partly to the noted appreciable promotion of biological activity. Mathan (1994) observed improvements in total porosity upto 67% in the sewage-irrigated soils over the control. In New Zealand, Magesan (2001) reported that the irrigation of primary-treated wastewater onto a sandy soil significantly decreased bulk density and increased total porosity.

Although spray irrigation of wastewater can generally increase porosity, some studies have shown that it did not have a consistent significant effect on the total or capillary soil porosities (Sopper and Richenderfer, 1979).

Macro-porosity is a measure of the volume of large pores (>0.06 mm diameter), normally air-filled except during short periods following heavy rainfall, and is calculated by subtracting the water content at –5 kPa from the total porosity value. Macro porosity is a useful indicator of saturated hydraulic conductivity and aeration. Pagliai and Antisari (1993) reported an increase in soil porosity following the application of pig slurry, and that most of the increases were due to the development of soil macro-pores (>50 μm). Magesan (2001) found that macro-porosity in the surface soils of wastewater-irrigated plots on sandy and volcanic soils significantly increased, whereas it decreased slightly in the lower layers of both soils. The significant increase in macro-porosity may possibly be due to increased biological activity, especially the increased burrowing activity of earthworms, following wastewater application.

Release Characteristics of Water

Water release characteristic gives the graphical relationship between volumetric water content and applied tensions (e.g. 0, –5, –10, –20, –40, –100, and –1500 kPa). This characteristic is one of the most important measurements for characterizing soil physical properties. It can indicate both the ability of the soil to store water and the aeration status of a drained soil, as well as be interpreted as a measure of pore-size distribution in non-swelling soils (Reeve and Carter, 1991). When organic matter is added to soils, more soil water is retained at various suctions. For example, after the addition of organic amendments, Mbagwu (1992) found a significant increase in soil moisture content at a range of matric potentials (the water retention). Khaleel et al. (1981) noted that the increase in water retention at low-tension range following the addition of poultry manure could be partly attributed to an increase in the number of water-storage pores. The magnitude of increase was more pronounced at field capacity (–10 kPa) rather than at wilting point (–1500 kPa). Mbagwu (1992) observed that organic waste application significantly increased soil water retention at all potentials except –1500 kPa. Magesan (2001) also reported that the water contents of wastewater-irrigated soils were generally greater than the control sites. Although irrigation on a volcanic soil decreased the water contents at lower suctions (<10 kPa), they were higher for the remaining higher suctions. As the suction increased, the difference in water contents between the control and irrigated plots decreased. At the lower depth, the water contents at different suction of the wastewater-irrigated plot were consistently higher than the control plot.

Total- and Readily-available Water Capacity

Total available water capacity (TAWC) is the proportion of soil volume drained between the pressure levels of –10 kPa (field capacity) and –1500 kPa (permanent wilting point). It is a measure of the volume available in a soil for storing water that can be potentially extracted by plants, and has become established as the chief measurement necessary to assess the drought resistance of a soil.

Soil water-holding capacity can be increased by the input of organic matter in the applied waste and by the improved soil physical conditions produced (Mbagwu, 1992). Obi and Ebo (1995) reported that land application of poultry manure on a sandy soil significantly improved the available water content, and showed that soil water capacity and organic matter are positively correlated. Magesan (2001) reported that while wastewater irrigation did not change TAWC of a sandy soil at the surface layer, it slightly increased in the lower depths.

Readily-available water capacity (RAWC) is the proportion of soil volume drained between the pressure levels of –10 and –100 kPa. It is a measure of the volume available in a soil for storing water that is easily accessible by plants. In practice, the proportion of RAWC in a soil may be as agronomically significant as TAWC in assessing draughtiness. It is generally calculated by subtracting water content at –100 kPa from water content at –10 kPa. Organic carbon content and particle-size class chiefly affects RAWC. The fall in available water with depth is a common feature of soil profiles, mainly due to decreased organic matter contents.

Saturated and Unsaturated Hydraulic Conductivity

In general, the addition of organic matter to soils increases their porosity and soil permeability (Khaleel et al., 1981). For example, Mbagwu (1992) reported that the addition of organic amendments to sandy clay loam soil significantly increased the porosity and saturated hydraulic conductivity. Obi and Ebo (1995) reported a fourfold increase in hydraulic conductivity following the addition of poultry manure, and also showed a significant positive correlation between organic matter and saturated hydraulic conductivity. This is probably due to an improvement in soil structure, coupled with high microbial activities. Mathan (1994) and Magesan (2001) reported increased porosity and soil permeability after wastewater irrigation. However, in many instances, decreases in permeability have been shown to occur as a result of wastewater addition (Lance et al., 1980; Clanton and Slack, 1987; Cook et al., 1994; Balks et al., 1997). Hinrichs et al. (1974) reported that the application of wastewater from beef feedlots to cropland over two years significantly decreased soil hydraulic conductivity. Sopper and Richenderfer (1979) also reported similar results after spray irrigation of wastewater onto silt loam soil in the pine forest areas. Cook et al. (1994) reported a 50% decline in infiltration rates at Whakarewarewa after 32 months of irrigation with the same wastewater, as measured by saturated hydraulic conductivity in a large in situ soil monolith, and suggested that particulate material from increased litter decomposition contributed to the clogging of soil pores.

Regarding unsaturated hydraulic conductivity, not many reports are available. Magesan (2001) reported that hydraulic conductivity decreased with an increase in suction (low matric potential). Also, the unsaturated hydraulic conductivity at a different rate of suction on a sandy soil decreased relative to the control plots.

NITRATE LEACHING

Nitrate leaching is a process by which excess nitrate is moved through the soil by moving water. Wastewater irrigation adds both water and nutrients to the soil system. The water stimulates plant growth and also nitrogen uptake, turnover of soil nitrogen, and denitrification. However, when wastewater is irrigated beyond the capacity of plant uptake, it then provides a source of readily-leachable nitrate. Leaching of nitrate and other solutes in wastewater-irrigated lands can lead to the degradation of both groundwater and surface water (Magesan et al., 1998; 1999).

Despite land application of wastes becoming a popular option, the processes controlling the fate of wastes in the soil and the effects on nitrate leaching and groundwater quality are poorly understood and not well documented (Cameron et al., 1997). Some studies on nitrate leaching, however, have been reported in North America (for example Cole et al., 1986), and in the Australasia-Pacific region (Polglase et al., 1995; Magesan et al., 1998). In this section, the effects of municipal wastewater application on nitrate concentrations in soil, soil water, and drainage water are discussed, together with some proposals to minimize nitrate leaching.

Effect of Nitrate Concentrations in Soil and Water

Increases in rates of municipal wastewater irrigation can increase nitrate concentrations in soil and water. For example, Magesan et al. (1998) reported that the amount of nitrate stored in the profile increased with the increase in the rate of wastewater irrigation to a forested soil over four years (Table 9.1). The amount of nitrate was greatest nearer the surface of the soil (0–5 cm).

It is generally expected that when the nitrogen demand of vegetation equals or exceeds the nitrogen supplied by wastewater, then nitrogen build-up in the soil and water would be minimal. However, when wastewater contains high nitrogen or nitrogen is applied throughout the year while vegetation nitrogen demand is seasonal, an increase in soil nitrogen could be expected. Vazquezmontiel et al. (1996) reported an accumulation in the soil nitrate when nitrogen was supplied continuously in the wastewater and exceeded crop requirements. Similarly, Sopper and Kerr (1979) reported that wastewater irrigation of a mixed hardwood forest at 50 mm per week increased nitrate concentrations in the soil solution. However, irrigating some forests at a lower rate (25 mm per week) during the growing season restricted nitrate concentrations to acceptable levels.

Table 9.1: Mean distribution (with standard errors) of nitrate-N (kg ha.$^{-1}$) in the soil before leaching, and the amount of nitrate-N (kg ha.$^{-1}$) leached (after Magesan et al., 1998)

Depth (mm)	*Wastewater application rate (mm per week)*		
	0 (control)	*29 (low)*	*88 (high)*
0–50	1.8 [0.1]	4.6 [0.1]	9.5 [1.4]
50–100	1.4 [0.0]	3.3 [0.1]	7.4 [1.3]
100–150	0.9 [0.0]	2.7 [0.2]	4.1 [0.3]
150–200	0.7 [0.0]	1.8 [0.1]	4.9 [0.3]
Total available (kg ha.$^{-1}$]	**4.8 [0.2]**	**12.4 [0.3]**	**25.9 [2.7]**
Total leached (kg ha.$^{-1}$)	**3.1**	**6.9**	**11.4**

Linden et al. (1984) studied nitrogen concentrations under two wastewater application rates, and found that nitrogen levels in the soil solution increased significantly with increased application rates, and sometimes exceeded 10 g m^{-3} when applications surpassed crop uptake demands.

Despite the large rates of nitrification, Polglase et al. (1995) found that nitrate leaching losses were relatively small, and that concentrations of nitrate in soil solution never exceeded 10 g m^{-3}. However, they reported that irrigation of treated wastewater at twice the rate of water use by trees could result in increased leaching losses. Application of wastewater at a rate that matched tree water use did not result in any adverse impact on groundwater quality.

Effect on Drainage Nitrate Concentrations

Nitrate losses in soil drainage water depend on the nitrate content of the soil profile, the volume of leaching water, and the efficiency with which the leaching water explores the pore volume (Magesan et al., 1994; 1996). Breakthrough curves (plots of relative concentration as a function of drainage) from the laboratory experiments are often used to reveal those processes that affect leaching.

Magesan et al. (1998) collected four undisturbed soil lysimeters (200 mm diameter × 200 mm depth) from each plot that had received an average of 0, 29 or 88 mm wastewater irrigation per week for four years. In the laboratory, these soil lysimeters were leached with continuous simulated rainfall at a rate of 11.4 mm per hour. The scientists reported that the amount and pattern of nitrate leached from the soils in the lysimeters (Figure 9.1) were affected by their wastewater irrigation history. In the control samples (no irrigation), the nitrate concentrations were less than 2 g m^{-3}, reflecting low background nitrate concentrations in the untreated soil. In the low-volume treatment samples (29 mm per week), the breakthrough curve was characterized by a rapid emergence of nitrate in the first drainage samples collected, followed by extensive tailing. In

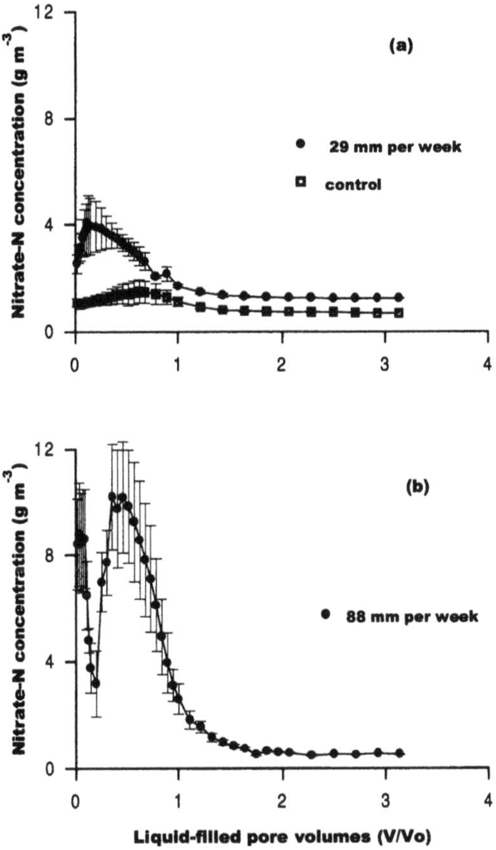

Fig. 9.1: Nitrate concentrations in drainage water of lysimeters receiving wastewater at (a) 0 and 29 mm per week, and (b) 88 mm per week. Error bars are standard errors of mean (n=4). (From Magesan et al., 1998)

contrast, the breakthrough curve for the high-volume treatment samples (88 mm per week) was bimodal, with peaks occurring after 0. 1 and 0.6 pore volumes of drainage, and then also followed by extensive tailing. Maximum nitrate concentrations in the leachate of the 'high' treatment were slightly below 10 g m^{-3}.

The rapid recovery of nitrate in leachate, with peak concentrations at drainage volumes substantially less than one pore volume, indicated that soil nitrate is readily transported beyond the topsoil. It has been suggested that the bimodal breakthrough curves, as found in the 'high' treatment plots, occur when solutes are leached via different pathways, for example via macro-pore flow followed by matrix flow (Magesan et al., 2002).

Lysimeter studies by Cameron et al. (1995) also showed small nitrate leaching losses when liquid wastes were applied to soils at relatively low rates, but considerably greater leaching losses at higher rates of application.

Strategies to Lowering Nitrate Concentrations in Drainage Water

Strategies are needed to minimize the amount of nitrate leaching from wastewater-irrigated soils (Magesan et al., 1998). Proposals for decreasing nitrate concentrations in the groundwater at land-treatment sites include decreasing the nitrate concentration in the waste stream (Tebbutt, 1983), encouraging plant nitrate uptake (Vazquezmontiel et al., 1996), enhancing denitrification in the soil (Monnett et al., 1995; Adelman and Tabidian, 1996), adding nitrification inhibitors to the waste-stream where most of the nitrogen is in organic or ammonium forms (Shepherd, 1996), and decreasing the amount of water and solute through-flow (Kluitenberg and Horton, 1990). Decreasing the amount of nitrate in the waste-stream (Tebbutt, 1983) is environmentally sound, but usually an expensive option.

Plant growth on the land treatment site is a key feature of the system as it removes nutrients and water from the waste-treated soil. Linden et al. (1984) studied the effect of two crops on nitrate concentrations under two wastewater application rates. The results indicated that high-demand crops such as forage grasses have the potential for better renovation of wastewater rather than corn at high application rates. Forests and tree plantations are different from agricultural crops. Trees generally store less nitrogen than crops, but add large quantities of carbon to the soil. Enhancing plant nitrogen uptake is difficult in an established forest, where the rotation frequency is some 25 years and nitrogen uptake rates usually decrease after the trees have reached 5–8 years of age. Increasingly, short-rotation forest trees are being considered suitable for the renovation of wastewater. Encouraging weed growth may increase short-term nitrogen plant uptake, but this option may not be supported for other forest management reasons.

Spray irrigation with domestic wastewater has the potential to pose as an effective treatment and disposal method for soils that have limited renovation capacity. Monnett et al. (1995) found that nitrogen removal via denitrification from spray irrigation fluctuated, owing to the alternating aerobic and anaerobic conditions associated with irrigation frequency. Maintaining wastewater in the upper, more microbially-active part of the soil profile through split applications, assists in nitrogen removal via denitrification. Adelman and Tabidian (1996) also found that soils with high carbon concentrations had minimal leaching relative to soils lower in carbon. This is because high-carbon soils support an active population of denitrifying bacteria.

In most wastewater, nitrogen is present as organic and ammoniacal forms. Both these forms undergo nitrification reactions in soils that create a buildup in nitrate that leads to leaching losses. A nitrification inhibitor could be used along with wastewater application to decrease the buildup in nitrate and subsequent leaching losses (Shepherd, 1996). Prasad and

Power (1995) have given examples of nitrification inhibitors for use in the environment. The cost of applying nitrification inhibitors would almost certainly be prohibitive in many wastewater schemes.

Several studies have shown that increasing the contact period of solutes with the soil can markedly decrease solute leaching losses (Kluitenberg and Horton, 1990). The best strategy for minimizing nitrate leaching losses, therefore, would be to decrease the amount of drainage and hence the flow of solutes in the soil, by applying the wastewater in instalments throughout the week instead of in a single application. This, in turn, would minimize the likelihood of deep drainage, and maximize the opportunity for plants and microorganisms to utilize the added nitrate.

CONCEPTUAL MODEL

One of the soil properties that controls nitrate movement in wastewater-irrigated soil is its hydraulic conductivity. However, the effect of wastewater application on conductivity is not clearly understood. While some studies have reported an increase in soil hydraulic conductivity after wastewater application (Mathan, 1994), most other studies have shown a decrease (Clanton and Slack, 1987; Cook et al., 1994; Balks et al., 1997). A possible mechanism for this decrease is biological clogging: blockage of soil pores by bacterial growth or its byproducts (extracellular carbohydrate, polysaccharides) that decreases the pore diameter (Kristiansen, 1981; Taylor and Jaffe, 1990; Vandevivere and Baveye, 1992).

Magesan et al. (1999) developed a conceptual model relating wastewater quality to hydraulic conductivity and nitrate leaching (Figure 9.2). If the applied wastewater C:N ratio is high, then N is immobilized in the soil due to the presence of excess C. This will result in low mineral N

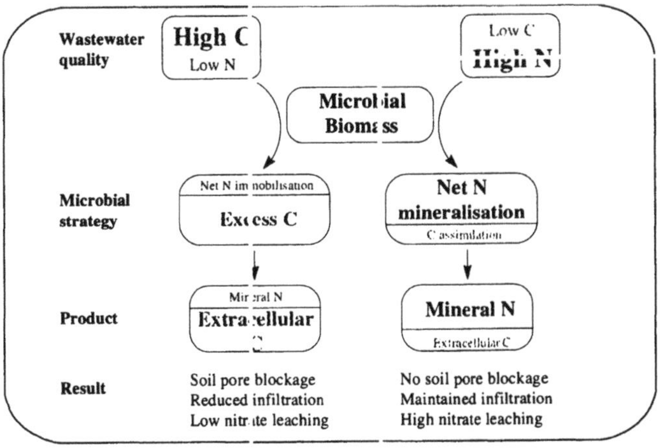

Fig. 9.2: Conceptual model relating wastewater C:N ratio with soil hydraulic conductivity and nitrate leaching. (From Magesan et al., 1999)

availability and high extracellular C deposition. Hence, soil pore blockage is to be expected; it will result in reduced infiltration and also a lower potential for nitrate leaching. Conversely, if the wastewater C:N ratio is low, then N will be mineralized and C will be assimilated. The result will be higher concentrations of mineral N with increased potential for nitrate leaching, but there will be little extracellular C production to reduce soil hydraulic conductivity.

To test this conceptual model, Magesan et al. (1999) used volcanic soil in a laboratory column experiment, repacked to a bulk density of 0.56 Mg m^{-3}. Water and wastewater with differing C:N ratios were irrigated onto these soil columns weekly for 14 weeks to determine whether wastewater characteristics increased microbial biomass and carbohydrate decreased soil hydraulic conductivity. Five treatments were employed:

I tap water
II secondary-treated municipal wastewater (C:N 2:1)
III dressing of 10 g dry weight *Pinus radiata* litter
IV synthetic wastewater (C:N 50:1)
V amended wastewater (C:N 50:1).

Leachate draining from the cores was collected each week and nitrate was determined. At the end of the experiment, some cores from each treatment were used to measure the soil properties, while other cores were used for measuring hydraulic conductivity.

Magesan et al. (1999) demonstrated that irrigation of wastewater with high C:N ratio resulted in an increase in microbial biomass C and soil carbohydrate (Table 9.2), which coincided with a marked decrease (75%) in hydraulic conductivity (Figure 9.3).

Unlike the high-C:N wastewater, the application of unamended wastewater resulted in high concentrations of soil-mineral N and total soluble N and substantial nitrate leaching (Figure 9.4), but there was no change in the other microbial parameters compared with the water

Table 9.2: Properties of soil cores irrigated on a weekly basis with different wastewaters for 14 weeks. All units are per g soil. Within rows, determinations followed by the same letter do not differ significantly ($p<0.05$) (after Magesan et al., 1999)

Soil properties		*Treatments*				
	Units	*I*	*II*	*III*	*IV*	*V*
Carbohydrate (H_2O)	mg glu eq	1.6 a	1.8 ab	1.8 ab	2.0 b	2.3 c
Carbohydrate (acid)	mg glu eq	18.3 a	18.4 a	18.9 ab	21.0 b	20.2 b
Microbial C	µg C	758 a	755 a	816 a	1176 b	1017 c
Mineral N	µg N	12 a	91 b	84 b	0.8 a	0.8 a
Total soluble N	µg N	29 a	103 b	94 b	16 a	19 a
Microbial N	µg N	130 a	90 a	112 a	155 a	168 a

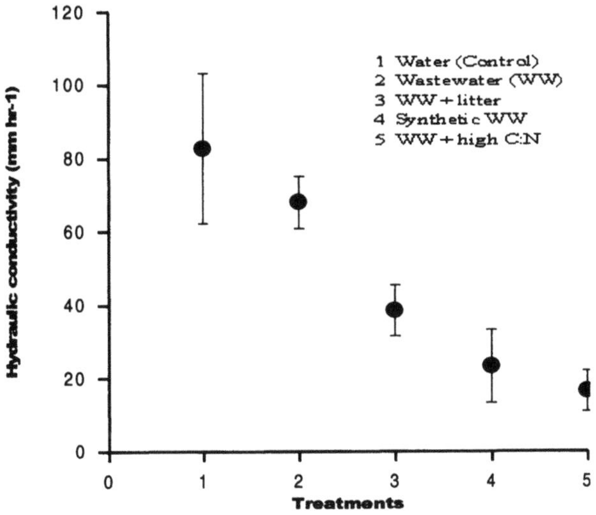

Fig. 9.3: Effect of wastewater quality on hydraulic conductivity

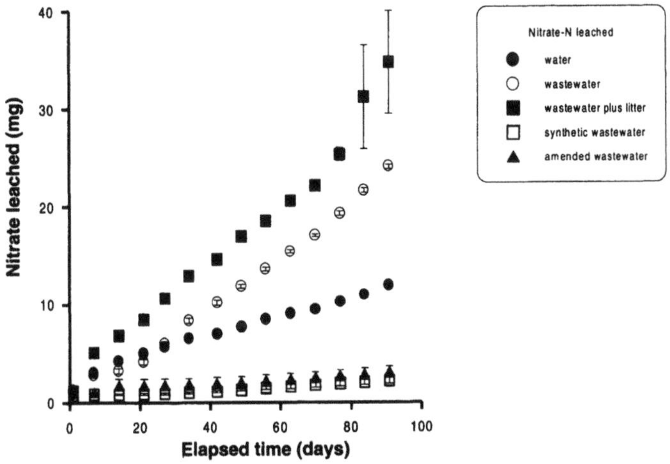

Fig. 9.4: Accumulated nitrate-N leached from soil cores treated with water (●), wastewater (○), wastewater plus litter (▪), synthetic wastewater (□), and amended wastewater with high C:N ratio (▲). Error bars represent standard error (SE) of mean. (From Magesan et al., 1999)

treatment. It, therefore, appears that there is a trade-off between nitrate leaching and soil hydraulic conductivity, depending on the wastewater quality.

The amended and synthetic wastewater treatments exhibited a decrease in mineral N and total soluble N concentrations, relative to the unamended wastewater and to wastewater-with-pine needle treatments. The result was large increases in mineral N and total soluble N concentrations, compared with the water-only application. Changes in microbial N with treatment were not significant.

Nitrate was the main form of N leached from the cores. The addition of litter, in conjunction with the wastewater application, resulted in the greatest amount of nitrate being leached. The amounts of nitrate leached from the cores receiving water, wastewater, and wastewater-plus-litter accumulated linearly over 0–100 days (Figure 9.4). After 20 days, the rates of nitrate leaching from cores receiving wastewater and those with wastewater and litter were similar. Negligible amounts of nitrate were leached from cores receiving the synthetic and amended wastewaters with high C:N ratio.

CONCLUSIONS

Land application is an effective management option for the treatment of wastewater. Such application, however, can affect soil physical properties such as bulk density, porosity, saturated hydraulic conductivity, aggregate stability, and water retention. While soil texture and particle density are not generally affected, they too can be affected under some situations.

Increases in rates of wastewater irrigation can increase nitrate concentrations in both soil and water. The amount and pattern of leached nitrate are generally affected by their wastewater irrigation history. One of the best strategies suggested to minimize the nitrate leaching losses from land treatment sites is to decrease the amount of drainage by applying the wastewater in instalments throughout the application period rather than as a single application. This would minimize the likelihood of deep drainage, and maximize the opportunity for plants and microorganisms to utilize the added N.

The results of the experiments conducted in this chapter suggest that more research on the impact of wastewater application on soil physical properties and nitrate leaching is needed, because information in the literature is minimal, and the fate of wastewater in the soil is complex and poorly understood.

REFERENCES

Adelman, D.D. and Tabidian, M.A., 1996, The potential impact of soil carbon content on ground water nitrate contamination. *Water Science and Technology* 33 (4–5): 227–232.

Abd Elnaim, E.M., Omran, M.S., Waly, T.M. and El Nashar, E.M.B., 1987, Effect of prolonged sewage irrigation on some physical properties of sandy soil. *Biological Wastes* 22: 269–274.

Balks, M.R., McLay, C.D.A. and Harfoot, C.G., 1997, Determination of the progression in soil microbial response, and changes in soil permeability, following application of meat processing effluent to soil. *Applied Soil Ecology* 6: 109–116.

Bond, W.J., 1998, Effluent irrigation—an environmental challenge for soil science. *Australian Journal of Soil Research* 36: 543–555.

Cameron, K.C., Rate, A.W., Carey, P.L. and Smith, N.P., 1995, Fate of nitrogen in pig effluent applied to a shallow stony pasture soil. *New Zealand Journal of Agricultural Research* 38: 533–542.

Cameron, K.C., Di, H.J. and McLaren, R.G., 1997, Is soil an appropriate dumping ground for our wastes? *Australian Journal of Soil Research* 35: 995–1035.

Clanton, C.J. and Slack, D.C., 1987, Hydraulic properties of soils as affected by surface application of wastewater. *Transactions of the American Society of Agricultural Engineers* 30(3): 683–687.

Cole, D.W., Henry, C.L. and Nutter, W.L. (Eds), 1986, *The Forest Alternative for Treatment and Utilization of Municipal and Industrial Wastes*. University of Washington Press, Seattle, US, 582 pp.

Cook, F.J., Kelliher, F.M. and McMahon, S.D., 1994, Changes in infiltration and drainage during wastewater irrigation of a highly permeable soil. *Journal of Environmental Quality* 23: 476–482.

Epstein, E., Taylor, J.M. and Chancy, R.L., 1976, Effects of sewage sludge and sludge compost applied to soil on some soil physical and chemical properties. *Journal of Environmental Quality* 5: 422–426.

Feigin, A., Ravina, I. and Shalhevet, J., 1991, Irrigation with treated sewage effluent. *Advanced Series in Agricultural Science*, Vol. 17. Springer-Verlag, Heidelberg, Germany, 224 pp.

Hillel, D., 1980, *Fundamentals of Soil Physics*. Academic Press, New York.

Hinrichs, D.G., Mazurak, A.P. and Swanson, N.P., 1974, Effect of effluent from beef feedlots on the physical and chemical properties of soil. *Soil Science Society of America Proceedings*, 38: 661–663.

Jury, W.A., Gardner, W.R. and Gardner, W.H., 1991, *Soil Physics*. John Wiley & Sons, New York.

Kardos, L.T. and Sopper, W.E., 1973, Renovation of municipal wastewater through land disposal by spray irrigation. In: Sopper, W.E. and Kardos, L.T. (Eds). *Recycling Treated Municipal Wastewater and Sludges Through Forest and Cropland*. The Pennsylvania State University Press, University Park.

Khaleel, R., Reddy, K.R. and Overcash, M.R., 1981, Changes in soil physical properties due to organic waste applications: A review. *Journal of Environmental Quality* 10: 133–141.

Kluitenberg, G.J. and Horton, R., 1990, Effect of solute application method on preferential transport of solutes in soil. *Geoderma* 46: 283–297.

Kristiansen, R., 1981, Sand-filter trenches for purification of septic tank effluent: 1. The clogging mechanism and soil physical environment. *Journal of Environmental Quality*, 10: 353–357.

Lance, J.C., Rice, R.C. and Gilbert, R.G., 1980, Renovation of wastewater by soil columns flooded with primary effluent. *Journal of Water Pollution Control Federation* 52: 381–388.

Lieffering, R.E. and McLay, C.D.A. 1996. The effect of strong hydroxide solutions on the stability of aggregates and hydraulic conductivity of soil. *European Journal of soil Science*, 47: 43–50.

Linden, D.R., Clapp, C.E. and Larson, W.E., 1984, Quality of percolate water after treatment of a municipal wastewater effluent by a crop irrigation system. *Journal of Environmental Quality* 13: 256–264.

Magesan, G.N., 2001, Changes in soil physical properties following irrigation of municipal wastewater onto two forested soils. *New Zealand Journal of Forestry Science*, 31: 188–195.

Magesan, G.N., White, R.E., Scotter, D.R. and Bolan, N.S., 1994, Estimating leaching losses from sub-surface drained soils. *Journal of Hydrology* 10: 87–93.

Magesan, G.N., White, R.E. and Scotter, D.R., 1996, Nitrate leaching from a drained, sheep-grazed pasture. 1. Experimental results and environmental implications. *Australian Journal of Soil Research* 34: 55–67.

Magesan, G.N., McLay, C.D.A. and Lal, V.V., 1998, Nitrate leaching from a freely-draining volcanic soil irrigated with municipal sewage effluent in New Zealand. *Agriculture, Ecosystems and Environment* 70: 181–187.

Magesan, G.N., Williamson, J.C., Sparling, G.P., Schipper, L.A. and Lloyd-Jones, A.Rh., 1999, Hydraulic conductivity of soils irrigated with wastewaters of differing strengths: field and laboratory studies. *Australian Journal of Soil Research* 37: 391–402.

Magesan, G.N., White, R.E., Scotter, D.R. and Bolan, N.S. 2002. Effect of Prolonged storage of soil lysimeters on nitrate leaching. *Agriculture, Ecosystems and Environment* 88: 73–77.

Mathan. K.K., 1994, Studies of the influence of long-term municipal sewage-effluent irrigation on soil physical properties. *Bioresource Technology* 48: 275–276.

Mbagwu, J.S.C., 1992, Improving the productivity of a degraded Ultisol in Nigeria using organic and inorganic amendments. Part 2. Changes in physical properties. *Bioresource Technology* 42: 167–175.

Monnett, G.T., Reneau, R.B. and Hagedorn, C., 1995, Effects of domestic wastewater spray irrigation on denitrification rates. *Journal of Environmental Quality* 24: 940–946.

Obi, M.E. and Ebo, P.O., 1995, The effects of organic and inorganic amendments on soil physical properties and maize production in a severely degraded sandy soil in Southern Nigeria. *Bioresource Technology* 51: 117–123.

Pagliai, M. and Antisari, L.V., 1993, Influence of waste organic matter on soil micro-and macrostructure. *Bioresource Technology* 43: 205–213.

Polglase, P.J., Tompkins, D., Stewart, L.G. and Falkiner, R.A., 1995, Mineralization and leaching of nitrogen in an effluent-irrigated pine plantation. *Journal of Environmental Quality* 24: 911–920.

Prasad, R. and Power, J.F., 1995, Nitrification inhibitors for agriculture, health and the environment. *Advances in Agronomy* 54: 233–281.

Reeve, M.J. and Carter, A.D., 1991, Water release characteristic. In: Smith, K.A. and Mullins, C.E. (Eds), *Soil Analysis: Physical Methods.* Marcel Dekker Inc., New York: 111–160.

Shepherd, M.A., 1996, Factors affecting nitrate leaching from sewage sludges applied to a sandy soil in arable agriculture. *Agriculture, Ecosystems and Environment* 58: 171–185.

Sopper, W.E. and Richenderfer, J.L., 1979, Effect of municipal wastewater irrigation on the physical properties of the soil. In: Sopper, W.E. and Kerr, S.N. (Eds), *Utilization of Municipal Sewage Effluent and Sludge on Forest and Disturbed Land.* The Pennsylvania State University Press, University Park: 179–195.

Sopper, W.E. and Kerr, S.N. (Eds), 1979, *Utilization of Municipal Sewage Effluent and Sludge on Forest and Disturbed Land.* Pennsylvania State University Press, University Park. 537 pp.

Taylor, S.W. and Jaffe, P.R., 1990, Biofilm growth and related changes in the physical properties of a porous medium. 1. Experimental investigation. *Water Resources Research* 26: 2153–2169.

Taylor, A. and Patrick, M., 1987, Looking at water through different eyes—the Maori perspective. *Soil and Water* (Summer edition): 22–24.

Tebbutt, T.H.Y., 1983, *Principles of Water Quality Control.* Pergamon Press, Oxford, UK.

UNEP, 1993, Environmental Data Report 1993–94. Blackwell, Oxford, UK.

Vandevivere, P. and Baveye, P., 1992, Saturated hydraulic conductivity reduction caused by aerobic bacteria in sand columns. *Soil Science Society of America Journal* 56: 1–13.

Vazquezmontiel, O., Horan, N.J. and. Mara, D.D., 1996, Management of domestic wastewater for reuse in irrigation. *Water Science and Technology* 33: 355–362.

Weil, R.R. and Kroontje, W., 1979, Physical condition of a Davidson clay loam after five years of heavy poultry manure application. *Journal of Environmental Quality* 8: 387–392.

10

Phosphorus and Nitrogen Availability in Agricultural Soil Irrigated with Recycled Water

B.L. Maheshwari, K. Sakadevan, and *H.J. Bavor*

INTRODUCTION

Throughout the world, recycled water is important in land and water resources management as it provides quality water for irrigation, industrial and urban water requirements (Asano and Levine, 1996; Bond, 1998). Land application of recycled water has been practised worldwide since the nineteenth century (Feigin et al., 1991). In addition to the water benefit to irrigated land, concurrent loading of nutrients (N and P) may improve the fertility of soils and increase plant productivity.

Interest in the application of recycled water to agricultural land is growing in many developing countries, particularly in arid regions which face severe water shortages (Ohgaki and Sato, 1991). In developed countries, however, the over-consumption of fresh water and strict government regulations on the discharge of pollutants, including nutrients, metals, pathogens, suspended solids and organic carbon, to water bodies has encouraged the application of recycled water to agricultural lands (Helena et al., 1996; Cameron et al., 1997).

Most studies on land application of recycled water have mainly examined the effect of recycled water application on plant biomass production, nutrient uptake and leaching of nutrients beyond the root zone (Loehr, 1984; Brechin and McDonald, 1994). Although such studies have generally provided information on the effectiveness of nutrient use and biomass production, they provide little information on the biogeochemical processes, the factors influencing them, and the availability of nutrients. An understanding of the processes and nutrient availability

School of Environment & Agriculture. Hawkesbury cumpus, Bldg J4, University of Western Sydney, Locked bag 1797 Penrith South DC, NSW 1797 Australia.

in soils irrigated with recycled water is needed to develop improved guidelines for sustainable use of recycled water.

In the soil-water-plant environment, there are many processes (for example, plant uptake, mineralization/immobilization, and adsorption/adsorption of N and P in the soil) which determine the interaction of N and P with various nutrient compartments (plant, soil and microorganisms), N and P availability, and removal of N and P from recycled water. Also, N in the soil-plant-water system may be removed by denitrification as nitrous oxide or volatilization as ammonia, or both. Many studies have been carried out to examine the leaching losses of N (both nitrate and ammonium) from agricultural lands (Mamo et al., 1999). Most studies dealing with P losses from soils have concentrated on erosion and surface runoff (Sharpley and Menzel, 1987; Gachter et al., 1998).

However, detailed studies have not been carried out on the transformations of N and P in soil-water-plant systems receiving recycled water or on influencing factors such as application methods or the rate and timing of application. Also, subsequent N and P availability in soil and losses to surface waters and groundwaters have not been studied in detail. The objectives of the present study were to examine the variation of N and P availability in plots irrigated with recycled water, and to identify factors which may influence the availability of N and P in soil receiving the recycled water.

MATERIALS AND METHODS

Experimental Site

Experiments were carried out at the University of Western Sydney—Hawkesbury Farm, situated within the University at Richmond, 63 km northwest of Sydney. The soil has been classified as sandy and well drained. Initial soil characterization indicated that the soil was mainly sandy (70% sand for the top 25 cm, and >70% sand between 25 cm and 1.0 m soil depths) to a depth of 1.0 m and sandy clay below 1.0 m. In the top 25 cm, the soil pH was 6.2 and total carbon content was 1.66 g C/kg soil. The values for soil bulk density, total N, total P, 2M KCl extractable ammonium and nitrate and 0.5M extractable phosphate at the beginning of the experiment are given in Table 10.1. Since the soil was very sandy down to a 1.0 m depth, total N and P concentrations were very low (Table 10.1). The site had never been irrigated with recycled water, although many other sites on the campus estate have been so over the past 15 years. The site was manually cleared of weeds and was sown with rye grass (*Lolium perenne*) to increase plant density.

Table 10.1. Bulk density, phosphorus and nitrogen content of soil at 0–25, 25–50, 50–75 and 75–100 cm depths (mg/kg soil)

Depth	*Parameter*						
(cm)	*Bulk density*	*Field capacity*[1]	*Total P*	*Total N*	*Ortho-phosphate*	*N-Nitrate*	*N-Ammonium*
0–25	1.55	13.3	248	527	115	7.6	8.5
25–50	1.63	15.1	174	105	81	2.8	6.9
50–75	1.68	20.0	115	98	46	2.1	7.4
75–100	1.86	22.4	60	92	48	1.9	7.3

[1] % volume basis

Treatments

The experimental design was a randomly constructed complete block with three replications of the following treatments.

1. Control: plots irrigated with fresh water once a week over the experimental period without fertilizer or recycled water application.
2. Single super phosphate fertilizer (SSP): plots treated with SSP at the rate of 375 kg/ha. at the beginning of the experiment and irrigated with fresh water once a week over the experimental period.
3. Recycled water and fresh water applied alternately: plots irrigated with recycled water for a week and followed by fresh water the next week, over the experimental period.
4. Continuous recycled water application: plots irrigated with recycled water once a week, over the experimental period.

Twelve plots were used in the study (4 treatments × 3 replicates). Each plot was 4 m × 4 m and was separated from other plots by a walkway of 1.0 m width.

Experimental Set-up

Porous suction cups were installed in all plots (two cups per plot) at one metre depth to collect the drainage water. The cups were acid-washed and rinsed with deionized water before installation to remove background phosphate, nitrate and ammonium, which might have been added to the cups during manufacture. Infiltrating interstitial soil water samples were collected by applying the suction to the cups. The recycled water used for irrigation was secondary-treated domestic effluent obtained from the Richmond Sewage Treatment Plant, situated adjacent to the experimental site. The first irrigation was carried out in September 1997 and was continued every week. The last irrigation was carried out on 1 May 1998. At the beginning of the experiment, 25 mm of water (fresh and recycled) was applied to all treatments; over time, this was increased to 50 mm per week. The plots were irrigated using a portable garden sprinkler system.

Soil, Water and Plant Measurements

Soil samples were collected from the site before the beginning of the experiment and total N, total P, mineral N (ammonium and nitrate) and 0.5M sulphuric acid extractable phosphate contents of the soil were measured separately for 0–0.25, 0.25–0.50, 0.50–0.75, and 0.75–1.0 m soil depths. The soil bulk density was also measured for these soil depths. Soil samples were collected at the end of the experiment from all 12 plots and the total N, total P, mineral N (ammonium and nitrate) and 0.5M sulphuric acid extractable phosphate contents were measured for each depth as above. The pH and total carbon for the soil were measured only for the top 0–0.25 m depth. The plant biomass was harvested three times during the experiment. The first harvest was carried out after two months, the second after four months, and the final after seven months of sowing. Pasture samples were dried at 55°C and dry matter weight and total N and P contents in the dry matter were determined. After each harvest, the plots were mown to clear the vegetation so that fresh plant growth occurred for the second and third rounds.

The total N and P contents of the soil were measured using an Alpkem Flow Solution 111 flow injection analyser (APHA, 1992) after digesting the soil with a mixture of sulphuric acid, potassium sulphate, and selenium (Kjeldahl mixture). The total N and P contents of the plant dry matter were also measured in this way.

The mineral N (nitrate and ammonium) content of the soil was measured using the Alpkem Flow Solution 111 after extracting the soil with 2M KCl (1:10 soil:solution ratio). The phosphate in the soil was measured using Alpkem Flow Solution 111 as described above after extracting the soil with 0.5M sulphuric acid (1:50 soil:solution ratio). The mineral N and phosphate contents of the drainage water were measured using the Alpkem Flow Solution 111 in the ratio mentioned above.

Estimation of Drainage Below Root Zone

Soil-water from all plots at 1.0 m soil depth was collected by applying suction to the porous cups. Rainfall and pan evaporation data were obtained from the University's weather station, located about one km away. The pan evaporation was converted into actual evapotranspiration for the experimental site. The amount of drainage (D) below the root zone was calculated using the following daily water balance equation:

$$SW_i = R_i + I_i - ET_i + SW_{i-1} + D_i \quad (1)$$

where

SW_i = soil water content (mm) of the root zone on i^{th} day

SW_{i-1} = soil water content (mm) of the root zone on the previous day (i.e. i–1th day)

R_i = rainfall depth (mm) on i^{th} day
I_i = irrigation depth (mm) on i^{th} day
ET_i = actual evapotranspiration (mm) on i^{th} day
D_i = depth (mm) of drainage below the root zone.

Drainage below the root zone will occur only when $SW_i > SW_{fc}$, where SW_{fc} is the soil water content (mm) of the root zone at field capacity. The actual evapotranspiration (ET) was calculated from the reference evapotranspiration (ET_r), using a daily crop coefficient (K_c):

$$ET = K_c ET_r \tag{2}$$

The value of K_c was estimated by the following equation (Harrington and Heermann, 1981):

$$K_c = K_{cb} K_a + K_s \tag{3}$$

where K_{cb} is the daily basal crop coefficient
K_a is a coefficient dependent upon the soil moisture
K_s is a coefficient to allow for increased evaporation from the soil surface occurring after rain or irrigation.

The generalized basal crop coefficient K_{cb} was defined to represent conditions where the soil surface was dry, with minimal evaporation from the soil but the soil water availability did not limit plant growth or transpiration. When available water within the root zone limits growth and ET, K_a in equation (3) will be less than 1.0 and can be approximated by the following equation (Burman et al., 1980):

$$K_a = \frac{[\ln(A_w + 1)]}{[\ln(101)]} \tag{4}$$

where A_w is the percentage of available water (100% when soil is at field capacity).

The reference evapotranspiration (ET_r) for the field conditions of the experimental site can be estimated from the pan evaporation by using the following relationship (Fleming, M., 1998, personal communication):

$$ET_r = 0.85E_p \tag{5}$$

where E_p is the pan evaporation.

RESULTS AND DISCUSSION

N and P Concentrations of Recycled Water

The concentrations of soluble phosphate, nitrate and ammonium in the recycled water applied to the soil varied throughout the study period (Figure 10.1). This variation generally resulted from:

- variation in the amount of water and sludge flowing through the sewage treatment plant;

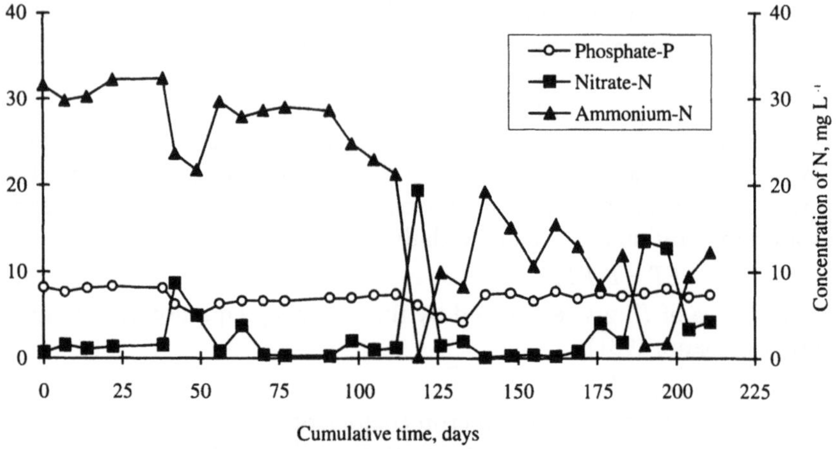

Fig. 10.1: Variation of nitrate, ammonium and phosphate concentrations in the recycled water used during the experiment.

- amount of water stored in the storage pond;
- water losses through evaporation and infiltration in the pond;
- P retention by sediments deposited in the pond;
- amount of rainfall; and
- losses of N through volatilization and denitrification within the pond.

The average phosphate, nitrate and ammonium concentrations were 6.97±1.00 mg P/L, 3.22±4.6 mg NO_3^-–N/L and 18.99±10.18 mg NH_4^+–N/L, respectively, during the study period.

The pH of the recycled water measured on two occasions during the study period did not exceed 8.0. Previous studies have shown that below pH 8.0, the loss of ammonium by way of volatilization may be less than two % of the total N applied to the system (Di et al., 1998). The variations in N and P concentrations in the recycled water over the study period may have influenced the transformations and availability of N and P in the soil-water-plant system.

The total amounts of phosphate, nitrate and ammonium added through fertilizer and recycled water to the plots are given in Table 10.2. As discussed below, N input to the control and SSP fertilised plots may have occurred through N fixation by the symbiotic association between white clover (*Trifoli repens*) and rhizobium bacteria. More than 99% of the P and N applied to the plots that received recycled water was in the form of orthophosphate and mineral N (nitrate + ammonium).

Table 10.2: Total amount of phosphate, nitrate and ammonium added to experimental plots through fertilizer and recycled water

Treatment	*Fertilizer added (kg N or P/ha.)*			
	Phosphate	*Nitrate*	*Ammonium*	*Nitrate + Ammonium*
Control	0	0	0	0
Superphosphate fertilizer	33.75	0	0	0
Alternate recycled water	45.20	16.40	119.60	136.00
Continuous recycled water	86.50	41.56	217.60	259.16

Water Input and Losses from Plots

A total of 1287.5 mm of fresh water was applied over a period of 246 days to the control and SSP fertilized plots, and the same amount of recycled water was applied to the treatment, which received recycled water continuously (Figure 10.2). The treatment that received recycled water alternately with fresh water was applied as 650 mm of recycled water (Figure 10.2) and 637.5 mm of fresh water. Apart from irrigation, the additional water applied was through rainfall. A total of 475.7 mm rainfall was recorded during the study period (Figure 10.2). The actual evapotranspiration during the period was 733.7 mm and the total drainage was 932.6 mm (Figure 10.2). Drainage occurred during both irrigation and rain events. Irrigation, rainfall, evapotranspiration and drainage, all these factors influence the water movement in soil-plant systems and, therefore, affect the transformations and availability of N and P within soil-plant-water systems.

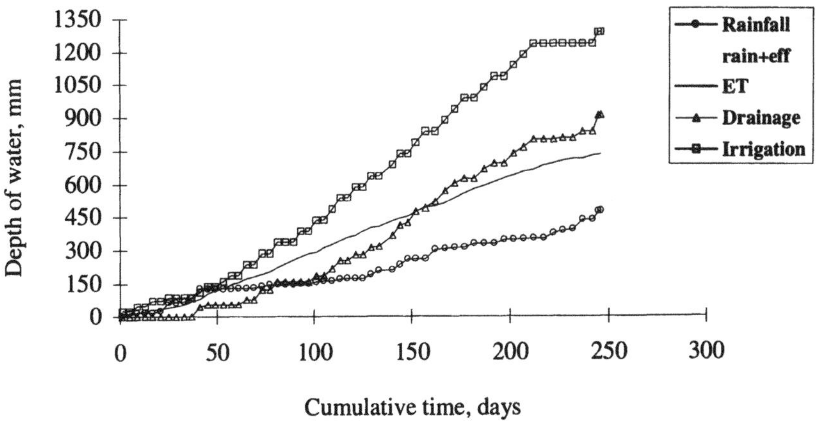

Fig. 10.2: Cumulative amount of rainfall, irrigation, evapotranspiration and drainage from the site during the experiment.

Dry Matter Production

The total amount of dry matter produced at each of the three individual harvests was significantly ($p<0.05$) greater in the recycled water treatments (both continuous and alternated) than in the control and SSP treatment (Table 10.3). There was no significant ($p<0.05$) difference in dry matter production between either of the recycled water treatments (Table 10.3). Similarly, there was no significant ($p<0.05$) difference in dry matter production between the control and the SSP treatment (Table 10.3).

Table 10.3: Dry matter production and N and P removed (±STD) in control, single superphosphate fertilizer and recycled water (continuously and alternately with fresh water) treatments. Values followed by same letter are not significantly different at $p<0.05$

Treatment	Characteristic (kg/ha.)		
	Total dry matter production	Phosphorus removed	Nitrogen removed
Control	3650±260a	11±1.2a	67±5a
Single superphosphate	3810±960a	12±2.4a	58±20a
Alternate recycled water	5990±1070b	19±4.3b	113±32b
Continuous recycled water	6020±660b	23±1.6b	160±11b

N and P Removal in Dry Matter

The total amounts of N and P removed in the dry matter for each of the three harvests were significantly ($p<0.01$) greater in the recycled water treatments (both continuous and alternated) than in the control and SSP treatment (Table 10.3). There were no significant differences ($p<0.05$) in the amounts of N and P removed in dry matter between either of the recycled water treatments (Table 10.3). Similarly, there were no significant differences in the amounts of N and P removed in dry matter between the control and the SSP treatment. Nitrogen was not applied to the control plots or the SSP-treated plots. The soil N (both total and mineral N) was measured at the beginning and at the end of the study period and showed no change for these two treatments (Tables 10.1 and 10.4).

The plant N in the control and SSP-treated plots (Table 10.3) may have been derived from N contributed through N fixation by white clover (Kahn and Yoshida, 1994), as it was observed that white clover contributed significantly to dry matter production in these two treatments. It has been shown in previous studies (Sakadevan et al., 1993) that mineralization of N from soil organic matter may also contribute to plant N. However, since the soil studied in this experiment is sandy and the soil N content is relatively low, the contribution through mineralization may be very low. The presence of greater levels of ammonium and nitrate in the soil of the recycled water plots may have reduced the symbiotic N fixation by white

Table 10.4: Phosphate, nitrate and ammonium present in the 1.0 m soil profile before and after the application of fertilizer and recycled water. Values followed by same letter are not significantly different at $p<0.05$

Treatment	*Parameter (kg N or P/ha.)*		
	Phosphate	*Nitrate*	*Ammonium*
Before application	1017±174a	55±10.8a	131±4.9a
After application			
Control	1038±147a	53±9.9a	125±3.9a
Single superphosphate	1177±154a	62±13.8a	127±3.2a
Alternate recycled water	1051±154a	56±11.1a	127±3.8a
Continuous recycled water	1277±181a	64±12.9a	125±2.6a

clover occurring in these plots. Previous studies on N fixation in legume crops have shown that the presence of ammonium and nitrate in soil may reduce the potential of legumes for N fixation (Rennie et al., 1982).

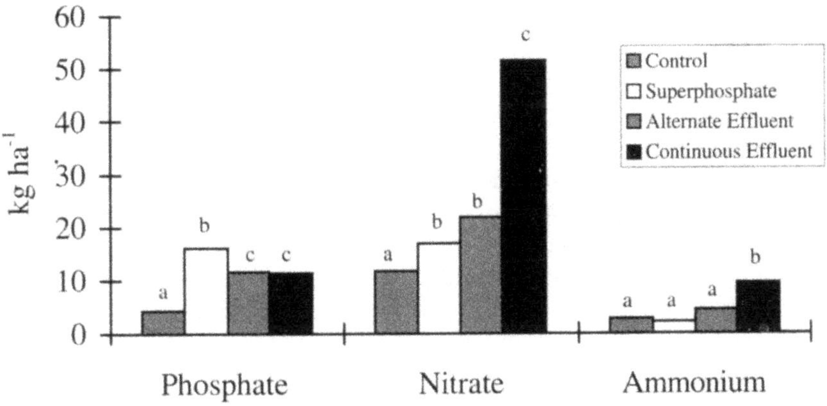

Fig. 10.3: Amount of phosphate, nitrate and ammonium leached from control, single superphosphate fertilised, recycled water applied alternately and recycled water applied continuously during the experiment. Treatments shown with same letters (a, b and c) are not significantly different from each other.

Leaching of N and P

Mineral N (nitrate and ammonium) and phosphate were leached below 1.0 m soil depth in all four treatments. The concentration of nitrate in the drainage water collected at 1.0 m soil depth from all the treatments ranged from low as 0.02 mg N/L for control to as high as 11 mg N/L for the continuous recycled water treatment. The total amount of nitrate leached at 1.0 m soil depth was significantly greater ($p<0.01$) in the continuous recycled water treatment than in the other three treatments. There was no significant difference ($p<0.05$) in the amount of nitrate leached between plots that received SSP and alternate recycled water. The amount of nitrate-N leached from the control was significantly ($p<0.01$) lower than that leached from the SSP and alternate recycled water plots. Although N fertilizer was not applied to the SSP treatment or the control, greater leaching of nitrate-N was observed in the SSP treatment than the control. This result indicated that the added phosphate may have influenced greater rhizobial N fixation in the SSP-treated plots, which would increase the soil nitrate and lead to a greater nitrate leaching from SSP treatment.

The concentration of ammonium-N in the drainage water in all the four treatments ranged from as low as 'below detection' for the control to as high as 6 mg N/L for the alternate recycled water treatment. Generally, the concentration of ammonium-N in the drainage water was lower than that in the effluent applied throughout the experiment. The total amount of ammonium-N leached below 1.0 m soil depth was lower than the amount of nitrate-N leached in all four treatments. The amount of ammonium-N leached was significantly ($p<0.01$) greater in the continuous recycled water treatments than in the other three treatments. There was no significant difference ($p<0.05$) between the amounts of ammonium-N leached from the control, the SSP treatment and the alternate recycled water treatment.

The total amount of mineral N (nitrate + ammonium) leached showed that the continuous recycled water treatment had lost the largest ($p<0.01$) amount of N by leaching (61.12 kg N/ha.) below 1.0 m soil depth, followed by the plots that received recycled water alternately (26.26 kg N/ha.), SSP fertilizer (19.18 kg N/ha.) and the control (14.63 kg N/ha.).

In the present study, about 8% and 4.3% of the applied N was leached below 1.0 m soil depth in the continuous and alternate recycled water treatments, respectively. Plots that did not receive mineral N (control and SSP treatments) also leached N below 1.0 m soil depth (4.7 and 4.2 kg N/ha.), but the losses were lower than those receiving recycled water. The greater leaching in the continuous recycled water treatment than in the alternate recycled water treatment may be attributed to the higher N loading in the continuous treatment. Almost double the amount of mineral N was applied in the continuous treatment than in the alternated treatment. Since nitrate was the dominant form of N leached, it may be considered that rapid nitrification of ammonium in this sandy soil may have occurred, which could lead to the availability of more nitrate in the soil solution than required by the plant mass. Nitrate is not strongly adsorbed by the soil and excess nitrate in the soil solution is leached down the soil profile. The results showed that leaching of N (mainly nitrate) is affected by the application method and frequency of application.

The concentration of phosphate in the drainage water in all the four treatments ranged from as low as 'below detection' for the control, to as high as 3.5 mg P/L in the plots that received recycled water continuously and was always lower than the concentration of P in the effluent irrigated to the soil. The total amount of P leached in the form of phosphate below 1.0 m soil depth was significantly ($p<0.01$) greater in the treatment that received SSP than the control and plots that received recycled water. There was no significant difference ($p<0.05$) in P leached below 1.0 m between the treatments that received recycled water continuously and alternately. The least amount of phosphate was leached from the control. In general, the total amount of phosphate leached during the study period was in the order: SSP > continuous recycled water = alternate recycled water > control. The greater amount of phosphate leached below 1.0 m soil depth in the SSP treatment compared to the other three treatments may be due to the greater availability of soluble phosphate in the soil immediately after SSP application (Harris et al., 1996; Sims et al., 1998). Single superphosphate was applied at 375 kg/ha. in one event at the beginning of the experiment compared to recycled water applied over a period of time. The single application of SSP at a higher application rate may have increased the available P (soluble ortho phosphate) in the soil and also increased the leaching of P. Recycled water, which was applied incrementally over a period of time, reduced the amount of readily available P (soluble orthophosphate) in the soil, decreased P leaching and increased P uptake by plants.

About 47.9% of applied P was lost by leaching from the SSP-treated soil as compared to soil irrigated with recycled water continuously (13.3%) and alternately with fresh water (26%) even though the amount of P applied through recycled water was much greater in the latter two treatments (Table 10.2). This indicated that the P application rate with an application interval may be important factors controlling P losses through sub-surface drainage. Phosphorus application rates, which balance or closely balance the system input and output may be likely to reduce P losses through subsurface drainage (Daniel et al., 1998).

N and P Mass Balance

The total amount of nitrate, ammonium and phosphate for the 1.0 m soil profile in all of the four treatments at the end of the experiment were not significantly different ($p<0.05$) from those present in the soil before the application of SSP or recycled water (Table 10.4).

Mineral N and phosphate applied through fertilizer and recycled water to the soil was not accounted for in plant uptake and leaching below 1.0 m. While mineral N was not applied to the control or the SSP treatment, mineralization of either soil organic matter or symbiotic nitrogen fixation, or both, may have contributed to plant uptake and leaching of N in these two treatments. About 13.7% N in recycled water was unaccounted for in the continuous recycled water treatment. The total amount of N removed in plant matter and lost by leaching (3.3 kg N/ha.) was more than the amount of N applied through recycled water in the alternate recycled water treatment.

The unaccountable N in the continuous recycled water treatment may have been lost by either denitrification or volatilization, or both, in addition to immobilization in the soil organic matter and ammonium adsorption on to cation exchange sites. As discussed by Di et al. (1998) ammonia volatilization from the soil may be very low since it accounts for less than two % loss of soil ammonium-N. In addition, since the soil is predominantly sandy with low organic matter content and has no larger cation exchange sites, it may be expected that immobilization to organic forms and fixation onto cation exchange sites may also be low. Nitrogen may have also been lost through denitrification in the soil. Denitrification occurs in field soils flooded with water during irrigation and during rainy periods (Ledgard et al., 1996; Vinten et al., 1998). In field soils receiving recycled water containing elevated levels of ammonium, conditions are favourable for denitrification.

About 5.61 kg P/ha. (6.7% of applied P), 52.0 kg P/ha. (60% of applied P) and 14.4 kg P/ha. (32% of applied P), respectively, were unaccounted for in the plots that received SSP, continuous recycled water and alternate recycled water. This unaccountable P might have been immobilized into soil organic matter or been adsorbed on to soil sesquioxide, or both. It is difficult to confirm that such immobilization or occurred in the soil due to the large amount of phosphate present (>1000 kg P/ha. for each depth) in the 1.0 m soil profile (Table 10.4) as compared to the actual amount being unaccounted for and the spatial variability of phosphate present in the soil.

CONCLUSIONS

Comparison of the efficiency of surface-applied recycled water, either intermittent or continuous, with SSP applied in a single event to a sandy soil, showed that N and P in recycled water can be used efficiently compared with fertilizer. The present study showed that high loading of mineral N (i.e. N applied as a component of recycled water) may lead to a greater leaching of nitrate from the root zone. The greater amount of nitrate leached from plots that received recycled water continuously than from plots that received recycled water alternately with fresh water, suggest that alternate application of recycled water with fresh water may reduce nitrate leaching to the groundwater.

REFERENCES

APHA, 1992, *Standard Methods for the Examination of Water and Wastewater* (18th ed.). American Public Health Association, Washington DC.

Asano, T. and Levine, A.D., 1996, Wastewater reclamation, recycling and reuse: past, present and future. *Water Science and Technology* 33(10-11): 1–14.

Bond, W.J., 1998, Effluent irrigation—An environmental challenge for soil science. *Australian Journal of Soil Research* 36: 543–555.

Brechin, J. and McDonald, G.K., 1994, Effect of form and rate of pig manure on the growth, nutrient uptake and yield of barley (cv. Galleon). *Australian Journal of Experimental Agriculture* 34: 505–510.

Burman, R.D., Nixon, P.R., Write, J.L. and Pruitt, W.O, 1980, Water requirements. In: Jensen, M.E. (Ed.), *Design and Operation of Farm Irrigation Systems*. American Society of Agricultural Engineers, St. Joseph, MI, US: 189–232.

Cameron, K.C., Di, H.J. and McLaren, R.G., 1997, Is soil an appropriate dumping ground for our wastes? *Australian Journal of Soil Research* 35: 995–1035.

Daniel, T.C., Sharpley, A.N. and Lemunyon, J.L., 1998, Agricultural phosphorus and eutrophication: A symposium overview. *Journal of Environmental Quality* 27: 251–257.

Di, H.J., Cameron, K.C., Moore, S. and Smith, N.P., 1998, Nitrate leaching from dairy shed effluent and ammonium fertilizer applied to a free-draining pasture soil under spray or flood irrigation. *New Zealand Journal of Agricultural Research* 41: 263–270.

Feigin, A., Ravina, I., and Shalhevet, J., 1991, *Irrigation with Treated Sewage Effluent. Management for Environmental Protection*. Springer-Verlag: Berlin.

Gachter, R., Ngatiah, J.M., and Stamm, C., 1998, Transport of phosphate from soil to surface waters by preferential flow. *Environmental Science and Technology* 32: 1865–1869.

Harrington, G.J. and Heerman, D.F., 1981, State of the art irrigation scheduling computer program. Proceedings of the ASAE irrigation scheduling conference, Irrigation scheduling for water and energy conservation in the 80s. American Society of Agricultural Engineers, St Joseph, MI, US: 171–178.

Harris, W.G., Rhue, R.D., Kidder, G., Brown, R.B. and Littell, R., 1996, Phosphorus retention as related morphology and taxonomy of sandy Coastal Plain soil materials. *Soil Science Society of America Journal* 60: 1513–1521.

Helena, M., Marecod, D.O., Monte, F., Angelakis, A.N. and Asano, T., 1996, Necessity and basis for establishment of European guidelines for reclaimed wastewater in the Mediterranean region. *Water Science and Technology* 33 (10-11): 303–316.

Kahn, M.K. and Yoshida, T., 1994, Nitrogen fixation in peanut determined by acetylene reduction method and ^{15}N isotope dilution technique. *Soil Science and Plant Nutrition* 40: 283–291.

Ledgard, S.F., Sprosen, M.S., Brier, G.J., Nemaia, E.K.K. and Clark, D.A., 1996, Nitrogen inputs and losses from New Zealand dairy farmlets as affected by nitrogen fertilizer application: year one. *Plant and Soil* 181: 65–69.

Loehr, R.C., 1984, *Pollution Control For Agriculture* (2nd ed.) Academic Press, Orlando, Florida.

Mamo, M., Rosen, C.J. and Halbach, T.R., 1999, Nitrogen availability and leaching from soil amended with municipal solid waste compost. *Journal Environmental Quality* 28: 1074–1082.

Ohgaki, S. and Sato, K., 1991, Use of reclaimed wastewater for ornamental and recreational purposes. *Water Science and Technology* 23: 2109–2117.

Rennie, R.J., Dubetz, S., Bole, J.B. and Muendel, H.H., 1982, Dinitrogen fixation measured by ^{15}N isotope in two Canadian soybean cultivars. *Agronomy Journal* 74: 725–730.

Sakadevan, K., Hedley, M.J. and Mackay, A.D., 1993, The mineralisation and fate of sulphur and nitrogen in hill country pastures. *New Zealand Journal Agricultural Research* 36: 271–281.

Sharpley, A.N. and Menzel, R.G., 1987, The impact of soil and fertilizer phosphorus on the environment. *Advances in Agronomy* 41: 297–324.

Sims, J.T., Simard, R.R. and Joern, B.C., 1998, Phosphorus loss in agricultural drainage: Historical perspective and current research. *Journal Environmental Quality* 27: 277–293.

Vinten, A.J.A., Davies, R., Castle, K. and Baggs, E.M., 1998, Control of nitrate leaching from a nitrate vulnerable zone using paper mill waste. *Soil Use and Management* 14: 44–51.

11

Kinetic Modelling of Leachate Migration in Soils

Jahangir Islam[1] and *Naresh Singhal*[2]

INTRODUCTION

Sanitary landfills have been, and continue to be, the most economical method for solid waste disposal. In New Zealand, until very recently, few municipal landfills were lined. The older landfills typically have no lining and great reliance is placed upon the natural attenuation of contaminants in the underlying unsaturated soils and the groundwater aquifer. Leakage of inorganic and organic pollutants from unlined landfills over a period of time can influence the groundwater quality and may pose a threat to drinking water resources. The transport of landfill leachate in soils is subject to various physical, chemical and biological processes that affect the eventual concentration of pollutants in soils and groundwater. Understanding the movement of leachate in the soil is essential for predicting the potential for groundwater pollution from landfills. In order to estimate the environmental risk and develop strategies for groundwater protection against contamination by landfills, an understanding of the chemical transport and behaviour of pollutants in soil-groundwater systems is required.

Mathematical models can serve as important tools to evaluate the effects of infiltrating leachate and to further design remedial options. Many mathematical models have been developed to simulate the migration of pollutants in soils. Most of these models simulate either geochemical processes (Engesgaard and Kip, 1992; Walter et al., 1994; Yeh and Tripathi, 1991) or biological transformations (Borden and Bedient, 1986; Kindred and Celia, 1989; Clement et al., 1996). Few models exist that include the interaction between biodegradation and inorganic geochemical reactions in soils (Lensing et al., 1994; McNab and Narasimhan, 1994; Schäfer et al.,

[1] Graduate student
[2] Lecturer, Department of Civil and Resource Engineering, University of Auckland, Private Bag 92019, Auckland, New Zealand

1998). Modelling geochemical interactions between organic biodegradation and inorganic species is a current research topic. Biochemical degradation of organic matter and geochemical transformations should be considered simultaneously as biodegradation reactions may exert an important control on the pH of pore water solutions in soils.

The migration of inorganic and organic contaminants of leachate in soil is influenced by the sequence of microbial redox processes. The effects have been described by several field studies (Baedecker and Back, 1979; Lyngkilde and Christensen, 1992a; Bjerg et al., 1995; Ludvigsen et al., 1998). This chapter presents a one-dimensional reactive multi-component landfill leachate transport model coupled to three modules (geochemical equilibrium, kinetic biodegradation, and kinetic precipitation-dissolution) to describe the interactions between microbial redox reactions and inorganic geochemical reactions. The model is applied to a hypothetical landfill site in order to simulate the extent of contaminant transport in soils under the landfill.

2. MICROBIAL REDOX PROCESSES IN THE SUBSURFACE

Once leachate migrates outside the confines of the landfill, its physico-chemical and biochemical character will be altered. The most active physico-chemical processes are likely to be dilution, adsorption, exchange reactions, and precipitation, with filtration for particulate matter (Bagchi, 1987; Christensen et al., 1994). The biochemical processes involve microbial degradation, either aerobic or anaerobic, with the latter expected to be most active, since most of the degradation takes place anaerobically (Lyngkilde and Christensen, 1992a). Depending on the composition of the leachate, a sequence of zones of increasing redox potential may develop the downgradient of a landfill. On the other hand, zones of methane production, sulfate reduction, ferric reduction, manganic reduction, nitrate reduction, and oxygen reduction will develop if the corresponding electron acceptors are present in the aquifer (Baedecker and Back, 1979; Lyngkilde and Christensen, 1992b; Christensen et al., 1994). Table 11.1 shows the sequence of different redox reaction downgradient of a landfill.

Redox zones develop in response to the decomposition of organic material and its effects on the behaviour of inorganic constituents. The boundaries and extent of these zones are controlled by competing rates of reactions and transport. There may be considerable overlap of redox reaction zones in both marine sediments (Wang and Van Cappellen, 1996) and leachate-contaminated groundwater systems (Baedecker and Back, 1979, Christensen et al., 1994). Simultaneous methane production, sulphate reduction and iron reduction was found in several samples from the strongly anaerobic part of the leachate plume at the Grindsted Landfill,

Table 11.1: Sequence of different redox reactions downgradient of a landfill (Christensen et al., 1994)

Reactants converted to products		ΔG^0 (kcal/mol) (pH = 7)
Methanogenic processes		
$2CH_2O \rightarrow CH_3COOH$	$CH_4 + CO_2$	-22
$4H_2 + CO_2$	$CH_4 + 2H_2O$	
Sulphate reduction		
$2CH_2O + SO_4^{2-} + H^+$	$2CO_2 + HS^- + 2H_2O$	-25
Iron (ferric) reduction		
$CH_2O + 4Fe(OH)_3 + 8H^+$	$CO_2 + 4Fe^{2+} + 11H_2O$	-28
Manganese (manganic) reduction		
$CH_2O + 2MnO_2 + 4H^+$	$CO_2 + 2Mn^{2+} + 3H_2O$	-81
Denitrification		
$5CH_2O + 4NO_3^- + 4H^+$	$5CO_2 + 7H_2O + 2N_2$	-114
Aerobic respiration, oxygen reduction		
$CH_2O + O_2$	$CO_2 + H_2O$	-120

Denmark (Bjerg et al., 1995; Ludvigsen et al., 1998). The sequence of redox processes strongly affects the migration of pollutants leaching from the landfill. The sequence of redox processes in the leachate plume is the main key to understanding the fate of reactive pollutants in the plume (Lyngkilde and Christensen, 1992b; Bjerg et al., 1995).

MODEL DESCRIPTION

Model simulations and field observations suggest the development of a biologically-active zone in which substantial organic degradation (90%) occurs within a very small distance (2 m) of the landfill (Sykes et al., 1982). The flow and transport of leachate leaking from the landfill in the unsaturated zone is assumed to be one-dimensional in a downward direction and the soil is assumed to be saturated. For unsaturated contaminant transport, Rowe (1987) recommends a simpler analysis that assumes the soil to be saturated in order to estimate the design contamination based on the worst case.

Reactive Transport Model

The transport equations for aqueous components can be derived on the principle of conservation of mass. The general form of contaminant transport equation for the mobile components in the aqueous phase (Bear, 1979) is:

$$\frac{\partial}{\partial t}(\theta C_k) - \frac{\partial}{\partial x_i}\left(\theta D_{ij}\frac{\partial C_k}{\partial x_j}\right) + \frac{\partial}{\partial x_i}(\theta v_i C_k) = \theta R_k \quad k = 1, \ldots, N_c \quad (1)$$

where

$$v_i = \frac{q_i}{\theta} = \frac{-K_{ij}}{\theta}\frac{\partial h}{\partial x_i} \quad (2)$$

and t stands for time (T), C_k is the total aqueous concentration of component k, x_i stands for Cartesian coordinates (L), v_i is seepage velocity or average pore fluid velocity (L/T), q_i is specific discharge or Darcy velocity (L/T), D_{ij} is the hydrodynamic dispersion tensor (L^2/T), θ is the porosity (dimensionless) and R_k is the chemical source/sink term (M/L^3/T) representing the changes in aqueous component concentrations.

Geochemical Reaction Module

The geochemical reaction module is based on the equilibrium speciation model MINTEQA2 (Allison et al., 1991), which includes a comprehensive set of chemical reactions including chemical speciation, acid-base reactions, mineral precipitation-dissolution, oxidation-reduction, and adsorption reactions. The ion association equilibrium-constant approach is used to represent the geochemical reactions. It is based on the solution of a set of nonl-inear algebraic equations obtained from the mass action and mass balance equations for the various components of the systems. The non-linear algebraic equations are solved by the Newton-Raphson approximation method to give the species distributions at equilibrium. Activity coefficients for all species are the functions of solution ionic strength and vary as species distributions alter the ionic strength. MINTEQA2 calculates the successive sets of activity coefficients for all solution species with each iteration. These sets are used to generate corrected values of the equilibrium constants that appear in the mole balance expressions.

Kinetic Biodegradation Module

The kinetic biodegradation module describes the biological degradation of organic compounds by different bacterial populations. Microbially-mediated redox sequences of reactions (aerobic oxidation, nitrate reduction, Mn(IV)-reduction, Fe(III)-reduction, sulfate reduction, and methanogenesis) can be modelled with different functional bacterial groups. The rates of reduction of terminal electron acceptor (TEA) and corresponding oxidation of organic substrate can be represented by multiple Monod-type expressions. Inhibition functions are used to model the sequential use of electron acceptors, depending on the presence of the dominant electron acceptors (Van Cappellen and Gaillard, 1996). The degree of overlap between successive organic matter degradation pathways would depend on the inhibition constants. Assuming a macroscopic approach (absence of diffusional resistance), biodegradation of the substrate and microbial growth can be represented by the following system of equations (Clement et al., 1996).

$$\frac{\partial S}{\partial t} = -K\frac{S}{K_s + S}\frac{C_{ea}}{K_{ea} + C_{ea}}\left(X^a + \frac{\rho_k X^s}{n}\right)\prod_i IF_i \tag{3}$$

$$\frac{\partial X^a}{\partial t} = YK\frac{S}{K_s + S}\frac{C_{ea}}{K_{ea} + C_{ea}}X^a\prod_i IF_i - K_{dec}X^a - K_{att}X^a + \frac{\rho_k K_{det} X^s}{n} \tag{4}$$

$$\frac{\partial X^s}{\partial t} = YK\frac{S}{K_s + S}\frac{C_{ea}}{K_{ea} + C_{ea}}X^s\prod_i IF_i - K_{dec}X^s - K_{det}X^s + \frac{n K_{att} X^a}{\rho_k} \tag{5}$$

where

S is the substrate concentration in the bulk (mobile) fluid (M/L^3)

X^a is the aqueous-phase biomass concentration (M/L^3)

X^s is the solid-phase biomass concentration (M/M)

C_{ea} is the concentration of terminal electron acceptor (M/L^3)

K_{dec} is the first-order endogenous decay coefficient (1/T)

K_{att} is the biomass attachment coefficient (1/T)

K_{det} is the biomass detachment coefficient (1/T)

K is the maximum specific substrate utilization rate (1/T)

Y is the yield coefficient for the biomass (cell mass produced per mass of substrate consumed)

n is the soil porosity

ρ_k is the bulk density of aquifer solids (M/L^3)

K_s and K_{ea} are the half-saturation constants for organic substrate and electron acceptors (M/L^3)

IF_i is the inhibition function for the inhibiting species i.

IF_i is used to model the sequential use of electron acceptors expressed as follows (Van Cappellen and Gaillard, 1996):

$$IF_i = \frac{IC_i}{IC_i + EA_i} \tag{6}$$

where IC_i is the inhibition constant for the inhibiting electron acceptors i, and EA_i is the concentration of the inhibiting electron acceptors.

Kinetic Precipitation-Dissolution Module

The leachate-polluted groundwater samples were supersaturated with carbonate minerals (Bjerg et al., 1995; Jensen et al., 1998), suggesting the fact that the processes are kinetically controlled. Supersaturation of carbonate minerals possibly persists due to the slow precipitation kinetics or complexation of cations with organic ligands in solution, or both. Kinetic rate laws are used to model the precipitation-dissolution processes of three common carbonate minerals, calcite ($CaCO_3$), siderite ($FeCO_3$), and rhodochrosite ($MnCO_3$). The precipitation-dissolution processes of iron sulfide (FeS) are modelled by the MINTEQA2 equilibrium speciation model. The kinetic rate laws and parameter values are given in Table 11.2.

Table 11.2: Kinetic rate equations for different minerals in an aqueous solution

Minerals	*Rate equations*	*Rate constant, k*	*Solubility product, K_{sp}*	*References*
Calcite ($CaCO_3$)	$R = k_1 [H^+] + k_2 [H_2CO_3^*] + k_3 [H_2O] - K_4 [Ca^{2+}][HCO_3^-]$ $K_4 = K_2/K_{sp} (k_1' + 1/[H^+]$ $(k_2 [H_2CO_3^*] + k_3 [H_2O]))$ $mmole/cm^2/s$	$k_1 = 0.051$ $k_2 = 3.45\times10-5$ $k_3 = 1.19\times10-7$ $k_1^- = 10\ k_1$	$1\times10^{-8.48}$	Plummer et al. (1978)
Siderite ($FeCO_3$)	$R = k (\{[Fe^{2+}][CO_3^{2-}]\}^{1/2} - K_{sp}^{1/2})^2$ mole/day	$k = 1.8 \pm 0.3\times10^6$ L/mole/day	$1\times10^{-10.40}$	Wajon et al. (1985)
Rhodochrosite ($MnCO_3$)	$R = k ([Mn^{2+}][CO_3^{2-}]/K_{sp} - 1)^n$ $\mu mole/hr/m^2$	$k = 1.56 \pm 0.39 \times 10^{-1}\ \mu mole/hr/m^2$ $n = 1.724 \pm 0.067$	5×10^{-11}	Sternbeck (1997)

[ ] = activity

Numerical Solution and Coupling of Modules

The one-dimensional advection-dispersion equation is solved by the finite-difference scheme with central weighting in space and time. The mathematical formulation of the kinetic biodegradation reactions leads to a system of ordinary differential equations, which is solved by a stiff ordinary differential equation system solver SFODE (Morris, 1993). A two-step sequential operator splitting method is used to solve the coupled transport model and the biogeochemical reaction equations. Two-step methods enable the use of appropriate numerical methods for the transport and the chemical model part separately. The different modules are coupled through the equations describing the biologically-mediated redox reactions. Change in concentrations of different terminal electron acceptors, hydrogen and inorganic carbon due to kinetic biodegradation is computed from the stoichiometry of the microbial-mediated redox reactions (Table 11.1).

Figure 11.1 shows the flow diagram of the sequential solution scheme used in the model. According to Smith and Jaffé (1998), redox potential (pE) is estimated at each location of the domain based on the concentrations of the dominant terminal electron acceptor and its corresponding reduced species from the transport and kinetic biodegradation model. Together with other computed species' concentrations, the estimated pE is then

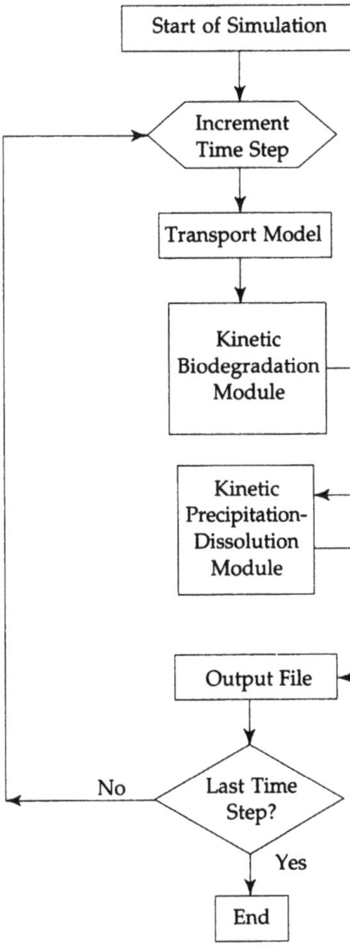

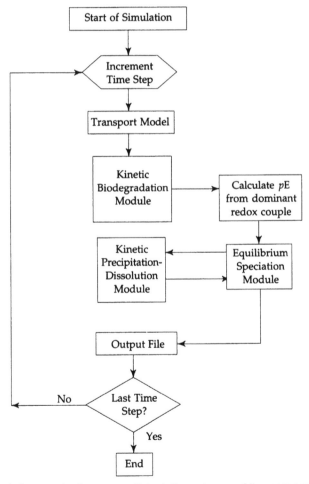

Fig. 11.1: Flow diagram for the sequential solution scheme of the composite model

used in the chemical equilibrium model to calculate the equilibrium state towards which the system should be driven (Smith and Jaffé, 1998). Precipitation can be allowed to take place at any location for which the chemical equilibrium model predicts the presence of that solid at equilibrium, and dissolution can be allowed for which the chemical equilibrium model predicts the absence of that solid at equilibrium. Based on the speciation calculation and specified rate expressions (Table 11.2), the rates of precipitation are calculated. Finally, the mass of each mineral phase of interest is updated and the aqueous solutions are re-equilibrated with the geochemical equilibrium model.

MODEL EVALUATION

The present model is evaluated by comparing the model results with the widely-used one-dimensional mixing-cell model PHREEQM (Appelo and Willemsen, 1987) for acidic mine tailings discharge in a carbonate aquifer. The details of the geochemical condition for this acid mine tailings simulation are given by Walter et al. (1994). Figure 11.2 shows the present model comparison results for this example problem, where the geochemistry is in equilibrium with respect to a series of mineral pH buffers. Good agreement is observed between the two model predictions. PHREEQM model simulation results show some oscillations, while the present model simulation results do not show any oscillation.

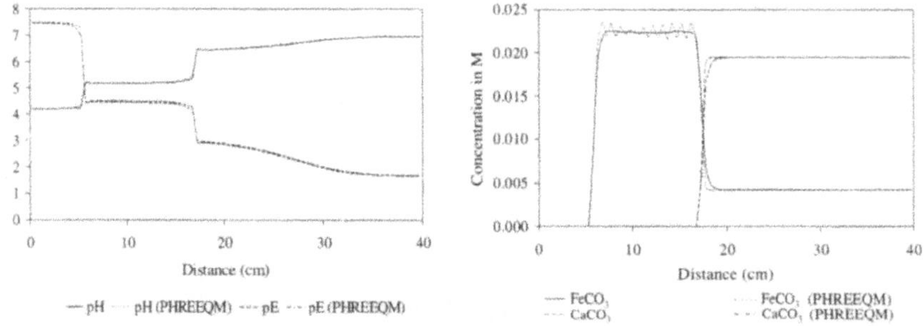

Fig. 11.2: Comparison of concentration profiles of pH, pE and solid phase minerals at 12 days between the present model and PHREEQM

EXAMPLE SIMULATIONS

The developed model is applied to a hypothetical landfill site to simulate the extent of contaminant transport in soils under the landfill. The simulation qualitatively shows the degradation patterns and their effect on groundwater geochemistry. The model is also applied to evaluate the effect of leachate pH, and soil buffering capacity on the retardation of contaminants in the leachate-contaminated soils. A one-dimensional hypothetical vertical column of soils under the landfill is assumed as the physical set-up of the model. The simulations are conducted for three test cases to simulate the behaviour of the contaminant transport in soils under the landfill:

Case 1 — pH of the leachate is considered to be 7.0

Case 2 — the pH of the lechate is 6.0

Case 3 — the pH of the leachate is 6.0 and the soil is considered to contain calcite mineral

The soil column selected is 3 m in length and the porosity 0.4. The groundwater velocity and the dispersivity are taken as 0.05 m/day and 0.05 m, respectively (Sykes et al., 1982). The model simulations are conducted with a grid spacing of 0.05 m and a time step of 0.05 day. Table 11.3 summarizes the kinetic parameters for the biodegradation module. The kinetic biological parameters are taken for methanogenic bacteria from Sykes et al. (1982) and for the aerobic bacteria from Metcalf and Eddy (1991). The biological parameters for the intermediate functional bacterial groups are linearly interpolated according to the different energy yield of the several dominant redox reactions (Table 11.1). Attachment and detachment coefficients are assumed to be 0.38 and 0.004 min^{-1} (Peyton et al., 1995). A value of 0.1 mg/l is used for inhibition constants against the electron acceptors O_2, NO_3, MnO_2, $Fe(OH)_3$, and SO_4.

Table 11.3: Summary of kinetic biological parameters used in model simulations

Bacterial group	*K (1/day)*	*K_s (mg/l COD)*	*K_{dec} (1/day)*	*Y*	*K_{ea} (mg/l)*
Methanogens	1.80	4000	0.015	0.040	-
Suphate reducers	1.93	3880	0.016	0.054	1.00
Fe(II)-reducers	2.06	3758	0.018	0.068	0.97
Mn(IV)-reducers	4.33	1616	0.042	0.317	0.47
Denitrification	5.74	282	0.057	0.472	0.16
Aerobes	6.00	40	0.060	0.500	0.10

Figure 11.3 shows the profile plots of oxygen, nitrate, MnO_2, $Fe(OH)_3$ at 15 and 30 days, and Figure 11.4 shows the profile plots of pE and dissolved organic carbon at 15, 30, and 360 days for Case 1. After around one year, the model reached steady state conditions. As the anaerobic leachate contaminated groundwater enters the aerobic aquifer, dissolved organic carbon is degraded by aerobic respiration, it continues until the concentration of oxygen has been depleted. As long as oxygen is present, the pE remains relatively constant. After the oxygen has been depleted, the pE decreases instantaneously to a lower value controlled by the second available terminal electron acceptor, nitrate. As long as the nitrate is

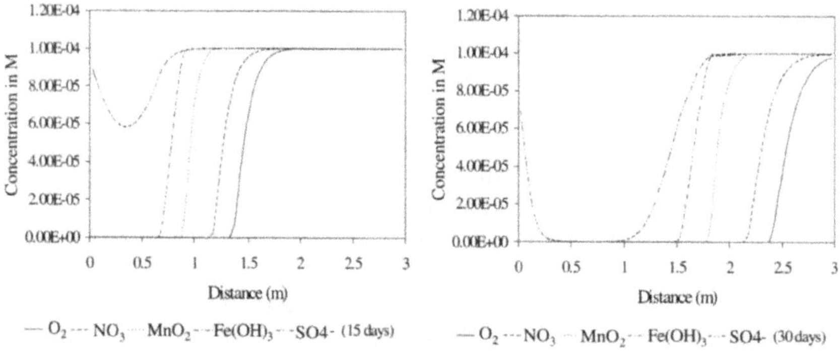

Fig. 11.3: Concentration profiles of terminal electron acceptors at 15 and 30 days

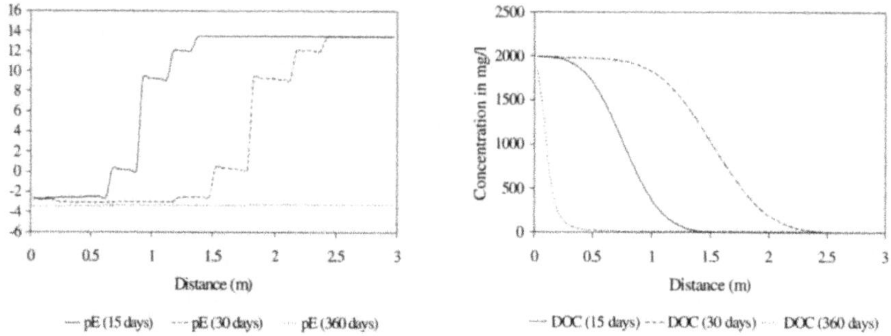

Fig. 11.4: Concentration profiles of pE and dissolved organic carbon at 15, 30 and 360 days

present, the pE remains constant and organic carbon is decomposed by denitrification process. The microbial degradation process continues and creates a sequence of redox zones downstream of the source. Oxygen reduction zone develops first, followed by nitrate reduction, manganic reduction, ferric reduction, sulphate reduction, methane producing and finally again sulphate reduction zone develops due to continuous input of sulfate from the leachate. The degree of overlap between different microbial redox zones can be controlled by the inhibition constants.

Figure 11.5 shows the concentration profiles of iron and pH for the three test cases. It shows that in all the cases, pH decrease along the length of the column due to precipitation of metals as sulfides and carbonates. The major minerals precipitated out are siderite ($FeCO_3$) and iron sulfide (FeS). Rhodochrosite ($MnCO_3$) is also precipitated in Case 1 and Case 3, but in a small amount as compared to siderite and iron sulphide. In Case 1, the pH is decreased from 7.0 to 6.5 at the first grid point due to precipitation of calcite, siderite and iron suphide. Calcite is precipitated out only at the first grid point. In Case 3, pH increases due to dissolution of calcite and it continues until the calcite is completely

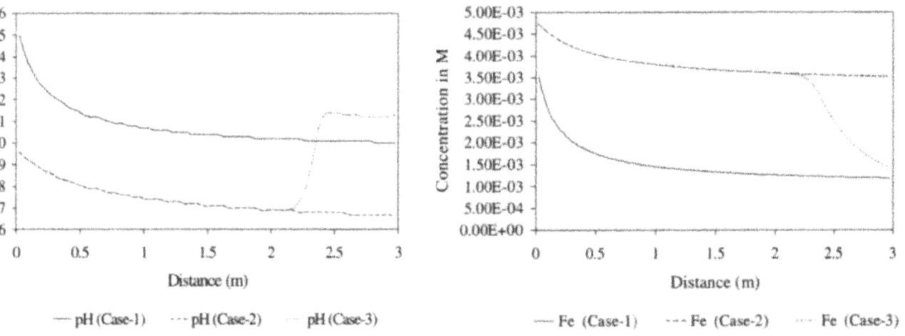

Fig. 11.5: Concentration profiles of pH and iron for the three test cases at 360 days

dissolved. The dissolution process moves downstream as indicated by the pH profile of Case 3. The decrease in pH of the influent leachate in Case 2 decreases due to mineral precipitation and thereby increases the concentration of iron in the aqueous phase. In Case 3, the concentration of iron decreases at the downstream of the column due to increase in pH.

CONCLUSIONS

A composite modelling approach is presented to simulate the complex interactions between microbial redox reactions and inorganic geochemical reactions in the landfill leachate contaminated soils. Simulations show the model results to be consistent with field and laboratory observations reported in the literature. The kinetic model presented here may be used to predict the migration of contaminants in soils under landfills based on physical transport, geochemical interactions and biological activity in soils. The composite leachate transport model can be used to assess the impact of contaminants leaking from landfills on the groundwater quality.

REFERENCES

Allison, J.D., Brown, D.S. and Nova-Gradac, K.J., 1991, MINTEQA2/PRODEFA2, A Geochemical Assessment Model for Environmental Systems: Version 3.0 User's Manual, Environmental Research Laboratory, Office of Research and Development, US Environmental Protection Agency, Athens, Ga.

Appelo, C.A.J. and Willemsen, A., 1987, Geochemical calculations and observations on saltwater intrusions, I. A combined geochemical/mixing cell model. *Journal of Hydrology* 94: 313–330.

Baedecker, M.J. and Back, W., 1979, Modern marine sediments as a natural analog to the chemically stressed environments of a landfill. *Journal of Hydrology* 43: 393–413.

Bagchi, Amalendu, 1987, Natural attenuation mechanisms of landfill leachate and effects of various factors on the mechanisms. *Waste Management Research* 5: 453–464.

Bear, J., 1979, *Hydraulics of Groundwater*. McGraw-Hill, New York.

Bjerg, P.L., Rügge K., Pedersen, J.K. and Christensen, T.H., 1995, Distribution of redox sensitive groundwater quality parameters downgradient of a landfill (Grindsted, Denmark). *Environmental Science and Technology* 29: 1387–1394.

Borden, R.C. and Bedient, P.B., 1986, Transport of dissolved hydrocarbon influenced by oxygen-limited biodegradation. 1. Theoretical development. *Water Resources Research* 22: 1973–1982.

Christensen, T.H., Kjeldsen, P., Albrechtsen, H., Heron, G., Nielsen, P.H., Bjerg, P.L. and Holm, P.E., 1994, Attenuation of landfill leachate pollutants in aquifers. *Critical Reviews in Environmental Science and Technology* 24: 119–202.

Clement, T.P., Hooker, B.S. and Skeen, R.S., 1996, Numerical modeling of biologically reactive transport near nutrient injection well. *Journal of Environmental Engineering* 122: 833–839.

Engesgaard, P. and Kip, K.L., 1992, A geochemical model for redox-controlled movement of mineral fronts in ground-water flow systems: A case of nitrate removal by oxidation of pyrite. *Water Resources Research* 28: 2829–2843.

Jensen, D.L., Boddum, J.K., Redemenn, S. and Christensen, T.H., 1998, Speciation of dissolved iron(II) and manganese(II) in a groundwater pollution plume. *Environmental Science and Technology* 32: 2657–2664.

Kindred, J.S. and Celia, M.A., 1989, Contaminant transport and biodegradation, 2. Conceptual model and test simulations. *Water Resources Research* 25: 1149–1159.

Lensing, H.J., Vogt, M. and Herrling, B., 1994, Modelling of biologically mediated redox processes in the subsurface. *Journal of Hydrology* 159: 125–143.

Ludvigsen, L., Albrechtsen, H.J., Bjerg, P.L. and Christensen, T.H., 1998, Anaerobic microbial processes in a leachate contaminated aquifer (Grindsted, Denmark). *Journal of Contaminant Hydrology* 33: 173–291.

Lyngkilde, J. and Christensen, T.H., 1992a, Fate of organic contaminants in the redox zones of a landfill leachate pollution plume (Vejen, Denmark). *Journal of Contaminant Hydrology* 10: 291–307.

Lyngkilde, J. and Christensen, T.H., 1992b, Redox zones of a landfill leachate pollution plume (Vejen, Denmark). *Journal of Contaminant Hydrology* 10: 273–289.

McNab, W.W. Jr. and Narasimahan, T.N., 1994, Modeling reactive transport of organic compounds in groundwater using a partial redox disequilibrium approach. *Water Resources Research* 30: 2619–2635.

Metcalf and Eddy, 1991, *Wastewater Engineering, Treatment, Disposal and Reuse* (3rd ed.). McGraw-Hill, New York.

Morris, A.H. Jr, 1993, NSWC Library of Mathematical Subroutines. Naval Surface Warfare Center, Dahlgren Division, Dahlgren, VA.

Peyton, B.M., Skeen, R.S., Hooker, B.S., Ludman, R.W. and Cunningham, A.B., 1995, Evaluation of bacterial detachment rates in porous media. *Applied Biochemistry and Biotechnology* 51: 785–797.

Plummer, L.N., Wigley, T.M.L. and Parkhurst, D.L., 1978, The kinetics of calcite dissolution in CO_2 water systems at 5 to 60°C and 0.0 to 1.0 atm CO_2. *American Journal of Science* 278: 179–216.

Rowe, R.K., 1987, Pollutant transport through barriers. In: Woods, R.D. (Ed.), *Geotechnical Practice for Waste Disposal.* Geotechnical Special Publication No. 13, ASCE: 159–181.

Schäfer, D., Schäfer, W. and Kinzelbach, W., 1998, Simulation of reactive processes relative to biodegradation in aquifers 1. Structure of the three-dimensional reactive transport model. *Journal of Contaminant Hydrology* 31: 167–186.

Smith, S.L. and Jaffé, P.R., 1998, Modeling the transport and reaction of trace metals in water-saturated soils and sediments. *Water Resources Research* 34: 3135–3147.

Sternbeck, J., 1997, Kinetics of rhodochrosite crystal growth at 25°C: The role of surface speciation. *Geochimica et Cosmochimica Acta* 61: 785–793.

Sykes, J.F., Soyupak, S. and Farquhar, G.J., 1982, Modeling of leachate organic migration and attenuation in groundwaters below sanitary landfills. *Water Resources Research* 18: 135–145.

Van Cappellen, P. and Gaillard, J.F., 1996, Biogeochemical dynamics in aquatic sediments. In: Lichtner, P.C., Steefel, C.I. and Oelkers, E.H. (eds). *Reactive Transport in Porous Media, Reviews in Mineralogy*, 34, Mineralogical Society of America, Washington, DC: 335–376.

Wajon, J.E., Ho, G.E. and Murphy, P.J., 1985, Rate of precipitation of ferrous iron and formation of mixed iron-calcium carbonates by naturally occurring carbonate materials. *Water Research* 19: 831–837.

Walter, A.L., Frind, E.O., Blowers, D.W., Ptacek, C.J. and Molson, J.W., 1994, Modeling of multicomponent reactive transport in groundwater—1. Model development and evaluation. *Water Resources Research* 30: 3137–3148.

Wang, Y. and Van Cappellen, P., 1996, A multicomponent reactive transport model of early diagenesis: application to redox cycling in coastal marine sediments. *Geochimica et Cosmochimica Acta* 60: 2993–3014.

Yeh, G.T. and. Tripathi, V.S., 1991, A model for simulating transport of reactive multispecies components: Model development and demonstration. *Water Resources Research* 27: 3075–3094.

Section B

WASTE MANAGEMENT TECHNOLOGIES: CASE STUDIES

12

A Review of the Land FILTER Technique for Treatment and Reuse of Sewage Effluent and Other Wastewater

N.S. Jayawardane[1], *T.K. Biswas*[2], *J. Blackwell*[2] and *F.J. Cook*[3]

INTRODUCTION

There is a widespread concern among various communities and governments in many countries about the treatment of wastewaters before their release to prevent pollution of natural waterbodies. Pollution of natural waterbodies can reduce the potential reuse of the water and pose risks to fisheries and public health in downstream locations (Pescod, 1992; Kirkham—this volume, Chapter 3). In many countries, urban expansion and industrialization has led to more water being diverted for domestic and industrial uses, thereby decreasing the future water supplies for agriculture and other uses. Therefore, wastewater reuse options need to be developed in order to meet the increasing demands on scarce freshwater supplies. The specific combination of wastewater treatment and release techniques, with wastewater reuse options adopted in any given river catchment will vary according to local factors, such as the need to maintain adequate environmental flows, land suitability and availability for irrigated cropping, and other potential reuses for wastewater (Jayawardane, 1995).

EXISTING SYSTEMS FOR WASTEWATER RENOVATION AND REUSE

High-tech engineering, biological and chemical wastewater treatment plants, which are expensive to build and operate, are widely used in large urban cities in developed countries. The less expensive land treatment/reuse techniques are more suitable for the developing countries and the rural areas in the developed countries.

[1] CSIRO Land and Water, GPO Box 1666, Canberra, ACT 2601
[2] CSIRO Land and Water, PMB 3, Griffith, NSW 2680
[3] CSIRO Land and Water, c/o QDNR, Meiers Road, Indooroopilly, QLD 4075

The existing land treatment and wastewater reuse techniques can be categorized into three groups: slow infiltration; rapid infiltration; and overland flow. The three types of land treatment differ in their capacity for hydraulic loading, adaptability to the site conditions, and the renovation processes involved (Iskander, 1981). The relative advantages and disadvantages of these different land application techniques are given in Table 12.1.

Table 12.1: Comparison of land application techniques

Technique	*Advantages and disadvantages*
Slow infiltration	Well suited for soils with moderate permeability that allows adequate leaching fractions to maintain the salt balance. Risk of water-logging and salinization on slowly-permeable soils. Requires capital expenditure for wet weather and winter storage.
Rapid infiltration	Only suited for coarser textured soils with high permeability to allow excess water to flow to aquifer storage for reuse. Pollutant retention on sandy soils could be limited.
Overland flow	Suited for soils with low permeability. Reduced nutrient removal capacity.
FILTER	Suited for soils with low permeability. Can be used to reclaim degraded saline and sodic soils. Does not require wet weather storage, but requires capital expenditure in installing subsurface drainage. Treated subsurface water released for reuse.

The slow infiltration or effluent irrigation technique involves the total reuse of wastewater for irrigated cropping and forestry. It can be used on either permeable or moderately permeable soils. Wastewater is applied at low rates to meet the evapotranspiration and salt-leaching requirements. The effluent can be applied as spray, drip or flood irrigation. Renovation of wastewater occurs by physical, chemical, and biological processes as wastewaters move slowly into and through the soil profile. Vegetation is an important component in this type of treatment, particularly for the removal of nitrogen, phosphorus as well as other nutrients from the wastewater (Iskander, 1978; Iskandar et al., 1976, 1977; US EPA, 1981). An adequate area of suitable land for irrigated cropping is needed. Effluent irrigation could save scarce water supplies during dry periods, while simultaneously preventing the pollution of waterbodies with pollutants present in the wastewater. However, during periods where rainfall exceeds evapotranspiration, the wastewater will need to be stored for reuse during the dry periods if pollution by untreated wastewater release is to be prevented. Provision for wastewater storage can be expensive, even where suitable wastewater storage sites are available close to the urban wastewater pollution sources, such as industrial and sewage works sites. Inadequate techniques for wastewater reuse through irrigation could also pose a risk to the valuable irrigated lands and their potential for producing edible crops, thus threatening their long-term use. For instance, where heavy metals are present in the wastewater, they can accumulate in the soil and contaminate edible crops. On soils with low permeability, the use of saline and sodic wastewaters could lead to soil degradation and crop yield decline through salinization and sodification of the irrigated lands. On highly-permeable soils, groundwater contamination with nutrients, chemicals and disease organisms could occur, threatening the underground drinking water supplies.

The rapid infiltration or the soil-aquifer treatment system (SAT) is a wastewater treatment technique combined with the storage of treated waters in aquifer systems. The recovery of treated water from underlying aquifers for subsequent reuse is an integral part of the system. It can be used on coarser textured and more permeable soils at high application rates of 10–250 cm/week (Bouwer et al., 1974; Aulenbach et al., 1975; Baillod et al., 1977; Satterwhite et al., 1976; Aulenbach and Clesceri, 1979). Commercial systems are found in the Southwest and the West Coast of the United States, where the main purpose is to recharge the groundwater for future reuse (Asano and Roberts, 1980). The wastewater is usually applied by flooding. Renovation occurs mainly through physical, chemical, and microbiological processes as the water passes into and through a deep soil profile. Vegetation or cropping is usually not a major component of this type of land treatment. The pre-treatment of the effluent applied varies from the primary (Baillod et al., 1977) to the secondary (Aulenbach and Clesceri, 1979). It has been reported that rapid infiltration systems can be operated year-round, with no storage requirement even under freezing conditions (Baillod et al., 1977; Satterwhite et al., 1976; Aulenbach et al., 1975) with suitable modifications at the soil surface.

The overland flow system involves partial wastewater reuse, combined with the release of excess treated wastewater. The soil often has low permeability, which limits deep water percolation. Raw or partially-treated wastewater is usually applied at a rate of 5–15 cm/week on sloping, vegetation-covered soil and recovered in collection ditches at the base of the slope for further treatment, reuse, or discharge. Renovation is achieved through physical, chemical, and biological processes as the wastewater flows down the slope and interacts with the vegetative layer and surface soil layers. The first overland flow system was built in Melbourne, Australia, in 1930. In the US, several small full-scale municipal systems have been constructed in Oklahoma, Mississippi, and South Carolina. Previous studies on system performance (Jenkins et al., 1978; Jenkins and Martel, 1979; Martel et al., 1980) demonstrated that overland flow could be used to obtain secondary or better effluent from screened primary-treated or lagoon-treated wastewater. In many instances, this method of waste treatment is not as good as the other two methods (slow and rapid infiltration) in removing phosphorus and faecal bacteria (Jenkins et al., 1978), due to limited contact with the soil. Treated wastewater from lagoon and overland flow systems at Werrribee, Melbourne, showed a high level of total P (6 mg/L) and total N (25 mg/L), as compared to land filtration systems where the treated wastewater had total P of 0.8 mg/L and total N of 10 mg/L (Melbourne Water, c1997).

THE FILTER TECHNIQUE

The FILTER (Filtration and Irrigated Cropping for Land Treatment and Effluent Reuse) technique involves wastewater reuse for agricultural cropping with the release of excess treated wastewaters for downstream reuse (Jayawardane, 1995). The technique combines using the nutrient-rich effluent to grow crops, with filtration of excess water through the soil to a subsurface drainage system. It aims at providing a sustainable land-treatment system on soils with restricted drainage by reducing the levels of nutrients and other pollutants below the stringent NSW EPA limits for discharge of the treated drainage water to sensitive inland waterbodies. The FILTER system is thus capable of handling high volumes of effluent during periods of low cropping activity or high rainfall, thereby eliminating the need for costly wet-weather storage on expensive urban lands. In this system, the rate of effluent application and drainage can be regulated as described below in order to ensure the required level of pollutant removal, thereby producing low-pollutant drainage water suitable for discharge to sensitive waterbodies.

For the FILTER system, the land at the effluent application site is prepared in the following manner. Physical loosening and chemical amelioration of the slowly-permeable soil to about one metre depth is used to increase soil macroporosity and hydraulic conductivity. A network of subsurface drains is installed at the bottom of this loosened layer, with control valves to allow for the regulation of leaching rates through the soil. Alternatively, controlled pumping from a drainage sump could be used to regulate the outflow, to approximately match the net hydraulic loading of the system (Figure 12.1).

The effluent is applied to each FILTER plot on a weekly or fortnightly rotation (Figure 12.2). The watertable resulting from the input of the effluent is drawn down in the soil by controlling the rate of outflow (via the subsurface drains), to approximately balance the net inflow. The controlled drainage system enables manipulation of the watertable and, hence, control of the depths of the aerated and anoxic soil layers above the drains. This facilitates management of the system to maximize pollutant removal and to provide adequate root-zone conditions for crop production. This regulated flow of effluent through the soil allows the nutrients to be sorbed on the soil particles, or be taken up by the crop, or

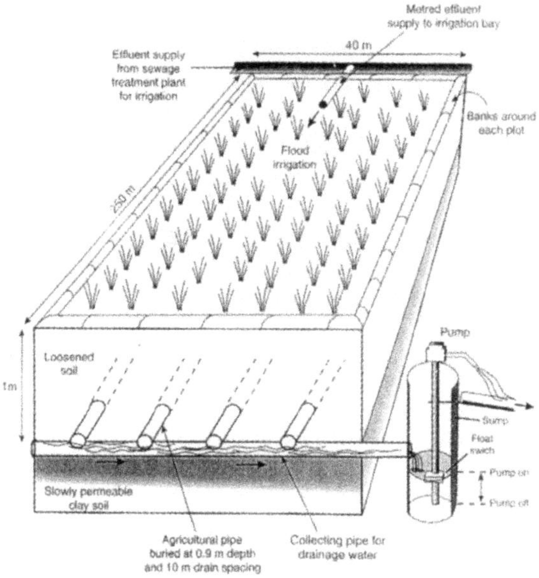

Fig. 12.1: Schematic diagram of FILTER plots.

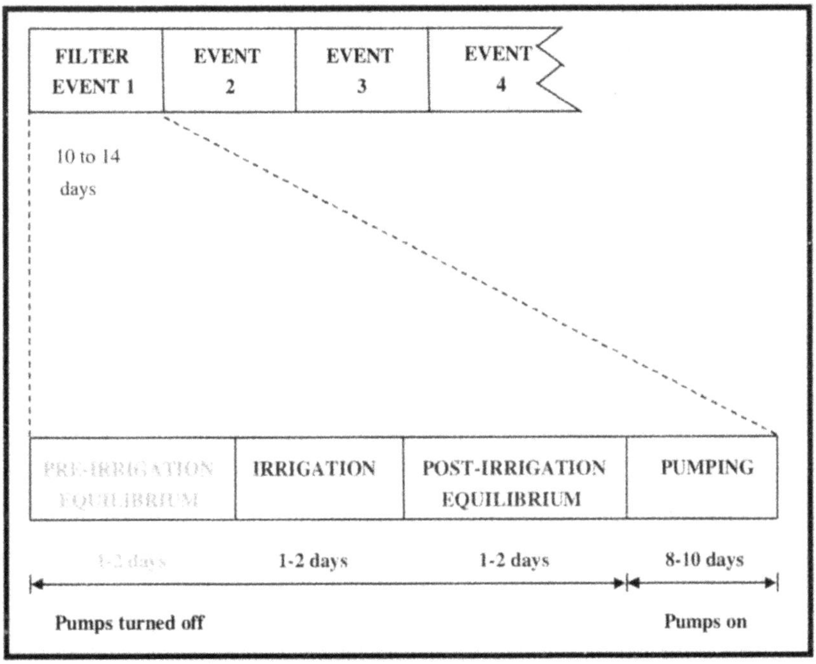

Fig. 12.2: FILTER operation procedures.

be lost through microbial degradation, ammonia volatilization and denitrification. These processes could thus be used to reduce the levels of phosphorus, nitrogen and other pollutants in the drainage outflow to levels less than those allowed by the EPA for discharge of effluent into inland waterways or ocean outfalls. The filtration processes could also be used to reduce the concentration of other contaminants in the sewage effluent to below EPA limits.

Each weekly or fortnightly effluent application cycle or filter event (Figure 12.2) consists of four consecutive stages, namely:

1. effluent application (irrigation);
2. post-irrigation equilibration period;
3. a pumping period (until drainage outflow approximately matches the net inflow); and
4. a no-pumping equilibration period (leading to flattening of the watertable).

The manipulation of these four-stage effluent application and drainage operations could be used to maximize the removal of nutrients, and to increase the uniformity in nutrient distribution and retention in the soil across the FILTER plots for future crop uptake.

The FILTER technique can be categorized as a rapid, controlled-flow system which combines some of the hydraulic flow characteristics and wastewater renovation processes of each of the slow infiltration, rapid infiltration and overland flow systems.

The success of the FILTER technique at a given site depends on achieving:

- concentration of the pollutants in the subsurface drainage waters reduced below the EPA specification for the receiving waters;
- macropore stability of the FILTER plots maintained so as to allow adequate flow rates to meet the hydraulic loading requirements; and
- long-term removal of pollutants through the FILTER system should balance the pollutant application rates, to ensure a sustainable system.

The following sections present data on a FILTER system designed to treat wastewater from the Griffith sewage works in NSW, Australia.

FIELD EVALUATION OF THE FILTER SYSTEM

Experimental Site

The field experiments were conducted at the Griffith City Council Sewage Treatment Works effluent disposal site (Jayawardane et al., 1997a,b). The experimental site is located on Chequers Road, off the Griffith-Hillston road, approximately 5 km west of Griffith (146°E 34°S) in NSW, Australia.

The soil at the site is a transitional red-brown earth, with a shallow saline ground watertable. The site was previously used for land application of saline sewage effluent for irrigation during summer months, without subsurface drainage. Consequently, the soils at the site are highly saline and sodic.

Installation of the FILTER System

Two FILTER experimental blocks were established within a 12 ha. area, which was laser-levelled to a 1:4000 slope. Field data is presented from Experimental Block A consisting of one-hectare FILTER plots 5, 6, 7 and 8 used for summer cropping and filtration during the period from November 1994 to May 1995 (the first cropping season at the trial site), followed by three further cropping seasons. Data are also presented from Experimental Block B consisting of FILTER plots 1, 2, 3 and 4 used for winter cropping during the period from May 1995 to November 1995 (the second cropping season at the trial site), followed by two further cropping seasons (Jayawardane et al., 1997a).

Each FILTER plot used for this study was 40 m wide by 250 m long (Figure 12.1), surrounded by a 0.4 m high bank. The plots were prepared as follows. To increase porosity and hydraulic conductivity, the plot was ripped to 0.90 m depth. The ripper tines were spaced 1.2 m apart. Two passes were made over the plots. During the second pass, the ripper blades were aligned between the rip lines of the first pass. Mined gypsum was broadcast at 8 t/ha., to reduce the potential for soil dispersion in the surface layers of this sodic soil after heavy rainfall.

The installation of the subsurface pipe drain system in each plot was started by digging a 1.5 m deep trench at the bottom ends of the plots. A commercial drainpipe layer was then used to install four parallel drainpipes at a spacing of 10 metres, at right angles to the trench. The 250 m 100 mm diameter pipes were laid at a slope of 0.1%. In Irrigation Block B, the subsurface drains were located at a mean depth of 1.2 m, while in Irrigation Block A, the drains were placed at 0.8 m depth. The diameter and slopes of pipes were selected to avoid any restrictions to flow within the pipes. The drainpipes were surrounded by a layer of fine gravel (less than 7 mm in diameter). The drainpipes were closed at the top end. At the bottom end, the four drainpipes at the 10 m spacing were connected to a collector pipe which opens into a 1.6 m deep sump to allow subsurface drainage outflow. Three other drainpipes were installed at the mid-point between the pipes spaced at 10 metres for the purpose of measuring the watertable depth in the plot. These three drainpipes were similarly connected to a second collector pipe, which opened into the sump. Inside the sump, a vertical tube was attached to the end of the second collector pipe, so that the watertable height inside this vertical tube indicated the watertable height in the plots at midway between the open drains. Trenches above the drainpipes were refilled. Each sump was fitted with a pump. The drainage water from the sumps was pumped into a drainage channel, which took the drainage water from the site.

Effluent treatment by FILTER was operated on approximately a fortnightly cycle, where about 1 ML of effluent was applied to each one-hectare plot, followed by a 1–2 day post-irrigation equilibration period. This was followed by an 8–10 day pumping period, when the pump was turned on and the effluent slowly passed through the soil and the subsurface drainpipes to a collection sump. After the pumping period, the pump was turned off for 1–2 days, allowing the watertable to reach equilibrium. The cycle was then repeated. The subsurface drainage system provided adequate drainage and soil aeration conditions for crop growth even during heavy rainfall and low evapotranspiration periods. During the regulated flow of effluent, nutrients and other pollutants were adsorbed on the surface of soil particles, taken up by the crop and weeds, or lost through volatilization and denitrification, thereby reducing the level of nutrients in the drainage outflow.

The cropping sequences in plots 1–8 are given in Table 12.2. In each Irrigation Block, two plots were used for perennial pasture and the other two plots were used for cereal crops. A detailed description of the agronomic management of each plot is given in Jayawardane et al. (1997a,b).

Table 12.2: The cropping sequence in the FILTER plots

Experimental block	*Plot*	*Cropping season*			
		1 (Summer 94–95)	*2 (Winter 95)*	*3 (Summer 95–96)*	*4 (Winter 96)*
B	1	-	Oats	Sorghum/Maize	Wheat
	2	-	Pasture	Pasture	Pasture
	3	-	Pasture	Pasture	Pasture
	4	-	Oats	Sorghum/Maize	Wheat
A	5	Pasture	Pasture	Pasture	Pasture
	6	Millet	Oats	Millet	Barley
	7	Pasture	Pasture	Pasture	Pasture
	8	Millet	Oats	Millet	Barley

During the first, second, third and fourth cropping and filtration seasons, effluent application rates of 7.7, 6.7, 5.6 and 6.2 mm/day, respectively, were used. During each of the four successive seasons, a total of 12, 12, 8 and 12 filter events, respectively, were carried out. During each filtration event, the following measurements were carried out on each plot. The volume of effluent applied to the plots was measured. The rate of drainage water running into the sumps was measured at regular intervals from the electricity consumption of the pumps used for emptying the sumps. The changes in the height of the watertable in the plots midway between subsurface drains were also measured periodically, using the vertical tube in the sump connected to the second collector pipe from the three drains at intermediate spacing, which were not used for drainage.

Soil samples were taken before effluent application commenced and at the end of the first, second, third and fourth filtration seasons. During soil sampling in each plot, soil cores were extracted and cut up into 0.10 m segments at soil depths 0–0.4 m, and into 0.20 m segments below 0.4 m depth, and retained for soil analysis after air drying. Soil measurements included pH, extractable phosphorus, and exchangeable cations (Ca, Mg, Na, K). Exchange sodium percentage (ESP) was calculated as the ratio of sodium to the total cations on the soil exchange complex. Extractable soil NO_3 and NH_4, was measured on fresh soil samples. SAR (sodium adsorption ratio) and electrical conductivity (EC) were measured in 1:5 soil:water extracts. SAR is a measure of potential sodicity hazard and is expressed as:

$$\text{SAR} = \frac{[\text{Na}]}{\sqrt{[\text{Ca}] + [\text{Mg}]}}$$

where [Na], [Ca] and [Mg] are the concentrations in millimoles per litre (mmol/L).

Continuous irrigation and drainage water samples were collected using a sample bleeding-tube arrangement. Samples were stored at 4°C before analysis for pH, EC, dissolved Na, Ca and Mg, electrical conductivity (EC), biochemical oxygen demand (BOD_5), total suspended solids (TSS), ammonium, oxides of nitrogen (NO_x), total Kjeldahl nitrogen (TKN), total phosphorus (TP), chlorophyll$_a$, *E. coli*, and oil and grease. The physical and chemical analyses of soil and water samples were carried out using the Griffith laboratory analytical procedures.

POLLUTANT REMOVAL THROUGH THE FILTER SYSTEM

Phosphorus

The concentration of total P in the effluent and the drainage waters over the four cropping seasons is given in Figure 12.3. In the first season, these values are a mean of the four plots in Experimental Block A and in the subsequent three seasons, the values are a mean of all eight plots in Experimental Blocks A and B. The mean total P levels in the effluent and the drainage waters from all the plots during the four cropping seasons were 4.47 and 0.41 mg/L, respectively. Thus, the total P value in the drainage waters was reduced and maintained below the NSW EPA limit of 0.5 mg/L for sensitive waters. In Experimental Block B, where the subsurface drains were located at 1.2 m depth and the drain-trench refilling

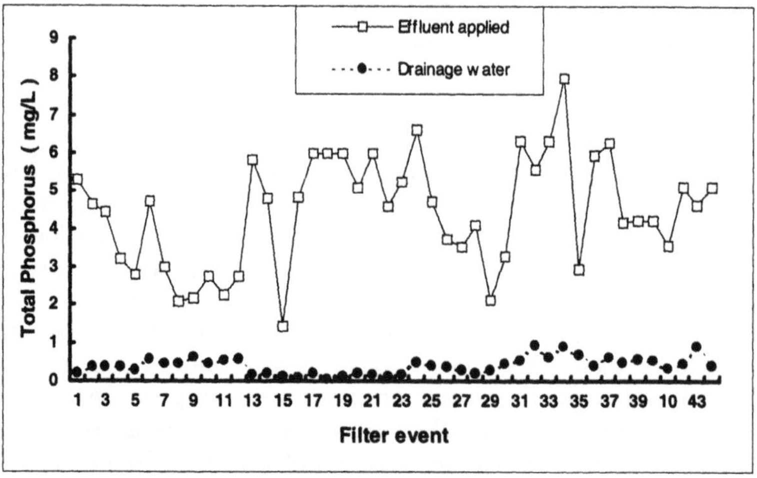

Fig. 12.3: Total-Phosphorus in the effluent applied and the drainage waters during cropping seasons 1 (filter events1-12), 2 (filter events 13-24), 3 (filter events 25-32), and 4 (filter events 32-44).

was done simultaneously with the deep-ripping operation to minimize the phosphorus-rich surface soil falling into this trench, the mean total P level in drainage waters over the second, third and fourth cropping seasons was 0.21 mg/L. The very low values of total P in filter events 13–24 of cropping season 2, is due to the low phosphorus concentrations in drainage waters from Experimental Block B.

Figure 12.3 shows that while total P in the effluent applied to the FILTER plots varied widely between 1.5 and 8 mg/L, the drainage waters had more uniform values at below 1 mg/L. This is due to the buffering effect of the clay soils at the site which have a high phosphorus sorption capacity. This buffering effect eliminates the need for regular monitoring of the total P in the effluent and drainage waters to maximize phosphorus removal. The mean reduction in the phosphorus load during flow through the FILTER system in all four seasons exceeded 96%.

Nitrogen

Total N in the effluent, soil and drainage waters occurs in three forms: NO_x-N; NH_4-N; and organic-N. The concentration of individual forms of N in the drainage waters depends on their interactions in the effluent and within the soil nitrogen pool.

During the first cropping season, the effluent was applied to plots 5 to 8 of Experimental Block A. The effluent applied had a concentration of NO_x-N below 1.0 mg/L. The mean concentration of NO_x-N in drainage waters in the four plots throughout the filtration phase is shown in Figure 12.4. The high concentration of NO_x-N in the drainage waters

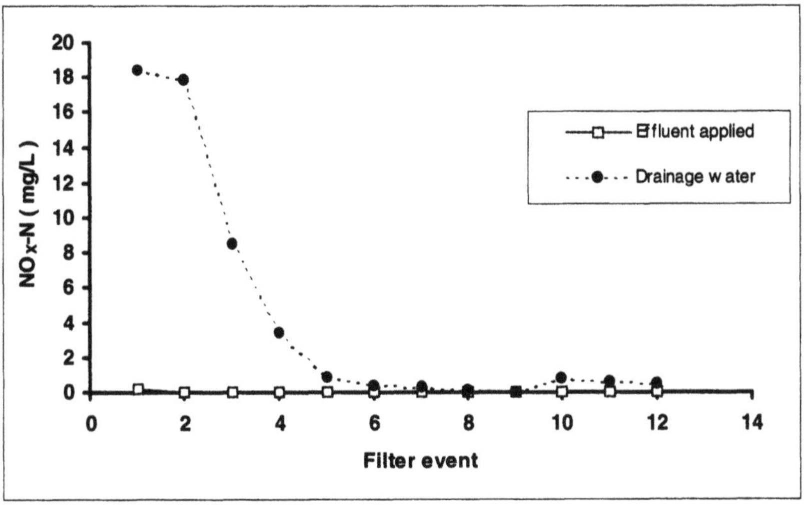

Fig. 12.4: NOx-Nitrogen in the effluent applied and the drainage waters during the first cropping season, in Experimental Block A.

during filter events 1 and 2 is related to the high values of soil NO_x-N before the start of the filtration phase (363 kg/ha. in the top 1 m of soil). However, the NO_x-N in drainage water fell to below 10 mg/L by filter event 3, and to below 5 mg/L in filter event 4 and thereafter. These reductions in NO_x-N concentrations during the operation of the FILTER technique is likely to be largely due to the high denitrification potential of the soil in the flooded soil layers, as was also shown in laboratory incubation studies. Crop uptake could also be a contributory factor in reducing the NO_x-N levels in drainage water. The mean NO_x-N content in all the plots at end of the summer cropping season was reduced to 7 kg/ha. in the top 1 m of soil.

The total-N concentration in the drainage waters on each plot during the first cropping season depends on the factors influencing the individual N components, as discussed previously. While the total-N concentration was high during filter events 1 and 2 due to leaching of soil NO_x-N accumulated before installation of the FILTER plots, it remained below 10 mg/L from filter events 3 to 12 (Figure 12.5). Thus, after removal of pre-FILTER soil N accumulations, the use of the FILTER technique resulted in maintenance of total-N levels well below the 15 mg/L limit specified by NSW EPA (and even below the future preferred target limit of 10 mg/L), under the effluent quality and FILTER operation conditions that prevailed during the 1994-95 summer cropping season. During the first summer cropping season, the lowering of the accumulated soil NO_x-N to a level which had only a small impact on the total-N in the drainage water, occurred quickly (within two filter events).

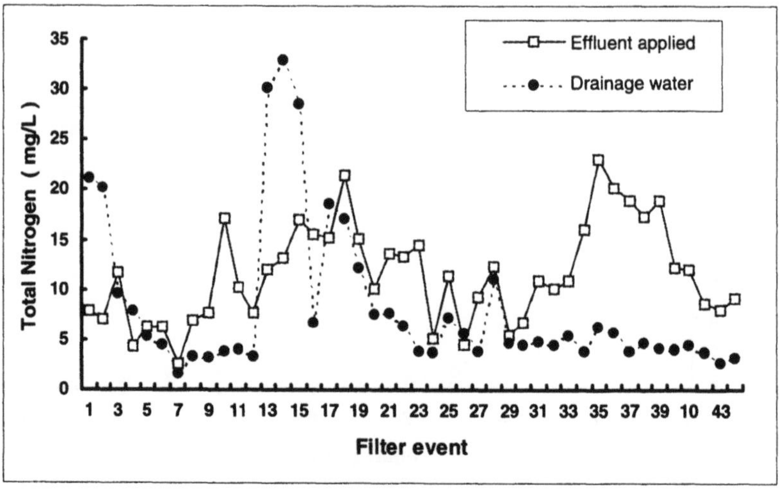

Fig. 12.5: Total-Nitrogen in the effluent applied and the drainage waters during cropping seasons 1 (filter events1-12), 2 (filter events 13-24), 3 (filter events 25-32), and 4 (filter events 32-44).

The mean total-N loads in the effluent and drainage waters in all the plots during the filter events 1–12 of the first cropping season were 104 kg/ha. and 16 kg/ha., respectively. This represents a total-N load reduction of 85%. The 15% balance includes a substantial amount of NO_x-N which had accumulated in the soil before installing the FILTER, and which was released into drainage water during filter events 1–3.

In the second cropping season (filter events 13–24), effluent was applied to the newly-installed plots 1–4 of Experimental Block B, in addition to plots 5–8 of Experimental Block A. As we observed in the first cropping season, the high soil NO_x-N of the newly-installed plots of Experimental Block B resulted in high NO_x-N and total-N concentration in the drainage waters from these four plots during the first few filter cycles of the second cropping season (Jayawardane et al., 1997a). The mean total-N concentration in the drainage waters from these four plots progressively dropped in the subsequent filter cycles. Consequently, the mean total-N concentration in the drainage waters from the eight plots progressively dropped in the subsequent filter cycles to values less than 10 mg/L (Figure 12.5). In the third cropping season, the mean total-N concentrations in the effluent applied and the drainage water from all eight plots were 8.9 and 4.9 mg/L, respectively. In the fourth cropping season, the corresponding values were 14.7 and 4.4 mg/L. Therefore, the concentrations of total-N in the drainage waters was reduced well below the EPA limits for discharge to sensitive surface waters (Figure 12.5).

The main nitrogen components in effluent from most sewage treatment plants are organic-N and NH_4-N. Table 12.3 shows the removal rates of these two nitrogen components during the four filtration seasons. The FILTER technique provides very efficient removal of these two components, especially NH_4-N.

Table 12.3: Pollutant load reduction during each filtration season

Cropping season	*Pollutant load reduction (%)*			
	NH_4-N	*Organic-N*	*Total-N*	*Total-P*
1	74.3	89.3	84.8	96.1
2	97.8	89.2	60.1	96.0
3	90.7	85.7	75.6	97.2
4	98.1	46.3	80.9	93.0
Means	90.2	77.6	75.3	95.5

In a commercial FILTER treatment system, several measures could be used to rapidly reduce the initial total-N concentration in the drainage water discharged from the site to values below the EPA limits. One option is to use sites on which the effluent has not been previously dumped. On sites where initial soil NO_x-N is high due to previous effluent applications, a non-irrigated crop could be grown before installing a FILTER system to deplete the soil NO_x-N. Another option is to install the major portion of the FILTER system in summer and to reuse the drainage water during the first few summer filter events on the balance area of land. As soon as the total-N in the drainage water falls below EPA limits, the drainage water could be discharged from the site. In the following winter, the drainage water from the plots installed in winter could be diluted with low total-N drainage water from those plots which had been installed in the previous summer (or in previous years), in order to get the total-N in drainage water discharged from the site below EPA limits. The combination of options employed at different sites will be determined by the site characteristics and the EPA regulations for the site.

N:P Ratio

During the first cropping season in Experimental Block A, the N:P ratio of the effluent was very low, ranging from 2 to 6, with a mean value of 3. Previous studies have indicated that as the N:P ratio is reduced below 12, the risk of occurrence of blue green algae progressively increases (SCM, 1992). In contrast to the effluent, the drainage waters showed a higher N:P ratio, ranging from 5 to 113, with a mean value of 23. This indicates a proportionally greater P removal rate compared to N removal in this heavy clay soil. The higher N:P ratio will lead to reductions in risk of downstream occurrences of blue-green algal blooms in the receiving waters.

Similar results were obtained in the later cropping seasons. Thus, during the second, third and fourth filtration seasons, the mean N:P ratio in the applied effluent was 2.6, 2.2 and 2.7, respectively, for all plots and filter events. The corresponding values in the drainage waters were 102, 33 and 11, respectively.

SALT

The initial mean soil EC1:5 of the four plots of Experimental Block B (Figure 12.6), shows that the site was extensively salinized by earlier applications of sewage effluent with EC-1:5 1.3 dS/m and SAR 7.2, because subsurface drainage was not provided. The initial ESP of the soil is also high (21–29%) at all depths (Jayawardane et al., 2001). Falkiner and Smith (1997) found that four years of irrigation with a slight to moderately saline and sodic effluent resulted in marked increases in soil salinity and increased soil sodicity to an ESP of 20–25. Smith et al. (1996) used modelling to show that it was essential to apply excess water at this effluent-irrigated plantation so as to promote the leaching of excess salt accumulating in the soil at the site.

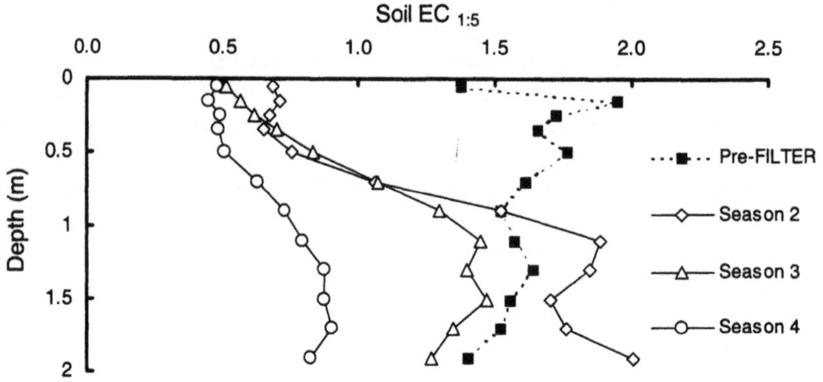

Fig. 12.6: Mean soil salinity in the Pre-Filter plots, and mean soil salinity at the end of the second, third and fourth cropping season in Experimental Block B.

With the application of moderately saline sewage effluent after establishment of the FILTER plots, initially, the salinity of the subsurface drainage water from the FILTER plots was high, as shown for data in Experimental Block B (Figure 12.7). This was due to leaching of the stored salt in the soil profile, and concentration of salts in the drainage water through crop evapotranspiration. The drainage water salinity progressively decreased during the second, third and fourth filtration seasons due to gradual leaching of the salt in the soil profile. Figure 12.6 shows that during the second filtration season, the reduction of soil salinity mainly occurred in the layers above the drains. During the third and fourth filtration seasons, the decrease in the soil salinity occurred above and

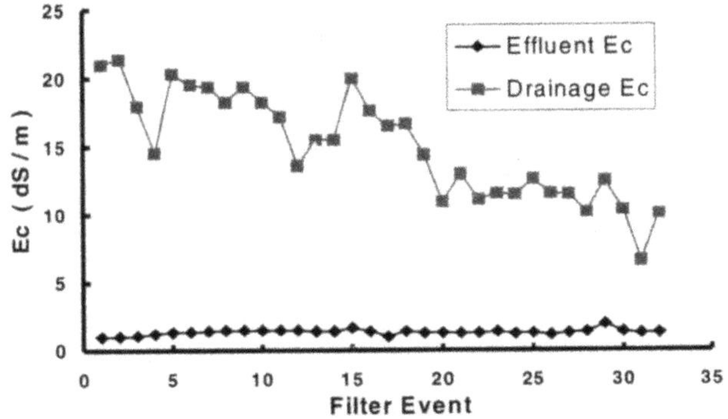

Fig. 12.7: The EC in the effluent applied and drainage water during successive filter events, in the second, third and fourth cropping season in Experimental Block B.

below the drain depth. Research studies are currently in progress to model the flow of water and salts through FILTER plots, to explain these progressive changes in soil salinity and drainage water salinity, using a modified HYDRUS-2D model.

The SAR of the composite drainage water sample during the second cropping season was higher than that of the effluent irrigation sample (Table 12.4). The high SAR of the drainage water could be related to the high initial soil ESP and the associated high SAR of the soil solution. The SAR of the composite drainage water sample during the fourth filtration season was slightly lower than the sample collected during the second filtration season. The soil ESP and the SAR-1:5 (Table 12.5) also decreased at the end of the fourth filtration season, as compared to the corresponding values before filtration. These changes could be attributed to the relatively low-value SAR wastewater used for irrigation and the provision of sub-surface drainage to leach the excess salts to prevent their accumulation in the soil. These results demonstrate that the provision of sub-surface drainage in the FILTER system not only prevented the development of salinity and sodicity in the soil, but also reversed these adverse soil degradation effects caused by previous pre-FILTER effluent applications.

Table 12.4: The mean SAR of composite effluent irrigation and composite subsurface drainage water samples in Experimental Block B

Sample type	*Cropping season*	*Mean SAR (meq/l)*			*SAR*
		Na	*Mg*	*Ca*	
Irrigation	2	7.6	1.1	1.0	7.4
	4	11.2	1.8	1.6	8.7
Drainage	2	112.6	40.6	9.5	22.5
	4	73.3	23.6	6.2	19.0

Table 12.5: Measured SAR and EC in 1:5 soil:water and calculated TEC from equations (1) and (2) proposed by Rengasamy et al. (1984) for pre-FILTER and at the end of the fourth cropping season, in Experimental Block B

Sampling time	*Depth (m)*	*SAR*	*EC (dS/m)*	*TEC*	
				Spontaneous	*Mechanical*
Pre-FILTER	0–0.2	13.9	1.46	0.24	4.28
	0.2–0.4	14.3	1.6	0.24	4.40
	0.4–0.8	15.2	1.52	0.26	4.68
	0.8–1.2	16.6	1.40	0.28	5.12
	1.2–1.6	14.1	1.43	0.24	4.32
	1.6–2.0	16.2	1.34	0.27	4.99
End of Season 4	0–0.2	6.2	0.39	0.11	1.82
	0.2–0.4	8.7	0.42	0.15	2.62
	0.4–0.8	11.0	0.49	0.19	3.33
	0.8–1.2	11.3	0.68	0.19	3.43
	1.2–1.6	13.5	0.79	0.23	4.14
	1.6–2.0	12.8	0.83	0.22	3.90

In the long term, after completion of the leaching of salts which had accumulated before field installation of the FILTER, the system could be expected to reach salt equilibrium, where the load of salt in the effluent applied equals the load in the drainage water. The ratio of EC of drainage water to that in the effluent will then reflect the concentration factor, depending on the evapotranspiration rate. During rainfall periods, a dilution factor will apply. The downstream effect of the salt load in the drainage water from any wastewater treatment site depends on the extent of mixing and dilution of the drainage waters with the receiving waters. If any alternative tertiary nutrient-removal treatment techniques which include chemical dosing are adopted to meet the EPA guidelines for nutrient removal, the salt loading of the receiving waters could be increased as compared to amount of salt added to the FILTER system.

pH

During the first filtration season in experimental block A, the pH of the drainage water was always less than the pH of the effluent with mean values decreasing from 9.4 (range 8.6–10.3) to 7.9 (range 6.8– 8.7). In the second, third and fourth filtration seasons, the mean pH of effluent applied was 8.1, 8.0 and 7.8, respectively. The corresponding values in the drainage waters were 8.1, 8.0 and 8.2.

Blue-Green Algae

The effluent from the Griffith Sewage Works showed the presence of blue-green algae with its characteristic greenish colour throughout the summer filtration period. These occurrences could be related to the high values of P and the low N:P ratio in the effluent ponds. In contrast, the drainage waters were largely colourless and did not show the presence of the blue-green algae. This reflects not only removal of the algae during effluent flow through the soil, but also the high N:P ratio achieved in the drainage waters.

Pollutant Removal in the 15-hectare Pilot FILTER Trial

Pollutant removal was also monitored in a 15-ha. pilot FILTER trial, constructed on a similar engineering design to the preliminary trial, with the subsurface drains placed at 1.2 m depth and 8 m spacing. Nutrient concentration and load reduction patterns during the first cropping season were similar to that obtained in the preliminary trials. In addition, significant reductions in total suspended solids, oil and grease, and *E. coli* were observed (Table 12.6).

Table 12.6: Reduction in concentration and loads of pollutants in sewage effluent during flow through pilot 15-ha. FILTER trial plots to the subsurface drains, during winter cropping

Pollutant	*Concentration (mg L)*		*Load (kg/ ha.)*		
	Effluent	*Drainage*	*Effluent*	*Drainage*	*% Reduction*
Total P	6.1	0.4	46.7	1.7	96
Total N	19.2	15.0	131.4	55.8	58
Organic N	6.3	1.2	46.3	4.9	90
NH_4-N	12.5	0.2	82.4	0.7	99
BODS	10	0.9	80.1	3.9	95
TSS	71	16.9	573.3	88.8	85
Chlorophyll a	0.07	0	0.01	0	100
Oil and Grease	1.8	0	15.9	0	100
E. coli counts	9	0	—	—	—

Note: E. coli is expressed as colony-forming units (CFU) per 100 mL of effluent.

MAINTAINING MACROPORE STABILITY AND FLOW RATES THROUGH THE FILTER PLOTS

An important feature of the FILTER system is the capacity to provide a rapid drainage of the excess water from the plots during the pumping stage of each filter cycle, especially during low evapotranspiration and high rainfall periods. Heavy clay soils with low permeability will require soil loosening by deep ripping to increase macroporosity and hydraulic conductivity properties, and thereby increase the flow rates to the subsurface drains. Stabilization and preservation of the macropores is required to maintain adequate flow rates. On all soils, loss of structural stability can occur through traffic-induced compaction and, hence, traffic management is important. In swelling clay soils, such as the soils at the Griffith sewage works site, structural stability can decline with an increase in exchangeable sodium percentage (ESP) and a corresponding decrease in EC (Quirk and Schofield, 1955; McNeal and Coleman, 1966; McNeal et al., 1966; Jayawardane, 1977, 1979, 1983, 1992; Jayawardane and Beattie, 1979; Jayawardane and Blackwell, 1991). Hence, maintaining the salinity-sodicity balance is important.

Rengasamy et al. (1984) proposed the use of equation (1) to calculate the threshold electrolyte concentration (TEC), which is the electrolyte concentration required to prevent spontaneous dispersion in surface and subsurface layers of a red-brown earth. They also proposed equation (2) to calculate the TEC for mechanical dispersion in the clayey subsurface layers of the red-brown earths, where the soil is subject to mechanical forces during cultivation.

$$\text{TEC} = 0.016\ \text{SAR} + 0.014 \tag{1}$$

$$\text{TEC} = 0.319\ \text{SAR} - 0.17 \tag{2}$$

where SAR is measured in 1:5 soil:water extracts.

The calculated TEC values for spontaneous and mechanical dispersion of soils collected before the establishment of the FILTER system and after the fourth filtration seasons in Experimental Block B are shown in Table 12.5. The soil EC values are greater than the TEC values for spontaneous dispersion, indicating that the pre-FILTER soils will be structurally stable during the flow of effluent through the soil to the subsurface drains. Although the soil EC is reduced due to leaching during the FILTER cycles, the associated reduction in SAR results in a reduction of TEC for spontaneous dispersion. Therefore, the soils will remain stable during effluent flow through the soil, even after three filtration seasons. However, as the EC of the soil sampled before filtration and after three filtration seasons is less than the TEC for mechanical dispersion, care is necessary to prevent structural breakdown due to factors such as rainfall impact on bare soil, mechanical disturbance and trafficking damage when the soil is wet.

The relationship between the subsurface drainage flow rate and the watertable height at the mid-point between drains was used to monitor the changes in the whole plot soil hydraulic properties (Jayawardane et al, 1997a,b). Figure 12.8 shows that the shift in this relationship from the second to the fourth filtration seasons was small.

For subsurface drains running normally without back-pressure, Youngs (1985) found that Hooghoudt's equation (Hooghoudt, 1940) could be simplified to:

$$\frac{H_m}{D} = \left(\frac{q_t}{K}\right)^{\frac{1}{a}} \tag{3}$$

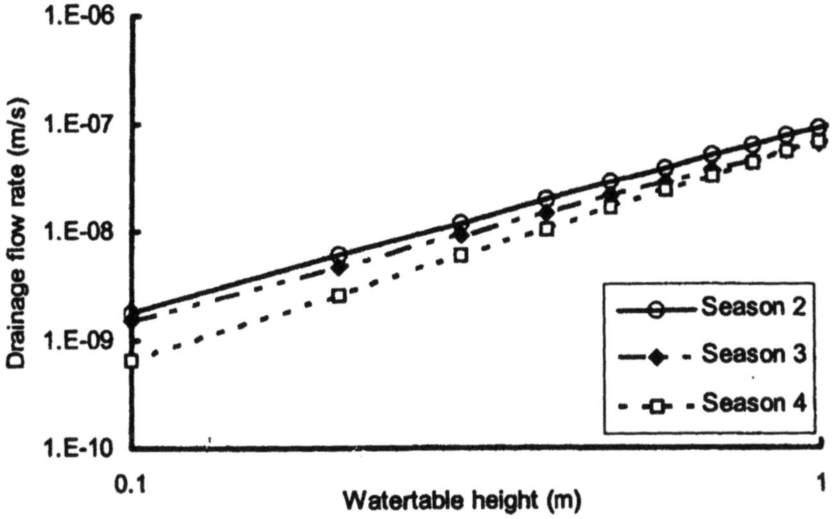

Fig. 12.8: The relationship between the flow rate (qt) to sub-surface drains and the watertable height (Ht) at the mid-point between two drains during the second, third and fourth cropping seasons, in Experimental Block B.

where

q_t is the drainage flow rate (m s^{-1})

H_m is the watertable height at the midway point between the drains (m)

K is the saturated hydraulic conductivity of the soil (m s^{-1})

$2D$ is the spacing between the drains (m)

a is a coefficient given by:

$$a = \left(\frac{d}{D}\right)^{\frac{d}{D}} \tag{4}$$

where

d is the depth below the drain to the impermeable layer (m), and $0 \le d/D \le 0.35$.

For the situation where $d/D \to \infty$, the value of a was found by Youngs (1985) to be approximately equal to 1.36 and where $d/D = 0$, the value of a is equal to 2.0.

Rearranging (3) and substituting the value of the watertable height at a given time (H_t) for H_m, we obtain

$$q_t = K\left[\frac{H_t}{D}\right]^a \tag{5}$$

$$\log q_t = a \log H_t + [\log K - a \log D] \tag{6}$$

By plotting log q_t against log H_t, we can derive the value of the slope a. By substituting in the bracketed term of Equation 6, the value of K can be calculated. The values of K and a for the initial and final filtration seasons were not significantly different (Table 12.7) in each of the eight plots (Jayawardane et al., 1997b). This indicates that the soil hydraulic properties of the subsurface layers were not markedly affected by the changes in soil salinity and sodicity during the successive filtration seasons. This observation is consistent with the soil EC remaining above the TEC during the filtration period. However, there was a tendency for surface ponding of water at the bottom end of the plots, especially during periods of prolonged rainfall during the third and fourth filtration seasons. This could be attributed to the low salinity of rainwaters and the reduced salinity of the leached surface soil layers, causing soil dispersion near the soil surface. This loss of soil structural stability is likely to have been aggravated by trafficking of the plots to carry out agronomic practices.

In swelling clay soils, careful long-term planning of the FILTER plot design and management and optimized routine FILTER operations could be used to minimize the risks of surface soil structural breakdown. For example surface soil dispersion problems could be countered by building up the surface soil organic matter by putting the area to be used for FILTER under a pasture crop for a few years before and after FILTER installation. Chemical amendment applications during and after this period

Table 12.7: Comparison of values of K and *a* between cropping seasons, for each plot (values within a plot followed by the same letter are not significantly different at the 5% level)

Plot	*Season*	*K* ($m\ s^{-1} \times 10^6$)	*a*	R^2
1	1			
	2	0.9[a]	1.66[a]	0.59
	3	4.6[b]	2.38[b]	0.80
	4	1.5[ab]	2.10[ab]	0.86
2	1			
	2	1.2[a]	1.68[a]	0.63
	3	1.2[a]	1.97[ab]	0.81
	4	1.5[a]	2.16[a]	0.91
3	1			
	2	1.8[a]	1.61[ab]	0.81
	3	0.89[a]	1.41[a]	0.69
	4	2.5[a]	1.91[b]	0.83

(Contd.)

Table 12.7: ***(Contd.)***

Plot	Season	K (m s^{-1} × 10^6)	a	R^2
4	1			
	2	5.3[a]	2.27[a]	0.82
	3	0.5[b]	1.50[b]	0.81
	4	5.1[a]	2.50[a]	0.86
5	1	34.9[a]	2.95[a]	0.88
	2	4.3[bc]	2.07[b]	0.91
	3	0.5[b]	1.64[b]	0.49
	4	15.6[ac]	2.92[a]	0.80
6	1	3.1[a]	2.04[a]	0.70
	2			
	3	0.2[b]	1.32[ab]	0.47
	4	1.4[ab]	2.08[ab]	0.70
7	1	7.0[ab]	2.38[ab]	0.85
	2	4.3[ab]	2.31[ab]	0.83
	3	0.9[a]	1.54[a]	0.44
	4	9.7[b]	2.63[b]	0.82
8	1	0.5[a]	1.20[a]	0.68
	2			
	3	0.7[a]	1.29[a]	0.65
	4	0.6[a]	1.13[a]	0.62

will help in increasing the surface soil stability during periods of application of mechanical forces. Several FILTER management options are available to reduce the TEC of the surface and subsurface soils and to increase the soil EC at different critical times in the FILTER operations, such as during exposure of the bare soil to rainfall, periods of land preparation and trafficking. These could include cutting off the drainage during summer to build up surface soil salinity, reapplying the highly saline drainage water to the soil surface as irrigation, and using chemical amendments during critical periods. Another approach could be to ensure that the surface soil layers are dry during the land preparation and trafficking periods.

An alternative approach is to minimize the adverse effects of surface soil structural breakdown on effluent infiltration and crop growth by using practices such as gypsum-slotting (Jayawardane and Blackwell, 1985) in the bottom sections of the FILTER plots. The highly calcium-enriched gypsum slots could provide routing of excess surface waters collecting at the bottom section of the FILTER plots to the drains during high rainfall periods. This approach illustrates how the FILTER engineering design can be modified to use different parts of the FILTER plots to perform the pollutant adsorption and hydraulic flow functions, thereby increasing the efficiency of the system on any given area of land.

The optimum combination of soil improvement and stabilization practices and subsurface drainage design will depend on specific site conditions.

MAINTAINING CROP YIELDS AND NUTRIENT UPTAKE

Provision of subsurface drainage in the FILTER systems should enable maintenance of adequate root-zone aeration in slowly-permeable soils to obtain high crop yields. Maintaining high yields will result in high rates of nutrient removal, allowing for the maintenance of nutrient balance on limited urban land areas. The high crop yields will also maximize economic returns to offset capital and operational costs.

The total dry-matter yields from the fodder and grain harvests obtained during the first cropping season are given in Table 12.8. The nutrient uptake and removal from the plots in Experimental Block A were calculated from the dry-matter yields and their nutrient concentrations (Table 12.8). The total dry-matter yields for the millet plots 6 and 8 include grain yields of 2.0 and 2.1 t/ha., respectively. Nitrogen removal in all plots was greater than the net additions (Table 12.8), indicating a depletion of the native N stored in the soil before FILTER installation. In the case of P, 75–100% of the net P added in the effluent applied at the relatively high loading rates used in the current experiment was removed.

Table 12.8: Dry matter yields and nutrient removal in a harvested crop as compared to net additions in the effluent to plots 5-8, during the first cropping season

Factor	*Plot 5 Pasture*	*Plot 6 Millet*	*Plot 7 Pasture*	*Plot 8 Millet*
Total dry matter (t/ha.)	14.2	10.8	10.1	11.2
Total-N removed (kg/ha.)	207	128	131	118
Total-N applied (kg/ha.)	90	89	88	85
Total-P removed (kg/ha.)	47	33	34	34
Total-P applied (kg/ha.)	42	42	41	43

Table 12.9 shows the crop dry-matter yields obtained during the first to fourth cropping seasons, measured in sample harvest areas at the top and bottom ends of each plot. These dry matter yields were comparable to estimates of whole plot forage yields from forage harvesters and straw bale counts. Due to selection of crops which were moderately tolerant to waterlogging and salinity, yields of crops and nutrient removal rates comparable to those obtained in well-managed land treatment systems (US EPA, 1981; NSW EPA, 1993) were achieved during the four filtration seasons (Table 12.9 and 12.10). Due to the poor establishment of the barley crop in plots 6 and 8 of Experimental Block A during the fourth filtration season, crop yields and nutrient uptake were low (Tables 12.9 and 12.10) during this season.

Table 12.9: Crop dry matter yields obtained during the four cropping seasons

Cropping season	*Experimental block*	*Crop*	*Dry matter (t/ha.)*
1	A	Pasture	12.2
	A	Millet	11.0
2	A	Pasture	6.6
	B	Pasture	5.0
	B	Oats	7.1
3	A	Pasture	5.1
	A	Millet	15.5
	B	Pasture	4.7
	B	Sorghum/Maize	7.6
4	A	Pasture	10.2
	A	Barley	3.9
	B	Pasture	10.0
	B	Wheat	12.2

Table 12.10: Nutrient uptake by crops during the four cropping seasons

Cropping season	*Nutrient uptake (kg/ha.)*			
	Nitrogen		*Phosphorus*	
	Pasture	*Cereal*	*Pasture*	*Cereal*
1	169	123	40	33
2	185	187	45	54
3	188	225	44	67
4	150	80	28	17
Mean	**173**	**154**	**39**	**43**

Table 12.10 shows the removal of N and P from the FILTER plots by the crops grown during four filtration seasons. The addition of N and P in the effluent from the Griffith Sewage Works to a 120 ha. commercial FILTER system during a six-month cropping season is estimated to be around 75 and 35 kg/ha., respectively. Hence, the data in Table 12.10 shows that a nutrient-balanced FILTER system could be maintained at the Griffith Sewage Works site by using a suitable cropping regime and cropping area, if comparable nutrient removal rates are obtained in large commercial plots. If the effluent applied to the commercial FILTER plots have the same chemical composition as the effluent being currently discharged from the Griffith Sewage Works, nitrogen fertilizer will need to be added to maintain nitrogen balance and maximum crop yields.

POTENTIAL USE OF THE FILTER SYSTEM FOR TREATING OTHER WASTEWATERS

Farms in irrigation areas face problems in managing their pesticide-contaminated farm drainage runoff (Korth, 1995). FILTER has the potential to clean up such a drainage before they are discharged to streams. Trials involving pesticide-spiking of sewage effluent were conducted on the one-hectare FILTER trial plots at the Griffith sewage works. The purpose of the trials was to evaluate the ability of the FILTER system in order to remove the range of pesticides commonly used in the Murrumbidgee Irrigation Area as the effluent flows through the soil into the subsurface drainage system (F. Pisan, personal communication). Pesticide load reductions exceeding 98% were observed with chlorpyrifos, molinate, malathion, bensulfuron, diuron, bromacil, atrazine, metalochlor and endosulfan (Table 12.11), and reductions in concentrations exceeded 92%. Studies are currently in progress to evaluate the long-term fate of the pesticides adsorbed onto the FILTER soils.

Table 12.11: Pesticide concentration and load reduction

Pesticide	*Effluent concentration (ug/L)*	*Drainage water concentration (ug/L)*	*Concentration Reduction %*	*Load Reduction %*
Chlorpyrifos	0.5	<0.05	100	100
Molinate	50	1.64	98.3	98.4
Malathion	3.0	0.06	99.3	99.4
Bensulfuron	30	0.40	99.5	99.5
Diuron	30	0.38	99.3	99.9
Bromacil	30	0.54	99.1	99.9
Atrazine	40.1	3.25	91.4	99.2
Metolachlor	40.9	1.25	97.0	99.7
Endosulfan	0.7	0.07	94.6	99.5

A modification of the FILTER systems—referred to as the Sequential Biological Concentration (SBC) technique—could be used to manage the saline drainage effluent from irrigation areas in a way that could be both economically and environmentally sustainable (Blackwell et al., 1999). This involves the use of a series of FILTER systems to sequentially concentrate the applied salt, combined with the use of progressively more salt-tolerant crops to reduce the volume of saline water requiring disposal in evaporation ponds.

CONCLUSIONS

The FILTER technique was developed for around-the-year treatment of wastewater at the Griffith sewage works site to meet stringent EPA criteria for the discharge of treated wastewaters to sensitive inland surface waterbodies. The technique was field-tested on a saline-sodic clay soil with impeded drainage. Field data shows that the FILTER system could reduce the nutrient, pesticide and other pollutant loads and concentrations below EPA specifications. Adequate effluent flow rates to the subsurface drains were obtained over four cropping seasons, through management of FILTER plot trafficking and soil salinity-sodicity interactions to maintain structural stability in these swelling clay soils. Substantial crop yields were obtained, which will provide the economic returns and nutrient removal rates required to maintain a sustainable system.

The use of the FILTER system on a saline-sodic soil led to the amelioration of the site. However, the associated increases in salt concentration and loads in the subsurface drainage water will need to meet the EPA salinity limits for discharge to surface waters at the specific treatment site.

Thus, the FILTER technique could potentially provide a sustainable system for around-the-year wastewater treatment and reuse on heavy clay soils with impeded drainage, provided that good FILTER design and management is used. The FILTER design and management requirements will vary widely, according factors such as the type of wastewater EPA specifications, and the local wastewater treatment site conditions.

ACKNOWLEDGEMENTS

We acknowledge funding for these studies from ACIAR, Griffith City Council, NHT, DLWC and DPIE. We thank G. Nicoll and D. Wallett for assistance in the field work to establish and operate the FILTER plots.

REFERENCES

Asano, T. and Roberts, P.V., 1980, *Proceedings Symposium on Reuse for Groundwater Recharge,* September 1979, Kellogg West Center, California State Polytechnic University, Pomona, California.

Aulenbach, D.B. and Clesceri, N.L., 1979, *Monitoring for Land Application of Wastewater.* Fresh Water Institute, Lake George, Troy, New York. Report 79-3. 34 pp.

Aulenbach, D.B., Clesceri, N.L. and Tofflemire, J.I., 1975, Water renovation by discharge into deep natural sand filters. *Proceedings of the 2nd National Conference on Complete Water Reuse.* Fresh Water Institute, Lake George, Troy, New York. Report 74-20.

Baillod, C.R., Waters, R.G., Iskandar, I.K. and Uiga, A., 1977, Preliminary evaluation of 88 years rapid infiltration of raw municipal sewage at Cabinet, Michigan. In: Loehr, R.C. (Ed.), *Land as a Waste Management Alternative.* Ann Arbor Science, Ann Arbor, Michigan: 489–510.

Blackwell, J., Biswas, T.K., Jayawardane, N.S. and Townsend, J.T., 1999, A novel method for treating effluent from rural industries for use in integrated aquaculture system. *Proceedings—Wastewater Treatment and Integrated Aquaculture Production.* 17–19 September. SARDI, PO Box 120, Henley Beach SA 5022, Australia. 11 pp.

Bouwer, H.R., Lance, J.C. and Riggs, M.S., 1974, High rate land treatment. II. Water quality and economic aspects of the Flushing Meadows Project. *Journal of the Water Pollution Control Federation* 46: 845–859.

Falkiner, R.A. and Smith, C.J., 1997, Changes in soil chemistry in effluent irrigated *Pinus radiata* and *Eucalyptus grandis* plantations. *Australian Journal of Soil Research* 35: 131–147.

Hooghoudt, S.B., 1940, Bijdrage tot de kennis van enige natuurkundige grootheden van de grond. *Verslagen van Landbouwkundige Onderzoekingen* 46(7): 515–707.

Iskandar, J.K., 1978, The effect of wastewater reuse in cold regions on land treatment systems. *Journal of Environmental Quality* 7: 361–368.

Iskander, J.K., 1981, *Modeling Wastewater Renovation.* U.S. Army Cold Regions Research and Engineering Laboratory. John Wiley and Sons, New York, USA.

Iskandar, J.K., Sletten, R.S., Leggett, D.C. and Jenkins, T.F., 1976, Wastewater Renovation by a Prototype Slow Infiltration Land Treatment System. CRREL Report 76-19, 44 pp.

Iskandar, J.K., Murrmann R.P. and Leggett, D.C., 1977, Evaluation of Existing Systems for Land Treatment of Wastewater at Manteca, California and Quincy, Washington. CRREL Report 77-24, 34 pp.

Jayawardane, N.S., 1977, The effect of salt composition of groundwaters on the rate of salinisation of soils from a watertable. PhD. Thesis. University of Tasmania, Hobart, Australia.

Jayawardane, N.S., 1979, An equivalent salt solutions method for predicting hydraulic conductivities of soils for different salt solutions. *Australian Journal of Soil Research* 17: 423–128.

Jayawardane, N.S., 1983, Further examination of the equivalent salt solution method for predicting hydraulic conductivity of soils for different salt solutions. *Australian Journal of Soil Research* 21: 105–108.

Jayawardane, N.S., 1992, Predicting unsaturated hydraulic conductivity changes of a loamy soil in different salt solutions using the equivalent salt solutions concept. *Australian Journal of Soil Research* 30: 565–571.

Jayawardane, N.S., 1995, Wastewater treatment and reuse through irrigation, with special reference to the Murray Darling Basin and adjacent coastal areas. CSIRO, Division of Water Resources, Griffith NSW, Divisional Report 95.1.

Jayawardane, N.S. and Beattie, J.A., 1979, Effect of salt solution composition on moisture release curves of soils. *Australian Journal of Soil Research* 17: 89–99.

Jayawardane, N.S. and Blackwell, J., 1985, The effects of gypsum enriched slots on moisture movement and aeration in an irrigated swelling clay soil. *Australian Journal of Soil Research* 23: 481–492.

Jayawardane, N.S and Blackwell, P.S., 1991, The relationship between equivalent salt solution series of different soils. *Journal of Soil Science* 42: 95–102.

Jayawardane, N.S., Blackwell, J., Cook, C.J., Nicoll, G. and Wallett, D.J., 1997a, The research project on land treatment of effluent from the Griffith City Council sewage work—Report 7. Final report on pollutant removal by the FILTER system, during the period from November 1994 to November 1996. CSIRO, Division of Water Resources—Consultancy Report No. 97–40.

Jayawardane, N.S., Cook, F.J., Ticehurst, J., Blackwell, J., Nicoll G. and Wallett, D.J., 1997b, The research project on land treatment of effluent from the Griffith City Council sewage work—Report 8. Final report on the hydraulic properties of the FILTER system, during the period from November 1994 to November 1996. CSIRO, Division of Water Resources—Consultancy Report No. 97–42.

Jayawardane, N.S., Biswas, T.K., Blackwell, J. and Cook, F.J., 2001, Management of salinity and sodicity in a land FILTER system, for treating saline wastewater on a saline-sodic soil. *Australian Journal of Soil Research* 39: 1247–1258.

Jenkins, T.F. and Martel, C.J., 1979, Pilot scale study of overland flow land treatment in cold climates. *Proceedings of Water Technology* 11: 207–214.

Jenkins, T.F., Martel, C.J., Gaskin, D.A., Fish, D.J. and McKim, H.L., 1978, Performance of overland flow land treatment in cold climates. In: *State of Knowledge in Land Treatment of Wastewater. Proceedings of an International Symposium.* Hanover, New Hampshire, August, 1978, Vol 2: 61–70.

Korth, W., 1995, Can pesticides used in irrigated agriculture get into drainage water? *Farmers Newsletter* No. 145, June 1995.

Martel, C.J., Jenkins, T.F. and Palazzo, A.J., 1980, Wastewater treatment in cold regions by overland flow. U.S. Army Cold Regions Research and Engineering Laboratory. CRREL Report 80-7, Hanover, New Hampshire.

Melbourne Water, c1997, *Wastewater Treatment*. Information booklet. Melbourne Water Corporation.

McNeal, B.L. and Coleman, N.T., 1966, Effect of solution composition on soil hydraulic conductivity. *Soil Science Society of America Proceedings* 30: 308–312.

McNeal, B.L., Norvell, W.A. and Coleman, N.T., 1966, Effect of solution composition on the swelling of extracted soil clays. *Soil Science Society of America Proceedings* 30: 313–317.

NSW EPA., 1993, *Guidelines for Utilisation of Treated Sewage Effluent by Irrigation*. May 1993. NSW Environment Protection Authority. 76 pp.

Pescod, M.D., 1992, Wastewater treatment and use in agriculture. FAO irrigation and drainage paper 47.

Quirk, J.P. and Schofield, R.K., 1955, The effect of electrolyte concentration on soil permeability. *Journal of Soil Science* 6: 163–178.

Rengasamy, P., Greene, R.S.B., Ford, G.W. and Mehanni, A.H., 1984, Identification of dispersive behaviour and management of red-brown earths. *Australian Journal of Soil Research* 22: 413–431.

Satterwhite, M.B., Stewart, G.L., Condike, B.J. and Vlach, E., 1976, Rapid Infiltration of Primary Sewage Effluent at Fort Devens, Massachusetts. CRREL Report 76-48. 48 pp.

SCM [Sinclair Knight, Consulting Environmental Engineers, Montgomery Consultants] 1992. Draft environment improvement plan, Lower Molonglo Water Quality Centre, Canberra, ACT, Australia. 35 pp.

Smith, C.J., Snow, V.O., Bond, W.J. and Falkiner, R.A. 1996. Salt dynamics in effluent irrigated soil. In: Land application of wastes in Australia and New Zealand: Research and Practice. In: Polglase P.J. and Tunningley, W.M. (eds). *New Zealand Land Treatment Collective—Proceedings of Technical Session 14 Australian Conference*, Canberra. September 1996: 175–180.

US EPA., 1981, Process Design Manual for Land Treatment of Municipal Wastewater. October 1981, U.S. EPA, U.S. Army Corps of Engineers, U.S. Dept. of Interior and U.S. Dept. of Agriculture.

Youngs, E.G., 1985, A simple drainage equation for predicting watertable drawdowns. *Journal of Agricultural Engineering Research* 31: 321–328.

13

Soil Physical Processes in Capping Systems of Landfills—Possibilities of Capillary Break Capping

T. Baumgartl[1], *R. Horn* and *B.G. Richards*[2]

INTRODUCTION

Increasing amounts of waste and the growing need for public awareness of possible pollution caused by landfills make it necessary to develop strategies to protect the environment from waste and its components. Several techniques and systems have been developed, which use either natural substances or technical material (fibre), or both, to achieve a high degree of security. The two main factors which must be considered when using these materials for lining landfills are the limitations of mechanical stability and the hydrological conductivity of fluids. Despite strict regulations regarding these materials, a number of cases have been reported which describe the problem of environmental pollution despite compliance within certain limits. Furthermore, the costs of planning and construction are often a point of discussion which may influence the standard of security. A major question that has to be dealt with is, therefore, not only the properties of capping systems but also their sustainability. This chapter summarizes the various aspects which must be considered when using natural substances or fibres for securing landfills, and proposes the capillary capping system as a possible design for long-term efficiency.

PROPERTIES OF LANDFILL CAPS

Certain requirements for the capping of a landfill must be met to prevent any environmental hazard. The following list summarizes the legal regulations of different countries concerning the properties and necessary requirements of landfill encapsulations.

[1] University of Kiel, Institute of Plant Nutrition and Soil Science, Olshausenstrasse 40, 24118 Kiel

[2] Geotech Research Pty Ltd, 1 Fleming Rd., Chapel Hill, QLD, 4064 Australia.

Germany

According to German Law, the requirements for waste material deposition are defined by:

- technical Instructions for waste TA-Abfall (1991). (This law defines the deposition for problematic waste such as heavy metal, organic waste, and xenobiotics);
- TA-Siedlungsabfall, (1993). (This law describes the requirements for municipal waste); and
- The requirements for the application of plastic (HDPE) coverage systems, as well as the application of asphalt or concrete, which are also regulated by supplementary laws.

The main components are listed below.

Bottom sealing system

This requires:

- drainage system: thickness 0.3 m, gravel layer 3 dm (diameter: 16–32 mm), $k_{sat} > 10^{-3}$ m/s, drainage pipes.
- mineral layer: 3 × 25 cm thickness, $k_{sat} < 5*10^{-10}$ m/s, proctor density, at 5% above optimal water content, clay >20%, clay minerals >10%, organic carbon <5%, $CaCO_3$<15%, air-filled pores <5%.
- equilibration layer: upto 3 dm thick prepared by ingredients such as gravel and coarse sand.

Capping system

This requires:

- mineral layer: thickness 5 dm = 2*2.5 dm (compacted to proctor density, $k_{sat} <5*10^{-9}$ m/s) clay >20%, clay minerals >10%, organic carbon <5%, $CaCO_3$<15%, air-filled pores <5% (sometimes plastic liner (HDPE): thickness 2.5 mm, completely dense, controlling system is not sufficient).
- drainage system: layer thickness: 3 dm; $k_{sat} >5*10^{-3}$ m/s.
- rehabilitation layer: at least 1 m thickness with a plant available water content of 200–250 mm, top 3 dm enriched in organic carbon.

Austria

Guidelines exist for the various waste landfill classes—industrial waste, municipal waste and organic waste. Limits are defined by parameters such as eluate concentration via factors such as CSB and heavy metals.

Bottom-sealing system

No filter layer is necessary, no geotextile, HDPE (thickness >2 mm), mineral layer: thickness 6 dm, k_{sat} $<10^{-9}$ m/s, natural soils are recommended.

Capping system

Drainage layer, mineral layer: 6 dm (compacted in 3 layers), no definition of k_{sat} values.

Denmark

No definitions are given for fly ash or inert material deposition.

Bottom-sealing system

This requires:

- drainage layer: thickness 3 dm, gravel and sand layer, drainage pipe distance: < 20 m.
- protection layer: geotextile, HDPE sheet.
- mineral layer: 5 dm thickness, compacted in 2 layers, k_{sat} $<10^{-10}$ m/s, bentonite addition recommended; only when clay is present in the deeper subsoil, can bentonite be excluded. Sometimes, the HDPE plastic sheet is also accepted as a complete sealing system.

Capping system

Apart from the drainage layer and HDPE layer, the mineral layer is primarily the recultivation layer. It should exceed a thickness of 17 dm.

USA

For state of knowledge, see EPA (1987, 1989, 1991). The criteria are nearly identical to those of Germany. However, more definitions are given for hazardous waste landfills.

Bottom-sealing system

Filter layer: geotextile application in addition to a mineral drainage filter.

- material: thickness 1.5 dm, leachate collection and removal systems
- protection layer: no geotextile, PEHD layer, bentonite layer (<2 mm), secondary leachate collection and removal system
- mineral layer: 9 dm thickness compacted into 6 layers to proctor density, thickness 9 dm, k_{sat} $<10^{-9}$ m/s.

Capping system

This requires:

- recultivation layer: thickness 6 dm, on top of a gravel layer: 3 dm
- filter layer: thickness 1 dm, geotextile species
- drainage system: thickness 3 dm, k_{sat}: 10^{-4} m/s, low density plastic material
- mineral layer: 6 dm thick, $k_{sat} < 10^{-9}$ m/s
- filter layer (2nd): geotextile layer
- equilibration layer: gravel with a depth of 3 dm

Japan

There are three types of waste deposit systems:

- Isolated-type landfill
- controlled-type landfill
- stable-type landfill for inert materials such as glass and combustion ashes

HYDRO-MECHANICAL PROCESSES IN SOIL SUBSTANCES USED FOR CAPPING

The sealing property of an encapsulation is mainly based on its physical aspects. In particular, good sealing requires low hydraulic conductivity. Low permeabilities through soil substrates are usually achieved by maximum compaction. As a guiding value, the Proctor-density is used for this purpose. It describes the maximum possible compaction, which depends for an applied amount of energy on the water content. The procedure of compaction, which is a result of shear processes, can be mathematically described by the characteristic of the Mohr-Coulomb-failure line. Stress transmission in soils is explained by different phases like the solid, liquid or gaseous phase, or a combination of all three.

The total stress equation for such a three-phase system (unsaturated soil) is described, according to Bishop and Blight (1963) by:

$$\sigma' = (\sigma - u_a) + \chi (u_a - u_w) \tag{1}$$

where

σ' = effective stress

σ = total stress

u_a = gas pressure

u_w = pore water pressure (= matric water potential under unsaturated conditions)

χ = factor which characterizes the amount of water-filled pores;

$\chi = 0$ for completely dry soils

$\chi = 1$ for saturated soils

In a two-phase system (solid-liquid), $\chi = 1$ and equation 1 can be reduced to the well-known Terzaghi equation:

$$\sigma' = \sigma - u_w \tag{2}$$

This mathematical theory clearly explains the important contribution of water, when the transmission of stress is considered.

With respect to friction forces between particles being the essential cause of stability, water is again a main factor. The angle of internal friction that describes the shearing property of a material is almost zero for liquids. The Mohr-Coulomb failure line at saturation relates to the frictional stability (shear stress) and total stress σ_n by:

$$\tau = c' + \sigma_n \tan\varphi' \tag{3}$$

where

τ = shear stress
c' = cohesion
σ_n = total stress
$\tan\varphi'$ = angle of internal friction.

For unsaturated conditions, equation 3 has to be extended by a term, which includes the effect of capillary forces of water between solids (see also Fredlund and Rahardjo (1993)). This term can be reduced to the factor tensile stress, which then gives:

$$\tau = c' + \sigma_\tau \tan\varphi' + \sigma_n \tan\varphi' \tag{4}$$

where

$\sigma_\tau = \chi\, u_w$ = tensile stress.

As the water phase is an important part of the equation, it is clear that any change in it affects shear stress. The water phase is able to stabilize solid particles by water meniscii forces between solids. The controlling parameter is the energy status of the water in the corresponding water-filled pore class, and hence, is a function of the water potential. Water has a destabilizing effect. However, when the soil is either saturated (or in the case of an unsaturated condition, when stresses cause a narrowing of particles) reduction of pore volume and a short-term change of the curvature of the water meniscii to the air interface lead to positive water potentials. This resembles in its consequence the conditions of a saturated soil.

In unsaturated water conditions, stresses caused by the matric water potential are defined as tensile stresses. They increase with decreasing matric water potential. Since the χ-factor decreases at the same time, the product of the χ-factor and matric water potential depends on the pore size distribution of the soil material. This, in turn, is influenced by the texture and degree of compaction.

Tensile stresses not only increase the stability between particles, but can also cause contraction of soil volume; for example when drying causes reduction of the water volume between pores. This leads to movement of particles. If the tensile stress exceeds the maximum tensile strength of the soil material, the solids detach from each other at the weakest point and cracking begins. The soil material will then not fulfil its purpose of sealing and protection any longer. Cracks will always be present even after an event of re-wetting and swelling of the soil (Nagel et al., 1995) as long as no dynamic energy is introduced into the aggregated soil and no remoulding of the soil takes place. Each drying is accompanied with a decrease of the matric water potential, which causes the particles to move into positions with a minimum inner energy that increases the stability by a higher coordination number (contact points of particles) and friction forces. Above all, in clay-rich substrates which have high swelling and shrinking potential, each cycle of desiccation causes an increase in the bulk density (decrease of void ratio = volume of pores/volume of solids). Eventually, cracking occurs, which is noticeable by an increase of the hydraulic conductivity in the saturated condition after re-wetting, even under constantly-applied stress (Figure 13.1).

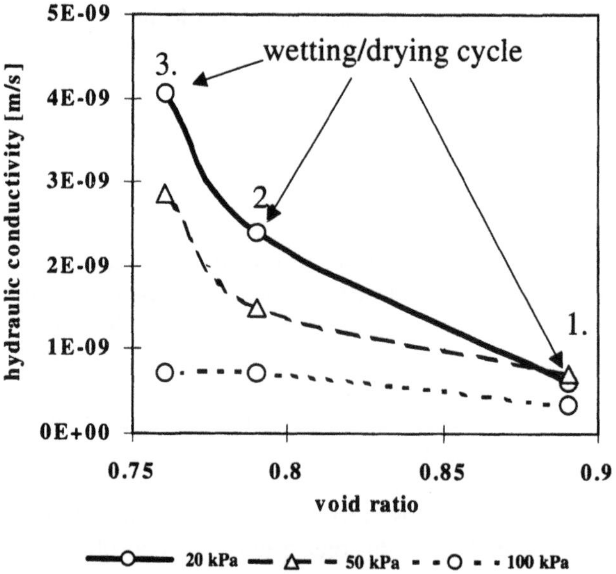

Fig. 13.1: Increase of the saturated hydraulic conductivity under constant stress with three drying and wetting cycles (after Nagel et al., 1995, 1996).

Shrinkage can occur even in highly compacted soil. It can be explained by the similar behaviour of a change in void ratio due to mechanical or hydrological stresses. Figure 13.2a illustrates this conceptual model (after Toll (1995)). The stress-strain relationship is well known, describing the decrease of void ratio with increasing applied mechanical stresses. A semilogarithmic plot ($e = f(\log\sigma_n)$) results in a straight line for a non-consolidated material. If a soil has already been exposed to stresses, the previous straight line is split into two parts. These two parts are named pre-compaction lines and virgin compaction lines, and are differentiated

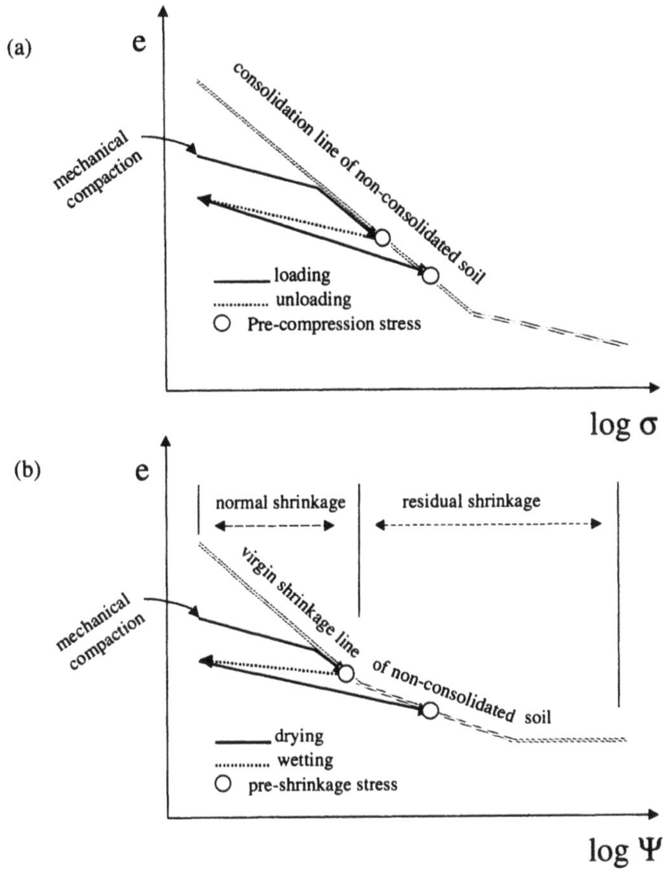

Fig. 13.2: Stress-strain relationship caused by mechanical stress (a) and hydrological stress (b) (Toll, 1995)

by the extent of pre-compaction stress. The same pattern can be found when soil material is undergoing hydrologically-caused tensile stresses when the soil is dried (Figure 13.2b). The first drying of a virgin material results in a straight line in the *e*-log(σ_n)-plot. Note, that σ_n and the matric water potential are both stresses that cause volume change. In reference to drying soils, the behaviour of the soil along this straight line is defined as normal shrinkage. If soil material has been exposed to hydrological or mechanical stresses, or both, in an earlier stage, a pre-shrinkage line can be defined, which is sometimes described as structural (Yong and Warkentin, 1975).

External mechanical stress only exerts pressure on the surface of a soil volume. Particles are moved when the maximum shear strength is exceeded. When unloading, the soil may rebound due to the elastic behaviour of the material. In the stress-strain-relationship, it will be the same loading and unloading line. Internal hydrological stress (tensile stress) caused by water menisci can not only move particles, but also reorientate them according to their physico-chemical attractions. This enables a higher packing density of particles. Each drying event can, therefore, lead to lower void ratios even when a former drying status (matric water potential) has not been exceeded.

Each volume change caused by wetting and drying or mechanical loading and unloading is due to a change in the pore volume and pore-size distribution. This relationship can be expressed as a dynamic water-retention curve, based on either mechanical stress (Figure 13.3) or hydrological stress. If the water flow through such soils is to be modelled, this change in void ratio and pore class has to be included. The variance depends on texture, and is most pronounced in clay-rich substrates. However, Gräsle (1999) found considerable deviations between neglected and considered void ratio dynamics on the water retention curve, even in sandy material.

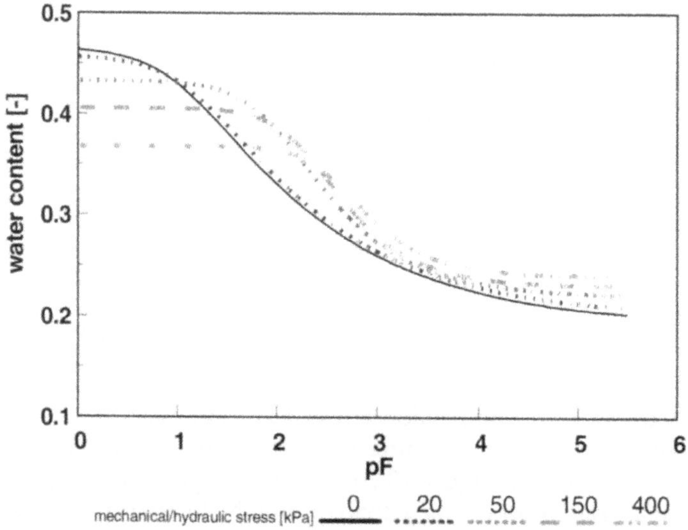

Fig. 13.3: Stress-dependent water retention curve of clay-enriched loess material (Hapludalf, derived from Loess); original bulk density = 1.35 g/cm^3

The above-mentioned mechanisms should be considered when using mineral substrates for sealing purposes. If the mechanical properties are known (stress-strain relationship; tensile strength), a system can be designed in which the risk of cracking is reduced. When capping a landfill, the main factor is climate, which controls how thick a reclaimed layer of soil on top of the sealing substrate has to be to avoid cracking due to drying. A study of 112 sites in Germany on the intensity of drying as a function of depth found that under moderate climatic conditions, the matric water potential did not significantly drop below a matric water potential of –40 kPa in depths of more than 1m (Baumgartl et al., 1998). Lilienfein et al. (1999) found even in Savanna Oxisols in Brazil, minimum matric water potentials were not less than –30 to –60 kPa at a depth range of 0.8–2.0 m. The variation depends on the usage of the soil. Combining a knowledge of the climate and hydromechanical behaviour of a substance allows for the design of a system in which the risk of cracking in a capping system is reduced.

CAPILLARY BREAK CAPPING

One possible way to reduce the risk of cracking is to design a landfill cap using materials that are not sensitive to tensile stresses and can, therefore, have a sealing effect because of their physical characteristics.

The principle of a capillary barrier is based on the characteristics of water flow in unsaturated soil substrate. Although such a material usually has a high saturated hydraulic conductivity, the resistance to water flow is sharply increased when the soil is non-saturated. This can be seen in the pore-size distribution, which is characteristic for these kinds of substances by a high percentage of coarse pores and a very small water-holding capacity. Therefore, very few pores are water-filled, which contributes to water flow once the soil is unsaturated. Combining two layers with different characteristics—one with coarse pores below a finer textured substance—water flow can be prevented under unsaturated conditions. In that context, the upper layer is called the capillary layer and the lower layer the capillary block: the combination is termed a capillary barrier. To maintain non-saturated conditions, the capillary barrier cannot be used in horizontal terrain, but only on slopes with a specific inclination. This is necessary to discharge water from the capillary layer along the 'break' surface or interface between the capillary block and capillary layer. Doing so prevents water saturation and avoids a breakthrough of water into the capillary block. For the capillary layer, an optimum texture has to be chosen which has a high hydraulic conductivity for both saturated and unsaturated conditions. At the same time, it should only have a small (or no) window of overlapping pores to the capillary block in order to avoid contact for water-flow into the capillary block. Several critical hydraulic parameters can be defined (Figure 13.4). The unsaturated hydraulic conductivity has a minimum (together with an equivalent suction = matric water potential), below which for a certain cap design (mainly slope length and inclination), the water flow will be too slow. This will cause accumulation of water until saturation or quasi-saturation, and eventually, a breakthrough of water into the capillary

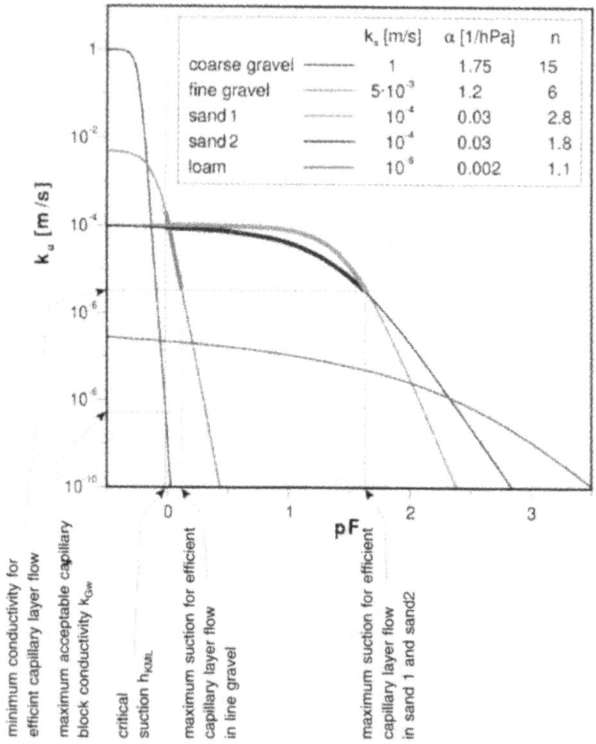

Fig. 13.4: Requirements of water retention curves of a capillary layer and capillary block length (after Gräsle and Horn, 1998)

block occurs. In addition, a critical suction has to be defined, which determines the maximum acceptable conductivity in the capillary block. The difference between the hydraulic conductivity in the capillary block and capillary layer should be as high and distinct as possible, subject to ensuring that a wide range of suctions with sufficient unsaturated hydraulic conductivity has to be provided in the capillary layer.

The slope of the capillary barrier determines how long a section can be, based on a risk scenario that no saturated conditions will appear in the capillary layer to avoid a failure of the system. If such a critical length is reached, drainage must be provided to discharge water from the capillary layer. This maximum length is described as the critical length (Figure 13.5). The modelling of risk scenarios mainly includes boundary conditions such as climate and vegetation. With this information, the infiltration rate and total infiltration can be calculated based on the knowledge of hydraulic properties (mainly saturated/unsaturated conductivity and field capacity) and dimension of the substance used for the capillary barrier. Models using the Finite Element technique (for

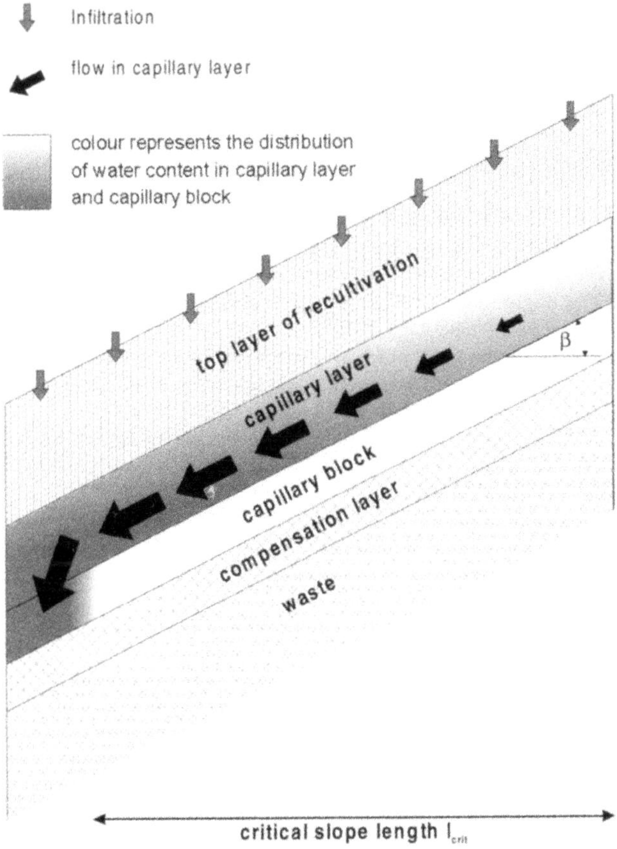

Fig. 13.5: Design of a capillary break capping; definition of critical slope length (after Gräsle and Horn, 1998)

example, Gräsle, 1999; Gräsle and Horn, 1998; Gräsle et al., 1995; Richards et al., 1997) are very suitable for the calculation of such relationships. With this knowledge, appropriate capping can be designed.

CONCLUSION

Capillary break capping has several important benefits. First, the use of substances with specific properties provides the chance to create a system which should be sustainable. The capillary break is able to follow any volume change within the landfill to settlement caused by gravity and decomposition of waste material. Irregularities in the settlement can be evened, whereas a geotextile liner might be exposed to high stresses with the risk of tearing. Furthermore, certain kinds of waste need small amounts of water to maintain microbial activity and subsequent decomposition of waste to reduce the contaminant's activity under long-term considerations. Last, but not the least, capillary break capping is more cost-effective than a system which combines the soil substance with geotextiles.

REFERENCES

Baumgartl, T., Taubner, H. and Horn, R., 1998, Matric water potentials in soils of different usage as a mean for the prognosis of crack formation of mineral substrates [German]. *Journal of Rural Engineering and Development*, 39: 215–219.

Bishop, A.W. and Blight, G.E., 1963, Some aspects of effective stress in saturated and partly saturated soils. *Geotechnique* 13: 177–197.

EPA, 1987, Minimum technology guidance on double liner systems for landfills and surface impoundments - design, construction and operation. EPA/530-SW-85-014, US EPA, Office of Solid Waste, Washington (D.C), 71pp.

EPA, 1989, Technical guidance document: Final covers on hazardous waste landfills and surface impoundments. EPA/530-SW-89-047, US EPA Risk Reduction Engineering Laboratory, Cincinnati (OH) 45268.

EPA, 1991, Design and construction of RCRA/CERCLA final covers (by Eastern Research Group, Inc., Arlington, MA). Seminar publication, EPA/625/4-91/025, US EPA, Center for Environmental Research Information, 26 West Martin Luther King Drive, Cincinnati (OH) 45268.

Fredlund, D.G. and Rahardjo, H., 1993, *Soil Mechanics for Unsaturated Soils*. Wiley, New York.

Gräsle, W. and Horn, R., 1998, Application of numerical simulations in the design of a capillary barrier as a capping system for the waste deposit 'Blockland' (Bremen, Germany) [German] *Journal of Rural Engineering and Development* 39(5): 199–205.

Gräsle, W., 1999, Numerische Simulation mechanischer, hydraulischer und gekoppelter Prozesse in Boden unter Verwendung der Finite Elemente Methode. Schriftenreihe des Inst. f. Pflanzenern. u. Bodenkunde, University of Kiel. Thesis, 400pp. ISSN 0933 680x.

Gräsle, W., Richards, B.G., Baumgartl, T. and Horn, R., 1995, Interaction between soil mechanical properties of structured soils and hydraulic processes—theoretical fundamentals of a model. *Proceedings of the First International Conference on Unsaturated Soils.* Balkema, Rotterdam/Brookfield: 719–725.

Lilienfein, J., Wilcke., W., Ayarza, M.A., Carmo Lima, S. do, Vilela, L. and Zedi, W., 1999, Annual course of matric potential in differently used Savanna Oxisols in Brazil. *Soil Science Society of America Journal* 63: 1778–1785.

Nagel, A., Baumgartl, T. and Horn, R., 1995, Change of the hydraulic conductivity in mineral base liners as a consequence of drainage cycles [German]. *Journal of Rural Engineering and Development* 37: 9–14.

Richards, B.G., Baumgartl, T., Horn, R. and Gräsle, W., 1997, Modelling the effects of repeated wheel loads on soil profiles. *International Agrophysics* 11; 177–187.

TA–Abfall, 1991, Gesamtfassung der zweiten Allgemeinen Verwaltungsvorschrift zum Abfallgesetz., 1. Auflage. Verlag Franz Rehm, München.

TA–Siedlungsabfall, 1993, Dritte Allgemeine Verwaltungsvorschrift zum Abfallgesetz. Technische Anleitung zur Vermeidung, Verwertung, Behandlung und sonstigen Entsorgung von Siedlungsabfällen, Bundesratsdrucksache 594/92.

Toll, D.G., 1995, A conceptual model for the drying and wetting of soil. In: Alonso. E.E. and Delage, P. (Eds), *First International Conference on Unsaturated Soils*. Balkema, Rotterdam, Paris/France: 805–810.

Yong, R.N. and Warkentin, B.P., 1975, *Soil Properties and Behaviour*. Elsevier, Amsterdam.

14

Waste Management in Central Queensland

P.F. Magnussen[1], *Nanjappa Ashwath*[1] and *C. Dunglison*[2]

INTRODUCTION

Waste management is now a widely accepted term, encompassing all the measures associated with waste avoidance, safe treatment, resource recovery, re-use, and waste disposal, while at the same time considering the environmental and economic factors (Bilitewski et al., 1997). Over the past several decades, many developed countries have become increasingly focused on the major issues relating to waste management: how and where to dispose of the waste generated by a modern society. This focus has fostered the concept of a revised waste hierarchy, a system in which emphasis is placed on waste avoidance rather than on waste disposal (Bilitewski et al., 1997).

In Australia, State and Territory Governments are primarily responsible for the instigation of policy and regulatory frameworks concerning waste-reduction programmes and their associated systems, as well as ensuring appropriate containment, treatment and disposal of hazardous wastes (Environment Australia, 1997). In Queensland, land disposal of wastes, as well as any resulting contamination, are dealt with under the following legislation DOEH 1996: the *Environmental Protection Act 1994;* the *Radioactive Substances Act 1958;* and the *Contaminated Land Act 1991*. Landfill operations are under the additional control of the Queensland Workplace Health and Safety Act (Sinclair Knight Merz, 1999).

As with other more developed economies (such as the United States, Europe and Japan), Australia is increasingly focusing on waste minimization and recycling (Krol et al., 1994). This is being achieved through the adoption of statewide approaches such as the Queensland Waste Management Strategy. This Strategy provides a framework within

[1] School of Biological and Environmental Sciences, Central Queensland, University, Rockhampton, Queensland 4702
[2] Rockhampton City Council, Rockhampton, Queensland 4700.

which efficient management of waste can operate to minimize or avoid adverse impacts on the environment (DOEH, 1996). It also promotes the concept of a waste management hierarchy—emphasizing prevention, rather than disposal—as well as consistent, preventative legislation to minimize the needless generation of wastes (DOEH, 1996).

Secondary objectives, which complement the primary objective of minimising harmful environmental effects, are:

- development of increased efficiency and conservation with respect to resource use, especially non-renewable resources;
- reduction in reliance on waste-treatment facilities;
- reduction in regulatory costs;
- creation of greater cost efficiency (e.g. reduced disposal costs) within the industry through reduction in the volume of raw materials; and
- consistency as well as harmony with national and international directives.

The strategy consists of two essential components:

- minimization, which through preventative and recycling options, will attempt to reduce the volumes and toxicity of wastes destined for destruction or disposal; and
- containment, the primary aim of which is the minimization of effects to the environment and public health through the entire handling, treatment and disposal stages.

Local and regional cooperation is, in turn, encouraged so that statewide waste management can be coordinated in an effective manner. Local Government Waste Management Groups (LGWMG) represent urban and rural local governments and regional groups; they initiate and coordinate waste management schemes for the collection, recycling, treatment and disposal of wastes (DOEH, 1996). Central Queensland Regional Organisation of Councils (CQ-ROC) is an advisory body, formed in 1996, consisting of seven local councils. Its aim is to identify areas such as waste management which might require a cooperative regional approach (Sinclair Knight Merz, 1999). Central Queensland Regroal Organisation of Councils (CQ-ROC) commissioned the collating of information on waste management in the region through a solid waste planning study (Stage 1 Waste Management Strategy), which was undertaken by Sinclair Knight Merz Pty Ltd.

This chapter provides an overview of solid waste management in Central Queensland, and an indication of whether the region is fulfilling its obligations under the Queensland Waste Management Strategy. This approach will be compared briefly with the United States (because it is a larger developed country also facing the problem of shrinking landfill capacity) with a view to providing some insight into the reasoning behind integrated waste management strategies.

CURRENT STATUS OF WASTE MANAGEMENT

The United States

Recent prioritization of waste treatment techniques by the US Environmental Protection Agency (EPA) has resulted in the current waste hierarchy being directed more towards changing attitudes at household, community and state levels rather than toward engineering solutions and geological parameters (Moon, 1994). This revised waste hierarchy is similar to that adopted by the 1996 Waste Management Strategy for Queensland. Its structure of this hierarchy is:

1. reuse;
2. waste reduction;
3. waste recycling;
4. energy recovery from waste;
5. incineration (e.g. the US); and
6. waste disposal (e.g. landfilling).

Approximately, 5300 municipal solid-waste landfills exist in the US. Due to increasingly stringent regulations demanding high standards of site-lining, monitoring of gas and leachate, and post-closure liability, higher costs have further dissuaded landfilling (Williams, 1998). The result has been a trend toward fewer and larger landfills, often located far from the source of waste production (Williams, 1998). According to Moon (1994), however, the movement of regulatory power to the states and then on to the districts is likely to result in the opening of more landfills. The reason for this potential reversal of the current trend might be high transportation costs and relative proximity of a county or district to a neighbouring jurisdiction which lacks waste disposal or storage facilities (Moon, 1994). The solid waste manager of the county or district mentioned above may then decide to take on the responsibility for waste generated outside their own borders (Moon, 1994).

Central Queensland

Stage 1 of the Waste Management Strategy (WMS) for Central Queensland has focused on collating the region's existing waste management information (Sinclair Knight Merz, 1999) to:

- determine the current status of existing waste disposal operations throughout the member councils;
- assess the potential of individual councils to meet future environmental requirements; and
- evaluate future options such as new regional landfills and transfer stations, as well as the potential for shared facilities by member councils.

Most importantly, the Stage 1 study identified a commitment by the member councils for improving the existing waste collection and disposal facilities, thereby minimizing potential environmental harm associated with waste disposal. The gathering of information was achieved through meetings with environmental representatives from each shire council, through inspections of landfill facilities and through community consultation. The key issues identified in Stage 1 of the WMS for Central Queensland (Sinclair Knight Merz, 1999) were:

- waste diversion options (e.g. only two out of seven councils found it viable to provide kerbside recycling pickup services) (Table 14.1);
- stockpiling of some regulated wastes, such as tyres and batteries) due to distances from disposal or treatment facilities; and
- existence of many small unmanned landfills in the region.

Table 14.1: Status of waste management systems amongst CQ-ROC member councils

Local authority	*Current waste facilities*	*Current waste collection systems*	*Waste statistics available*	*Future known landfill sites*
Duaringa Shire Council	Four landfills (one major landfill)	240 L bin system. Contract. No collection of recyclable items	Minimal	Nil
Emerald Shire Council	Six landfills, three transfer stations	240 L bin system. Contract. No collection of recyclable items	Medium level	Nil
Fitzroy Shire Council	Four landfills, 7 transfer stations	240 L bin system. Contract. No collection of recyclable items	Medium level	Two
Livingstone Shire Council	One landfill, 6 transfer stations	240 L bin system. Contract. No kerbside collection of recyclable items	Minimal	Nil
Mount Morgan Shire Council	One landfill	240 L bin system. No collection of recyclable items	Minimal	One
Peak Downs Shire Council	Two landfills, one transfer station	240 L bin system. Contract. Kerbside collection of recyclable items	Minimal	Nil
Rockhampton City Council	One major landfill, one Materials Recovery Facility	240 L bin system and a bag recycling system. Day labour	High	Nil

Source: Adapted from CQ-ROC, 1998.

Kerbside recycling services provided by Rockhampton City Council (RCC, 1999) had, until March 1999, contributed to a 60% reduction in the landfill waste stream. The main reason for the success of the Council's recycling programme was the provision of part of the labour force by the local Corrective Services facility (Sinclair Knight Merz, 1999). However, the cost of recycling in Rockhampton—about $400,000 per annum—is still believed by RCC to be an economical alternative to the establishment of a new landfill (Sinclair Knight Merz, 1999). Upon consideration of the external costs of landfilling ($13.10–$33.20 per tonne in Sydney) and the potential for future requirements which may see regulations requiring the stabilisation of waste prior to landfilling (possibly $50–150 per tonne), NSW EPA deemed the real costs of recycling to be significantly less than faced value (Butler, 2000).

Although used tyres collected by RCC are sent to Townsville for recycling (Rutherford, 1997), many tyres are believed to be disposed of via landfill within the region (Sinclair Knight Merz, 1999), thereby contributing to a reduction in available landfill volume. Future initiatives likely to assist in addressing this and other waste management concerns are discussed later in this chapter.

According to the Queensland EPA (QEPA, 1999) past problems with Queensland's management of regulated and hazardous wastes have been due to a lack of integration, a lack of adequate facilities for the disposal of industrial waste, and a general inconsistency in the standard of municipal landfills. The small unmanned landfills throughout the region are of the trench-type and many have been operating without surface water or groundwater monitoring (Table 14.2) as well as without knowledge of the depth to groundwater (Sinclair Knight Merz, 1999). Adequate knowledge of the depth to groundwater remains the main criterion for choosing either the open or trench-type waste disposal options (Davis and Cornwell, 1998). The above assessment may, therefore, be partly justified where Central Queensland is concerned. However, a commitment by CQ-ROC members to upgrading the existing facilities forms a major part of the proposed waste management strategy across the Central Queensland region (Sinclair Knight Merz, 1999).

MODIFICATIONS TO PRACTICES

As a result of site visits by environmental specialists and the holding of discussions with officers from other shire councils, many improvements were implemented by member councils before the Stage 1 study had concluded. These improvements included groundwater monitoring, exchange of knowledge on waste management between the various councils and the upgrading of small landfills (Sinclair Knight Merz, 1999).

It is proposed that the region's unmanned landfills, as mentioned above, be either manned or converted to transfer stations, thereby, enabling policing of any regulated wastes as well as providing better knowledge of waste streams for auditing purposes (Table 14.1) (Sinclair Knight Merz, 1999). According to Capricorn Conservation Council, many potentially contaminated sites already in existence in Central Queensland are not on

Table 14.2: Categories and monitoring status of landfill sites in the CQ-ROC region

Monitoring parameter	*Nature of facility [and number in the region]*				
	Relatively large, technically adequate [1]	*Relatively large, poorly sited [1]*	*Moderately sized, with limited environmental controls [4]*	*Very small, isolated [2]*	*Very small unmanned trench' type [10]*
Groundwater	1	1	1	0	3
Surface water	0 (required)	0 (required)	0 (required at 1)	0	0
Dust	*	*	*	*	*
Noise	#	#	#	#	#
Fly	ν	ν	ν at 1 required at 1	0 required at 1	0 required at 4
Rodent	ν	ν	ν	0	0 • at 1
Bird	ν	ν	ν • at 2	0	0 • at 1
Erosion control	ω	ω	ω	ω	ω
Vegetative cover	∇	∇	∇	∇	∇

Source: Sinclair Knight Merz, 1999
• Inspected visually—no obvious problems
* Visual or complaint basis
ν Adequately controlled through compaction and daily cover
∇ Not specified in regulating authorities' operating licence
ω Visual basis
Complaint basis

the local council registers (Fitzgerald, 1990). Consequently, investigations into these sites have, to a large extent, been reliant on public recollections of past waste disposal activities (Fitzgerald, 1990). Future improvements to the monitoring of waste streams should, therefore, help reduce problems associated with the generation of unknown, and potentially contaminated, dumping sites.

Three of the seven CQ-ROC shires, as well as RCC, have prepared site management plans (Table 14.3) as part of their integrated approach to waste management. This accounts for the two largest landfills in the region, in addition to one of the four moderately-sized landfills which previously had employed limited environmental controls (Sinclair Knight Merz, 1999).

From the summary of operational and environmental monitoring conducted at Lakes Creek Road landfill (Table 14.3) it can be seen that additional monitoring parameters have been included in the site management plan. For example, a comprehensive surface water monitoring regime is now in place, and the leachate is being disposed off the Landfills (Rockhampton City Council, 1999). A study conducted by Kinhill Pty Ltd in August 1999 for RCC revealed that the high volume and low risk of the leachate produced allowed for:

Table 14.3: Summary of operational and environmental monitoring at the Lakes Creek Road landfill, Rockhampton

Management parameter	*Application frequency*	*Additional information*
Environmental Auditing	• Internally every 6 months • Externally every 5 years	• Internal audit by trained staff • External audit by Certified Auditor
Waste Audits	• Every 5 years	• Wheelie Bin and Landfill audits
Particulate	• 12-month collection and analysis programme • Repeat dust collection every 3–5 years • Complaint monitoring	• Dust monitoring to Australian Standards • Complaint monitoring to be performed by qualified persons
Landfill Noise	• Equipment monitored every 12 months • Noise complaint monitoring	• Noise testing to Australian Standards
Landfill Odour	• No set frequency	• Landfill operating procedures to limit odour generation
Pest and Vermin Control	• Site inspections to identify increase or decrease in numbers • Spray at selected times	• Part of landfill operations visual inspections
Flora and Fauna	• Every 5 years. Next study in 2005	• Study to be conducted by qualified persons
Meteorological	• Installation of a weather station at the landfill	• Information on wind speed and direction, rainfall and barometric pressure
Surface Water	• Quarterly monitoring within 24 hours of rainfall event exceeding 50 mm	• Analysis for parameters indicated • Reporting of significant changes in trends
Groundwater	• Quarterly monitoring of groundwater wells	• Analysis for specified parameters under licence • Reporting of significant changes in trends
Landfill Gas	• Monthly for first 12 months, then quarterly using a Natural Gas Detector	• Surface monitoring (to identify offsite migration sources) and facility structure monitoring

Source: Adapted from Rockhampton City Council, 1999.

- pumping to the sewer main—the main method
- on-site irrigation and evaporation—conditions permitting
- additive in biodegradable (newspaper-based) spray-on landfill cover.

Background information provided by the councils has shown that two of the seven councils are certain of having future landfill sites available to them (Table 14.1) (Sinclair Knight Merz, 1999). It has, therefore, become necessary to consider the available capacity of existing main landfills: Lakes Creek Road (Rockhampton); Blackwater (Duaringa Shire); and Lochlees (Emerald Shire). Without any modification of the current practices, these landfills would require closure within ten years (Sinclair

Knight Merz, 1999). Further, due to the poor siting of the Lakes Creek Road landfill, environmental problems are likely to demand closure soon. This option could, however, be avoided through the implementation of an effective site plan, thereby extending the life of this site to 30–50 years. Such a plan is likely to involve (Sinclair Knight Merz, 1999) the following measures:

- grading the current landfill footprint areas to shed surface water;
- designing a long-term water-shedding surface (post-settlement) to top the existing landfill footprint; and
- development of a programme to progressively raise the landfill to a final grade.

REGIONAL INITIATIVES

The Waste Management Strategy for Queensland places great emphasis on education regarding the waste minimization, the basic approach being to inform the public of the options and the procedures regarding waste management (DOEH, 1996). In line with this approach, RCC has initiated a television advertising campaign aimed at enhancing community awareness of environmental issues associated with packaging of products and produce (Sinclair Knight Merz, 1999). The result is expected to be, for example, a reduction in the amount of packaging going to landfill as a result of domestic waste and better knowledge of recyclable plastics (Sinclair Knight Merz, 1999).

Agriculture is the main industry in the region under study and, therefore, it is likely that any regulated wastes generated, especially in the more western shires, would comprise a fair proportion of farm chemicals. As farm chemicals are often packaged in plastic containers, the additional problem arises of having to physically deal with these at the landfills. To encourage reuse of chemical waste drums and hence reduce the amount going to landfill, CQ-ROC has implemented the 'drumMuster' system—an industry incentive developed by the National Farmers Federation (NFF), the National Association for Crop Protection and Animal Health (AVCARE), the Veterinary Manufacturers and Distributors Association (VMDA) and the Australian Local Government Association (ALGA) (Sinclair Knight Merz, 1999). A 4 cent per litre or per kilogram levy on various farm-related products, imposed from 1 February 1999 onwards, aims at funding a national collection and recycling scheme (Sinclair Knight Merz, 1999).

A number of potential options exist for the region which may assist in reducing the problems associated with disposal of tyres. For instance, a trial conducted at Queensland Cement Limited (QCL) Gladstone plant, in February 1999, involved the use of discarded tyres (70,000) as a source of fuel in the plant's kiln (QCL, 1999). The trial was designed and conducted in accordance with the most stringent health, safety and environmental guidelines. This was evident from the emissions readings at the plant (conducted by an independent emissions monitor), which were found to be amongst the lowest monitored for any cement or thermal-firing process in Australia and New Zealand (QCL, 1999, cited in Sinclair Knight Merz, 1999). Factors such as high transport costs and the small number of tyre recyclers in Central Queensland have so far reduced the scope for minimizing tyre waste in the region (Sinclair Knight Merz, 1999). However, according to Rutherford (1997), there is a potential for establishing a processing plant in the Rockhampton area.

CONCLUSIONS

Through the formation of CQ-ROC, the basic components of the Queensland Waste Management Strategy—waste minimization and waste containment—are being addressed. A commitment by CQ-ROC members to address the strategy priorities has been evident. For example, site visits by environmental specialists during the course of the solid waste planning study resulted in a range of prompt improvements to the waste disposal facilities.

Distances from disposal and treatment facilities in some areas of the region have resulted in both reduced recycling opportunities and stockpiling of some regulated wastes such as tyres. However, rural industry initiatives in the region have addressed both a potential source of regulated waste, in the form of agricultural waste-chemicals, as well as the problems associated with reduced volume in landfills caused by discarded containers.

Addressing the disparities in the standards of landfills across the region, as well as future capacity requirements of existing main landfills, are the primary priorities. Varying degrees of upgrading have been proposed for the region's smaller landfills (in cases where closure is not an option), and long-term closure plans proposed to extend the limited filling capacities of the larger landfills. Extension to the operating life of Rockhampton City Council's Lakes Creek Road landfill is largely dependent on the continued success of the Council's recycling initiative. This may, therefore, be complemented through the undertaking of a regional education initiative aimed at encouraging waste reduction.

ACKNOWLEDGEMENTS

The authors would like to thank the organizers of the conference for the opportunity to present this paper. Appreciation is also extended to Sinclair Knight Merz (Environmental Consultants, Rockhampton, Queensland) for giving permission to use their comprehensive report. We would also like to acknowledge the encouragement given by the following people: Prof. David Midmore, Director, Plant Sciences Group, Central Queensland University and Richard Yeates, Phytolink Australia, Pty Led., Brisbane.

REFERENCES

Bilitewski, B., Härdtle, G. and Marek, K., 1997, *Waste Management*. Springer-Verlag, Berlin.

Butler, M., 2000, Recycling. Is it worth it? *Waste Management and Environment* 11(4): 24–28.

CQ-ROC (Central Queensland Regional Organisation of Councils), 1998, Preparation of a Waste Management Strategy—Tender No. CQ-ROC1.

Davis, M.L. and Cornwell, D.A., 1998, *Introduction to Environmental Engineering* (3rd ed.), WCB/McGraw-Hill, Boston.

DOEH [Queensland State Department of Environment and Heritage], 1996, Waste Management Strategy for Queensland. DOEH, Brisbane.

Environment Australia, 1997, *Waste Minimisation for Local Government*. Environment Australia. Deakin, Queensland.

Fitzgerald, S., 1990, 'Dumpsites cause query', *The Morning Bulletin* (2 August): p 1.

Kinhill Pty. Ltd, 1999, *Lakes Creek Road Landfill Leachate Disposal and Sampling*. Kinhill Pty Ltd., Rockhampton.

Krol, A., Rudolph, V., Hart, K. and Swarbrick, G., 1994, Landfills, a containment facility or a process operation? http://www.civeng.unsw.edu.au/staff/Swarbrick.G/NHSWC94?NHSWC94.html. Accessed April 2000

QEPA (Environmental Protection Agency, Queensland), 1999, *State of the Environment*, Queensland, Environmental Protection Agency, Brisbane.

QEPA, Queensland Environmental Protection Agency, Moon, H., 1994, Solid waste management in Ohio. *Professional Geographer* 46(2): 191–198.

Queensland Cement Limited, 1999, QCL trials recycling programme for Gladstone kiln line. (Media Release, 24 February 1999) URL: http://www.qcl.com.au/news/trialrecyc.htm. Accessed April 2000

Rockhampton City Council, 1999, Lakes Creek Road Landfill, Rockhampton, Queensland, Site Management Plan. Rockhampton City Council, Rockhampton.

Rutherford, M., 1997, Rocky on road to tyre plant. *The Morning Bulletin* (11 November): p 8.

Sinclair Knight Merz, 1999, Waste Management Strategy Report. Central Queensland Regional Organisation of Councils. Sinclair Knight Merz, Pty. Ltd., Rockhampton, Queensland, Australia.

Williams, P.T., 1998, *Waste Treatment and Disposal*. John Wiley and Sons, New York.

15

Waste Management Practices in Sri Lanka

Azeez M. Mubarak

INTRODUCTION

Sri Lanka, with 18.7 million people of whom 24% are urban, ranks fourth among the SAARC countries in the level of urbanization. With some two million people, the Colombo district is the country's largest and most important urban area with an urban population of 62% and a population density of 3253 per square km (Central Bank, 1999). Some 80% of the country's industrialization is in Colombo and its neighbouring district Gampaha (Figure 15.1). While the environment in this area has not been degraded as is the case in larger Asian metropolitan areas, there is clear evidence that growth and economic development are proving to be problematic. Colombo and its peripheral districts are feeling the pressure of development and waste generation, and waste management has become one of the major environmental issues of the country.

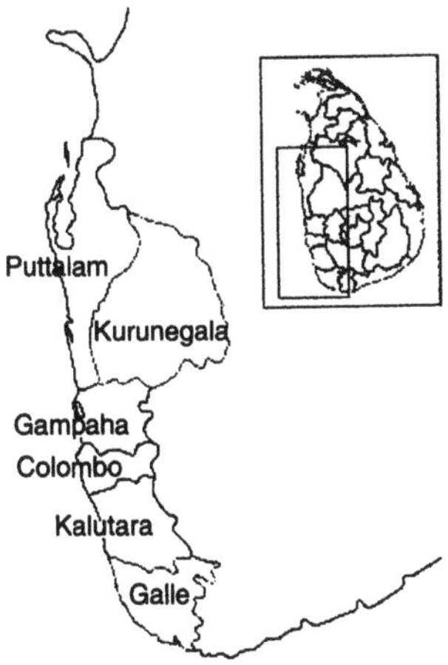

Fig. 15.1 Location of industrialized areas in Sri Lanka

Director, Industrial Technology Institute, Colombo, Sri Lanka

WASTES GENERATION—CURRENT STATUS

Municipal Solid Waste

Solid waste in urban areas arises from a wide variety of sources including household, industry, business and commercial enterprises. The total quantity of solid wastes collected in Sri Lanka is estimated at 2694 tonnes per day (Table 15.1). The bulk of the collection, about 1170 tonnes, is from the Colombo District of which nearly 64% is from the Colombo Municipal Council (Ministry of Forestry and Environment, 1999). Estimated per capita waste generation per day varies from 0.85 kg in the Colombo Municipal Council to 0.25 kg in *Pradeshiya Sabhas* (rural areas) (ERM, 1997). About half the municipal solid wastes generated in the Colombo Metropolitan Region remain uncollected. The composition of municipal solid waste collected in the Colombo District and other information on solid waste collection and disposal are shown in Table 15.2. Some 80% of municipal waste is organic and biodegradable, which is amenable for conversion to useful products such as compost and methane.

Table 15.1: Waste collection by the Local Authorities of Sri Lanka

Province	*District*	*Waste (tonnes/day)*
Western Province	Colombo District	1170
	Gampaha District	197
	Kalutara District	80
Southern Province	Galle District	76
	Hambantota District	22
	Matara District	46
Central Province	Kandy District	162
	Matale District	16
	Nuwaraeliya District	45
Sabaragamuwa Province	Kegalle District	39
	Ratnapura District	78
North-Western Province	Kurunegala District	66
	Puttalam District	23
Uva Province	Badulla District	147
	Monaragala District	23
North-Central Province	Anuradhapura District	23
	Polonnaruwa District	10
North and Eastern Province	Ampara District	46
	Batticaloa District	125
	Jaffna District	194
	Trincomalle District	87
	Others	19
Total		**2694**

Source: Ministry of Forestry and Environment, 1999

Table 15.2: Composition of solid wastes collected in the Colombo District

Local authority	*Area (km^2)*	*Pop. (× 1000)*	*Disposal method efficiency %*	*Availability of landfill facilities*	*Gross weight per day (tons)*	*Waste composition(%)*						
						Plastic	*Biodeg. (Short-term)*	*Biodeg. (Long-term)**	*Metal*	*Wood*	*Glass*	*Paper*
Colombo M.C.	37.0	800.0	95	Temp.	680.0	6.0	83.0	0	2.0	2.5	0.5	6.0
Dehiwala-Mt Lavinia M.C.	21.6	247.0	50-75	Private	150.0	2.0	93.0	0.5	0.5	1.7	0.5	1.0
Homagama P.S.	140.8	200.0	50-75	Private	6.0	8.0	71.0	10.0	2.0	1.0	1.0	7.0
Maharagama P.S.	21.9	118.0	25-50	Yes	20.0	5.0	50.0	20.0	4.0	8.0	3.0	10.0
Moratuwa M.C.	23.3	200.0	50.75	Yes	135.0	10.0	70.0	8.0	0.0	0.0	0.0	12.0
Sri Ja'pura-Kotte M.C.	43.8	137.4	>75	Yes	95.2	28.2	35.4	14.2	0.7	14.2	0.3	7.0
Kaduwela P.S.	87.7	172.0	>75	Yes	28.0	3.6	57.1	17.9	7.1	7.1	0.0	7.0
Seethawakapura P.S.	212.8	115.3	50-75	Yes	1.5	0.4	80.0	10.0	0.4	2.0	0.4	6.8
Kotikawatte-Mulleriyawa P.S.	29.6	98.0	25-50	Marsh	6.0	1.7	83.0	6.7	0.0	5.0	1.7	1.7
Seethawakapura U.C.	19.4	24.5	>75	Yes	6.0	0.4	87.0	6.7	0.4	1.7	0.3	3.3
Kesbewa P.S.	55.0	186.0	25-50	Yes	40.0	10.0	40.0	30.0	5.0	7.0	3.0	5.0
Kolonnawa U.C.	23.1	65.0	50-75	Yes	3.0	5.0	67.0	4.0	0.0	10.0	9.0	5.0
Total	**716.#**	**2363.#**			**1170.7**							

*Biodegradable (long-term): coconut shells, king coconut shells

Source: Ministry of Forestry and Environment, 1999.

Collection, transport and disposal of municipal solid wastes are the responsibility of the local authorities. The collection service provided by the local authorities (LA) includes either one or a combination of:

- Door-to-door service—this is for larger local authorities who have skips; Refuse Collection Vehicles (RCV) provide this service.
- Primary and secondary collection service—The local authorities collect the waste from households by handcarts, which is then deposited at temporary waste-collection points. The secondary collection system, using equipment such as tractor-trailers and RCVs, delivers the waste from primary collection points to the disposal sites.
- Secondary collection service only—Residents bring the waste to a collection point, from where the local authorities transport it to the disposal site.

In recent years, some local authorities have privatized the waste collection services, causing a marked improvement in street cleaning and door-to-door waste collection. With the move towards privatization, it is anticipated that total waste collection will increase significantly. The disposal of garbage by the use of open dumping is common among most of the local authorities. The main reason for this has been that it is the simplest and, in terms of disposal methods, the cheapest. Most of these sites are in low-lying areas such as marshy lands and abandoned paddy fields. Over 60% of these sites are privately owned and are being filled with waste primarily as a means of land reclamation. Some 70% of the sites in the Colombo District are less than a hectare, and 46% have a remaining lifetime of less than three years (ERM, 2000).

Garbage at open dumps is rarely covered with layers of soil which often leads to the contamination of surface-waste bodies and pollutants leaching into the groundwater. Furthermore, such dumping sites are frequented by scavenging animals, or provide and ideal breeding ground for parasitic insects, or both. Blocked drains and waterways, a result of uncontrolled dumping of solid wastes, is one of the major contributors to the mosquito problem in the urban cities.

Industrial Solid Waste

Most solid waste generated by industries is sent to municipal sites or dumped within the zone. Some wastes from packaging, garment, and rubber industries are incinerated but with little or no control.

Industrial wastewater treatment plants (physico-chemical and/or biological) produce sludge that requires safe disposal. Since a significant proportion of industry—particularly those established before the 1990s—do not have proper treatment plants, no realistic estimate could be made on the amount of sludge generated by the industry. However, the sludge-generation potential could be gauged from the total quantity and quality

Table 15.3: Profile of industrial sectors generating pollution in Sri Lanka

Factor	*Textile*	*Desiccated coconut*	*Rubber processing*	*Food and beverage*	*Tanning*	*Metal finishing*	*Paint and chemicals*
No. of medium- and high-polluting establishments	41	53	229	47	15	76	33
Total wastewater volume (m^3 day^{-1})	7100	1200	4840	4111	1614	6692	928
BOD (kg day^{-1})	4970	4200	9670	6166	3229	3229	
COD (kg day^{-1})	11,360	7200	29,040	12,333	8070		
Total toxic metals (kg day^{-1})*	NA	NA	NA	NA	161	669	93
Location	Colombo and Gampaha Districts	Rural areas within the 'coconut triangle'	South-western Sri Lanka	Greater Colombo and Kalutara	Colombo and Gampaha Districts	Widely distri-buted within 4 subareas of Colombo	50% in Ratmalana; the rest mainly in Ekala and North Colombo

*Based on an assumed average concentration of 100 mg/litre

Source: ERM (1994)

of wastewater generated by various industry sectors in the country (Table 15.3). Since viable outlets for the disposal of such sludge is lacking in the country, the industry has adopted a range of uncontrolled practices for sludge disposal, including spreading on land within factory premises or disposal by private contractors at unknown sites with associated risk of groundwater pollution.

Hazardous Waste

It is estimated that at least 40,000 tonnes of hazardous wastes, both liquid and solid, are generated in the country annually (ERM, 1996). This comprises about 10,000 tonnes of inorganic wastes, 15,000 tonnes of organic wastes, and 14,000 tonnes of oil wastes from motor vehicles (Table 15.4). Of the inorganic wastes, acids and alkali constitute about 7000 tonnes, while heavy metal wastes account for 27 tonnes per annum. Oil wastes (3600 tonnes, solvent wastes) 3030 tonnes and agrochemical wastes (2657

Table 15.4: Sri Lanka's adjusted hazardous waste inventory

Waste type		*Tonnes p.a.*
Inorganic wastes		
Inorganic acids		2744.00
Inorganic alkalis		4396.00
Zinc-bearing wastes		8.75
Heavy metal wastes		18.75
Waste-treatment sludges		271.75
Containers contaminated with inorganic metals		1.25
Solid waste contaminated with inorganic material		2837.50
	Total	10278.00
Organic wastes		
Oil wastes (liquid)		2371.25
Oil wastes (semi-solid)		1237.50
Solvent wastes (non-halogenated)		1533.75
Solvent wastes (halogenated)		1497.50
Waste paints, lacquers, varnish and related material		255.00
Waste agrochemicals		2857.50
Waste pharmaceuticals		210.00
Wood preservative wastes		38.75
PCB, PBB, PCT wastes		6.25
Containers contaminated with organic material		8.75
Solid waste contaminated with organic material		4722.50
	Total	14739.00
Other wastes		
Asbestos wastes		117.50
Plastic/resin wastes		1482.50
	Total	1600.00
Oil wastes from motor vehicles		
	Total	14000.00
Total hazardous wastes		40617.00

tons) are the major categories in the organic class. As there is no system in the country to dispose of these hazardous waste, they are either dumped within the site or transported outside for recycling or dumped in marshy areas by contractors. These practices can lead to the contamination of surface-water bodies and underground aquifers.

Integrated Solid Waste Management Strategy

A comprehensive national management system for the control of wastes would comprise a mix of:

- legislation;
- implementation and enforcement;
- facilities; and
- infrastructure and support services.

For an effective management system, all of these four elements must be in place, for example, legislation without facilities and support services cannot be enforced satisfactorily, while in the absence of suitable legislation, attempting to implement a waste management system nationally would be futile.

POLICY AND LEGISLATON

The waste management strategy that is being currently adopted by the Sri Lankan Government includes short-term actions for immediate implementation and a phased approach to implement longer-term actions. Successful waste management is a cooperative as well as coercive exercise, which ultimately relies on contributions from a range of Ministries, industrial bodies, individual companies, and consumers. Implementation and enforcement programme must be practical within the limits of resources and skills available.

A National Industrial Pollution Management Policy Statement, signed jointly by the Ministers of industry, Environment and Science & Technology (Ministry of Environment, 1998) advocates:

- pollution prevention at source;
- a 'polluter pays' principle;
- clustering industrial units in estates or parks;
- incentives and enforcement; and
- community, private sector and government interaction.

A national solid waste management strategy proclaimed recently also encourages cleaner production, including waste minimization, reuse and recycling.

In terms of the National Environmental Act, the Central Environmental Authority (CEA) was created in 1981 as a policy-making and coordinating body. The 1988 amendments transformed the CEA as an enforcement and implementing agency. Until recently, all industries, whether they were classified as low-, medium- or high-polluting, were required to obtain an Environmental Protection Licence (EPL). In view of the thousands of industries scattered around the country and the regulatory burden placed on the limited CEA staff, EPL issuance to all industries had become very difficult. In order to relieve this constraint, licensing of low-polluting industries was delegated to the relevant local authorities in 1994.

It is very difficult to build a control system for industrial pollution without strengthening the institutional capacity for monitoring pollution and enforcement. Accordingly, some of the powers of the CEA have been delegated to the Board of Investment (BOI) and other local authorities to help in the implementation and enforcement of the legislation.

A national definition on hazardous wastes and regulations to control the collection, storage, transport and disposal of hazardous wastes has been published (Gazette, 1996) and a hazardous waste management system is also being formulated to handle priority wastes. Legal framework required for solid waste management is provided under the Local Government Acts, and the local authorities are responsible for the collection and disposal of solid waste (Local Government Acts). The provisions relating to solid waste management are as follows.

All street refuse, house refuse, night-soil or other similar matter collected by local authorities under the provisions of this part shall be the property of the Council, and the Council shall have full power to sell or dispose of all such matter.

Every Pradeshiya Sabha, Urban Council and Municipal Council shall, from time to time, provide places convenient for the proper disposal of all street refuse, house refuse, night-soil, and similar matter removed in accordance with the provisions of the Law, and for keeping all vehicles, animals, implements and other things required for that purpose and shall take all such measures and precautions as may be necessary to ensure that no such refuse, night-soil, or similar matter removed in accordance with the provisions of the law is disposed of in such a way as to cause a nuisance.

Further, under the National Environmental Act, the CEA with concurrence from the Ministry can give necessary directions to any local authority for safeguarding and protecting the environment. In addition, no person shall discharge, deposited or emit waste which will cause pollution into the environment, except under the authority of a licence issued by the authority. The National Environmental Act also requires landfill sites to undergo environmental impact assessment when the capacity of the site is over 100 tonnes per day.

As in many developing countries, waste disposal has been given a relatively low priority by the state. As a result, implementation of the legal and regulatory requirements has not been effective. For example, local authorities continue to operate existing open-dumping sites without an environmental protection licence, which is a violation of the law and a clear threat to the environment. The CEA does not, and is unable to, enforce the regulations and prohibit the use of sites, as there is no alternative system in place. Without proper waste disposal facilities, regulation of solid waste disposal cannot achieve the desired results.

Facilities

To minimize the environmental degradation of waterbodies in Ratmalana and Ekala industrial areas, where there is a high concentration of industries interspersed with houses, common wastewater treatment systems are being planned to treat combined industrial and domestic wastewater. Since Ratmalana is located close to the coastline, wastewater will be disposed of directly via a new ocean outlet with primary treatment only. In the case of Ekala, since there is no ocean outlet nearby, the combined wastewater will be sent through a secondary treatment plant prior to discharge to inland watercourses. At present, Sri Lanka does not have an engineered sanitary landfill site at which to dispose of solid waste in an environmentally sound manner. A sanitary landfill with a receiving capacity of 1000 tonnes a day, including a composting plant with a minimum capacity of 100 tonnes a day to treat market garbage, has been proposed for municipal and non-hazardous industrial solid waste collected in the Colombo region. However, due to objections by some NGOs representing the local residents, the project has been stalled.

The problem of hazardous waste generation and its safe disposal has also become a major concern in the country. A low-technology central treatment facility for chemical treatment and stabilization of hazardous wastes, as well as a landfill site for stabilized wastes, has been proposed. An existing cement kiln will be used to burn pumpable organic wastes. There are technical, commercial and economic benefits to be gained from locating industries, particularly industries in the high-polluting category, in estates or zones that have been provided with common waste management facilities. The Ministry of Industrial Development has short-listed, on a regional basis, several areas suitable to be developed as industrial estates (Ministry of Industrial Development, 1998). Construction of an industrial estate at Sithawaka is now complete, while EIA is being performed on other sites in different parts of the country.

Treatment of waste from an individual industrial sector, particularly high-polluting industries such as tanneries, in a combined facility results in a number of technical and economical advantages. Therefore, the government is proposing to move all tanneries to a new location with the provision of a combined wastewater treatment system. A site at Bataatha in the Hambantota District has been identified for this purpose. A team of UNIDO consultants is now assisting these tanneries in cleaner production practices and processes before relocation.

Land- and water-based disposal systems may not be possible in the existing industrial areas due to limited land availability, high price of land and high concentration of residential and commercial activities. However, these systems may be introduced in new industrial estates to be developed in less populated regions if technically and economically feasible.

Cleaner Production

One of the more important actions of the government strategy is to actively promote cleaner production. Presently, there is little evidence of application by any industry of source management and control procedures. The reasons are not only technical and financial, but also lack of awareness among industrialists of the advantages. The potential benefits both to industry and the environment are substantial when disciplined procedures are adopted for implementation of cleaner production. Reduction of waste and emissions at source can eliminate some 20–40% of pollution. Several initiatives are already underway to promote the cleaner production concept in Sri Lanka. The Pollution Control and Abatement Fund (PCAF) and E-Friends are two financial assistance schemes available for any industry engaged in cleaner production activities. Upto 1998, a total sum of US$3 m has been disbursed as grants and soft loans to 50 projects under the PCAF programme.

The UNIDO-sponsored Industrial Pollution Reduction Programme (IPRP) implemented by the CEA from 1994 to 1998 was aimed at identifying and implementing financially-viable waste minimization options in textile, metal finishing, distillery and tannery industries and setting up site-specific demonstration projects to promote clean technologies. Some of the successful demonstration projects are: chrome recovery units for tanneries, low liquor ratio jet dyers for textile dyeing, and an improved distillation column for a toddy distillery. A pollution prevention information clearinghouse named CleaNet (www.cleanet.lk) has also been set up by the Ceylon Chamber of Commerce and Industrial Technology Institute in order to assist industry on cleaner production activities.

The solid waste management strategy promoted by the government also includes several cleaner production initiatives such as waste avoidance and reduction and reuse and recycling.

Waste Avoidance and Reduction

Excessive packaging causes a rapid increase in solid waste generation. Reducing the demand for such packaging systems by consumers through considering their own disposal problems at the source, and the problems associated with the final disposal, can reduce generation of such wastes to a considerable extent. Manufacturers should, therefore, be encouraged to reduce unnecessary packaging so as to reduce the generation of solid waste. Generation of market garbage remains very high at present due to unsuitable transportation, packaging, handing and storage of fruit and vegetables and other perishables. Post-harvest marketing chains with appropriate technology will avoid or minimize such wastes and, therefore, should be actively promoted. Haphazard disposal of plastic waste has become a serious environmental and health problem in the country. In the absence of a well-established plastic recycling facility in the country, the use of plastics is being discouraged as much as possible in favour of paper, glass, cloth and biodegradable packaging materials. Production of long-life products and multi-use packaging is being promoted instead of throw-away packaging.

Reuse and Recycling

Reuse and recycling of waste is being encouraged to reduce the overall generation of waste. At present, the Ceylon Glass Company, one of the major glass manufacturers in the country, uses 40% of cullet (waste glass) in their process. The use of cullet has helped to significantly reduce energy costs, and has increased the lifetime of their equipment. The state-controlled paper corporation uses wastepaper in paper production. In addition, there are several small-and medium-scale units that produce recycled paper to eco-friendly hotels, a small but growing niche market in Sri Lanka. At present, wastepaper, glass, metal scrap and plastic are collected directly from households and offices, during collection and transport (by LA), or at the final disposal site by rag-pickers and municipal workers. These are sold to collection shops where they are cleaned and supplied to local recyclers or exported for recycling overseas. India is the main export market for such materials. Strengthening this informal market-driven system by providing incentives and infrastructure facilities to widen its scope and coverage will be more effective than setting up an entirely new system for this purpose.

Sorting at source would improve the economic viability of recycling operations. Infrastructure facilities could be established in order to motivate the sorting of waste at source, which in turn, would facilitate the establishment of formal and informal waste collection systems. Some local authorities have established collection points where residents bring their recyclable material for sale, but the success rate for this venture is low.

Composting

As discussed previously, most solid waste generated in Sri Lanka contains biodegradable waste, with a high moisture content and, therefore, is suitable for composting. Several initiatives are underway to promote compositing of household and market garbage. Both local authorities and NGOs are promoting composting barrels or bins at the household level. Kesbewa PS is also operating a centralized composting project. An estimated 0.2% of the current waste collected in the Greater Colombo Area is composted (ERM, 1999). The National Engineering Research and Development Centre has developed a 40-tonne capacity anaerobic batch digester to convert market garbage into organic fertilizer. Currently, trials are being run to use the methane gas produced during the anaerobic digestion to generate electricity. Experiments are also being conducted using a mixture of water hyacinth, a weed commonly found in waterways, with market garbage as a feedstock for these digesters.

The quality of the compost produced is likely to depend mostly on the degree of contamination and type of organic waste present in the market garbage. Plants and animals tend to bio-magnify heavy metals; so if commercial use of compost is to be actively promoted, a standard for compost must be formulated. This will ensure that the product is devoid of harmful constituents and will guarantee that food-chain crops grown with such compost are not at risk.

Infrastructure and Support Services

As noted above, it is very difficult to build a control system for waste without simultaneously strengthening the institutional capacity for monitoring pollution and enforcing the regulations. The CEA is now being strengthened with a new laboratory and personnel to effectively implement the regulations. It is also necessary to build up the local capacity to develop cost-effective waste management systems, to select and assimilate imported technology, and to provide access to foreign technical information to the industry. Several state organizations such as the Industrial Technology Institute (ITI), National Building Research Organization (NBRO), Rubber Research Institute (RRI), National Engineering Research & Development Centre (NERD) and universities offer testing and consulting services to the industries to help address their pollution-related problems. In addition, many private sector companies and consultancy firms have appeared to meet the growing demands to control pollution. However, to provide a more effective and reliable service to the industry, expertise and facilities available in these laboratories need strengthening and accreditation to international standards.

Education and Research

Several environment-related courses at post-graduate levels have been initiated by the Universities of Colombo, Moratuwa, Peradeniya and Kelaniya to cater to the growing demand for environmental engineers and scientists. The Ministry of Science and Technology is implementing a comprehensive S&T manpower development programme sponsored by ADB to enhance the quality and R&D capability of the environmental S&T manpower in the country. Environmental education at pre-school levels and a programme to train master teachers to strengthen teaching in schools have been initiated by the CEA and the National Institute of Education (NIE). But these efforts would need to be expanded and sustained if they are to have any meaningful impact among school children. Training programmes and workshops on issues and programmes such as cleaner production, waste audits, EMS, and ISO 14000 are also being organized regularly by several oganizations with the aim of promoting waste prevention awareness in the country.

Community Participation

Cooperation and participation of the community in waste management, particularly for solid waste, is essential in implementing any solid waste management strategy. Each individual generates solid waste. On the other hand, the public displays strong emotions on the environmental and health impacts of mismangement of solid waste. In general, the public looks at solid waste management as a function of local authorities without considering the important role that can be played by themselves in order to ensure solid waste management in an environmentally-sound manner, which in turn, reduces their health costs. Community involvement in the decision-making process in developing solid waste management strategies should be encouraged at its inception to make the implementation of any solid waste management programme a success.

Key Issues for Waste Management in Sri Lanka

One of the major problems for effective waste management in Sri Lanka is the lack or slow progress of infrastructure programmes such as industrial estates, common waste treatment and solid waste disposal facilities. There is no single landfill in the country to dispose of solid waste, but all efforts to construct a landfill for this purpose are being objected to by environmentalists and the public mainly due to lack of understanding, misinformation and unfounded fear among the residents regarding the environmental implications of the project. A conciliatory approach by all parties concerned is vital to arrive at a rational and cost-effective solution to the solid waste problem.

The lack of a hazardous waste disposal facility is also a major issue for industries generating such wastes. Although there is a plan to establish such a facility with a proper management strategy, the progress is extremely slow. Most of the technologies used, particularly by the small and medium industries, are outdated and inefficient, and as such themselves generate much waste and consume much energy. Although there are financial incentives to encourage industries to replace such technologies, the response from industries is poor. The possible reasons for this are: the package is too costly to the SMI, industry is unaware of the environmental problems or is unable to access new technologies, and there is no commitment to improving the environment.

CONCLUSIONS

The following actions are suggested to improve the current situation:

- an awareness campaign on programmes like cleaner production and EMS;
- setting up of mechanisms to assist industries to source, access, and transfer clean technologies; and
- educating the public on the importance of the environment, and through them, exerting pressure on the industries to act in an environmentally-responsible manner.

Several commercial composting projects have failed for various reasons such as (ERM, 1999): presence of small quantities of plastics and other low-biodegradable items like coconut husk and banana stalk in market waste, resulting in poor quality compost; inability to market compost due to variation in quality and consistency; and public opposition to siting composting facilities due to inherent visual and odour problems. However, recently, a private composting facility has been established in Colombo to convert some of the municipal waste to bio-fertilizer. The success of this venture is yet to be seen.

Two other key issues that need to be addressed are:

- lack of a National Zoning plan, leading to conflict between industry, business and residence, and
- insufficient information on the state of the environment and trend analysis to formulate the best practicable plans and programmes.

REFERENCES

Central Bank, 1999, Economic and Social Statistics of Sri Lanka, 1999. Central Bank of Sri Lanka, 1999.

ERM [Environmental Resources Management], 1999, MEIP/SMI-IV Strategy, Guidelines and Institutional Strengthening for Industrial Pollution Management: Final Report, (February).

ERM [Environmental Resources Management], 1996, Consulting Services for Pre-feasibility Study on Hazardous Waste Management and Disposal in Sri Lanka, Final Draft Report (November).

ERM [Environmental Resources Management], 1997, Environmental Impact Assessment for a proposed Sanitary Landfill, Alupotha Division, Salawa Estate.

ERM [Environmental Resources Management], 1999, Strategic overview of potential solutions to the crisis of disposal of Municipal Solid Waste, the Local Authorities of the Greater Colombo Area, Western Province, Sri Lanka.

ERM [Environmental Resources Management], 2000, Review of Current Municipal Solid Waste Dump Sites in the Greater Colombo Area, Western Province, Sri Lanka.

Gazette, 1996, The Gazette of the Democratic Socialist Republic of Sri Lanka, Extraordinary, No. 924/13 (May 23).

Local Government Acts. The Municipal Council Ordinance of 1980, the Urban Councils Ordinance 1987 and the *Pradeshiya Sabha Act 1987.*

Ministry of Industrial Development, 1998, Review of Status.

Ministry of Environment, 1998, National Industrial Pollution Management, Policy Statement 1998, Ministry of Transport, Environment & Women's Affairs, Ministry of Industrial Development and Ministry of Science & Technology and Human Resources Development.

Ministry of Forestry and Environment, 1999, Municipal Waste. Database of Municipal Waste in Sri Lanka.

ABBREVIATIONS

ADB	Asian Development Bank
BOI	Board of Investment
CEA	Central Environmental Authority
EIA	Environmental Impact Assessment
EMS	Environmental Management System
EPL	Environmental Protection Licence
IPRP	Industrial Pollution Reduction Programme
ITI	Industrial Technology Institute
LA	Local Authority
NBRO	National Building Research Organization
NERD	National Engineering Research & Development Centre
NGO	Non Governmental Organization
NIE	National Institute of Education
PCAF	Pollution Control Abatement Fund
PS	Pradeshiya Sabha
RCV	Refuse Collection Vehicle
UNIDO	United Nations Industrial Development Organization

16

The Current Status and Future Outlook for Waste Management in Thailand

Somtip Danteravanich

INTRODUCTION

Located in the centre of the Southeast Asian mainland, bordered by Malaysia to the south, Laos and Cambodia to the east and Burma (Myanmar) to the west, Thailand has an area of 513, 115 km^2 and is divided into five regions. The north of Thailand is covered with thick tropical forest and fertile agricultural land. By contrast, the northeast is dry, arid and rich with mineral deposits, notably potash. The Central Plain is also very fertile, and the south is endowed with mineral deposits, tropical forest and rubber plantations. The western region is also rich in mineral deposits, and the land is generally suitable for farming and cattle ranching. Thailand's main natural resource is agriculture, while other significant resources are natural gas reserves in the Gulf of Thailand, tin deposits in the South, and small quantities of petroleum.

The economically-important industries of Thailand are all agriculturally based. Examples are rubber, palm oil, tapioca, sugarcane processing, pulp and paper, tannery, fish-mill and frozen seafood industries. The petrochemical industry is becoming more significant. Industrial activities are not all located as agglomerations in urban centres. Manufacturing industries in Thailand are generally located either along riverbanks or in coastal regions; agro-industries, a significant fraction of the industrial base of Thailand, are distributed in remote areas.

Although Bangkok remains Thailand's industrial core, promotion of industry in the provinces has been encouraged and industrial estates have been established. These industrial estates may be operated by the government's Industrial Estates Authority of Thailand (IEAT) or run entirely by private enterprises. Presently, Thailand has 41 designed

Faculty of Technology and Management, Prince of Songkla University, Surat Thani Campus, Surat Thani, 84100 Thailand industrial estates and industrial areas, divided almost equally between the public and private sectors. Currently, the government is focusing on development and promotion of small- and medium-sized enterprises (SMEs) to help stimulate the industrial sector.

In 1995, the law in Thailand established new local organizations named Tambon Administrative Organizations (TAOs), which upgraded the old Tambon Councils. These organizations have become important stakeholders in development, environmental protection and management of the country.

CURRENT WASTE GENERATION AND MANAGEMENT

Municipal Solid Waste

In 1998, 38,000 tonnes of municipal solid waste was estimated to have been generated in Thailand per day. This value reflects an increasing rate of solid waste generation from 1992 of approximately 4.4% (Rotchpaiwong, 1999). The proportion of solid waste generated in each region was: central (27%), southern (20%), northeastern (19%), northern (17%), eastern (12%), and western (5%) An average urban community in Thailand generates between 0.5 and 1.5 kg of solid waste per capita per day. The greater the extent of urbanization, the more is the solid waste generated. For example, it is estimated that the average solid waste generation rate in Bangkok is 1.5 kg per capita per day, while the average in municipal areas outside Bangkok is 0.8–1.0 kg; in rural areas, it is 0.5 kg (Yoosook, 1999).

Previous research on solid waste composition in several Thai cities has shown that solid waste usually comprises more than 30% recyclable content. The typical recyclable proportion of combustible solid waste, which is high in moisture and organic matter content, is 70–80%. Moisture content has been recorded as high as 50%, and the dry solid calorific value of solid waste is 3300–4700 kcal kg^{-1}, the actual figure depending on the physical composition. The calorific value seems to be increasing as a result of a proportionate increase in plastic and paper components of waste (Danteravanich and Siriwong, 1998). Due to the co-disposal of solid wastes, municipal solid waste normally comprises 0.1–1.0% of domestic hazardous waste.

The Bangkok Metropolis produced the highest amount of solid waste in Thailand with an average of 8500 tonnes per day in 1997-98. The Bangkok Metropolitan Authority was able to collect and transport 95% of the waste produced. Approximately, 76% was disposed in landfills and 13% was composted to produce organic fertilizer. The remaining 11% was left to decompose by itself at the treatment plants. The total mass of solid waste generated in all urban municipalities in Thailand was calculated to be more than 11,000 tonnes per day. The remaining 18,500 tonnes per day was solid waste produced outside municipal areas, which was dealt with by TAOs (Danteravanich and Siriwong, 1998 and Pollution Control Department, 1997).

Disposal of Municipal Solid Waste

Currently, municipal solid waste in Thailand is disposed either to landfills or treated by incineration. Although urban municipal authorities have the capacity to collect and transport large quantities of solid waste, sanitary landfills are not often used as a disposal method. At present, there are 23 landfill sites under construction and 16 landfill sites in operation in Thailand. The remaining municipalities use open land disposal and open burning as disposal methods. The sanitary landfill option is preferred for public health, environmental sanitation and environmental pollution control reasons. However, most municipalities do not employ this method due to limitations in technical experience and operational budgets. Operational costs associated with solid waste disposal to landfill in Thailand ranged from US$1.46 to 3.06 per tonne, with an average cost of US$2.30. The general costs of solid waste management facilities are relatively invariant, and the cost of operating solid waste disposal vary with the scale of operation (Sayamon, 1997). At present, there is no charge to the population residing in the municipal service areas for solid waste disposal, highlighting another major issue that needs to be resolved.

TAOs operate outside municipal areas and, therefore, have a very low capacity to collect, transport and dispose of solid waste, especially since they have only recently been established. Solid waste generated in TAO areas is usually dumped in open areas, which has lead to the increase in unsanitary landfill sites. Waste is left to decompose naturally, which is unsightly, a cause of nuisance, and could cause severe health and environmental problems.

The alternative to landfill disposal of municipal solid waste is incineration. Two solid-waste incineration plants have been constructed in southern Thailand. The island of Phuket has a plant with a capacity of 250 tonnes per day, while Samui Island has two small incinerators each with a capacity of 70 tonnes per day (Phuket and Samui are popular tourist destinations). The incineration plant in Phuket can produce upto 2.5 MW of electricity from waste heat. Since the operating costs of these plants and the time required for technology transfer are high, the Public Works Department (PWD) of the Ministry of Interior, who originated the incineration projects, is supporting the two local government authorities. At present, the PWD has contracted a private company to operate the incineration plant at Phuket for the first two years of operation. Although incineration is effective at destroying many types of wastes, emission of pollutants to the environment is a major concern. In the light of this potential environmental problem, the Ministry of Science Technology and Environment established air emission standards for incineration operations in 1997 (Table 16.1).

Table 16.1: Thai air emission standards for the incineration of domestic solid waste

Parameter	*Incinerator capacity (tonnes per day)*	
	> 50	*< 50*
Dust (mg m^{-3})	120	400
HCl (ppm)	25	136
SO_2 (ppm)	30	30
NO_x (ppm)	180	250
Dioxin as PCDD and PCDF (ng m^{-3})	30	30

Key ministries concerned with general solid waste management in Thailand are the Ministry of Interior and the Ministry of Science Technology and Environment. In general, the solid waste management framework of the whole country of Thailand has been centrally developed and is administered by the Ministry of Interior. The Ministry of Science, Technology and Environment provides some technical assistance.

Hazardous Waste

Hazardous waste is a serious problem in Thailand that is expected to worsen in the future. Its generation is increasing rapidly from various sources, including industries, agriculture, communities, businesses, services, ports, hospitals and laboratories. Hazardous waste generated in Thailand was estimated by the Pollution Control Department (PCD) to be 1.6 million tonnes in 1996. The main portion (73.3%) was from industry, and the remainder was from services and businesses (8.4%), hospitals and laboratories (8.2%), ports and navigation (8.1%), domestic areas (1.3%), and agriculture (0.7%). Hazardous waste can be classified into two categories: locally-generated hazardous waste and industrial hazardous waste.

A 1998 Pollution Control Department report revealed that locally-generated hazardous waste throughout Thailand currently exceeded 304,000 tonnes per year. The types of waste were: used oil; waste oil from gas stations; solid and liquid heavy metals; toxic wastes from commercial printing shops; chemicals from photographic processing; lead acid batteries, toxic chemicals from automotive maintenance and repair shops; household wastes such as dry cell batteries and fluorescent bulbs; biohazardous wastes from hospitals and clinics; laboratory waste; waste oil from sea ports; chemicals from military installations; and chemicals used in agricultural processes. Locally-generated hazardous wastes are not segregated from general community waste and so contain a broad range of hazardous wastes generated from commercial and domestic activities. Currently, these wastes are co-disposed with municipal solid waste at designated sites. Urban centres produce the greatest amount of hazardous waste per capita per year (for example, Bangkok generates 13.28 kg), while regional areas generate much smaller quantities (3.25–4.77 in the central, southern, northern and northeastern regions).

Infectious waste produced in hospitals is disposed by incineration. Incinerators have been installed at each of the 852 hospitals throughout the country under the control of the Ministry of Public Health by the end of 1997, in order to comply with policy on management of infectious wastes. However, these incinerators have not been equipped with appropriate air pollution control devices. At present, only two infectious waste incineration plants (Bangkok and Hat Yai) have been designed and constructed to guarantee compliance with minimum standards. The facility in Bangkok has two incinerators each with capacity of 10 tonnes per day and the Hat Yai incinerator has a capacity of 5 tonnes per day. There are no regulations to control emissions from infectious waste incinerators (Yoosook, 1999).

Hazardous industrial wastes, such as heavy metal sludges, strong spent acids, pesticide ingredients and 'off-spec' dry batteries, are generated as byproducts of various industrial processes. The majority of hazardous wastes are not properly handled, causing potential hazards to health and the environment. Under the Factory Act in 1992, the Ministry of Industry established standards for hazardous waste identification and introduced regulations requiring proper treatment and disposal of such wastes. In 1997 and 1998, the Ministry introduced further standards for industrial hazardous waste identification with general guidelines on treatment and disposal methods; these laws require waste generators to report the type and amount of wastes they generated to the Department of Industrial Works together with information on treatment and disposal procedures.

At present, facilities for the treatment and disposal of industrial hazardous waste are insufficient: only certain types of hazardous waste can be treated, and waste production exceeds treatment capabilities. Hazardous wastes can only be treated at three locations in Thailand: the Samae Dum Centre in Bangkok; at a landfill site in Ratchburi province; and at the GENCO (General Environmental Conservation Public Company Limited) facility in Map Ta Phut Industrial Estate, Chonburi province.

The GENCO operated Samae Dum Center was commissioned in the western suburbs of Bangkok in 1998. The plant has a maximum treatment capacity of 200 tonnes per day of inorganic acid and alkaline aqueous waste, 800 tonnes of textile dyeing wastewater, and 200 tonnes of inorganic heavy metal sludge and solids. Currently, the plant is treating wastes from some 500 factories and is handling 8500 tonnes per year of spent acids, 60,000 tonnes of acid or alkaline waste contaminated with heavy metals, and 14,200 tonnes of heavy metals sludge and solids. The Ratchburi Landfill Centre, Thailand's first secure landfill, was built shortly after the Samae-Dum plant. It has an area of 48 hectares and is located in Ratchburi province. It complies with international standards (built with a double high-density polyethylene (HDPE) lining) and receives approximately 30,000 tonnes of waste per year from the Samae-Dum plant.

GENCO also operates a ten-hectare waste management site in the Map Ta Phut Industrial Estate in Rayong province. The site has the capacity to receive about 650 tonnes per day of hazardous waste, of which 550 tonnes is treated by waste stabilization and landfill, while the remaining 100 tonnes is disposed of by using a fuel blending system. Second and third phases for the treatment plant are planned; they will include facilities for chemical-physical treatment and incineration of wastes. The second phase chemical-physical treatment unit will have a capacity of 250 tonnes per day, while the third phase incineration unit will have a capacity of 100 tonnes per day.

Sites Contaminated with Solid and Hazardous Waste

In addition to domestic and industrial waste management concerns, contamination of land and groundwater with solid and hazardous wastes is becoming a serious concern. Some examples of such sites in Thailand are arsenic contamination caused from tin mining at Ronpiboon district in Nakorn Srithammarat province (Southern Thailand), herbicide-contaminated sites at Boo-Phai, Prachuap Khiri Khan province and municipal landfill sites. The movement of contaminated leachates from the landfill sites to the surface and groundwater sources is a serious issue which potentially has serious risks to human health. As mentioned above, the process of co-disposal of locally-produced hazardous waste with domestic solid waste is common in Thailand and can be expected to result in landfill contamination and toxic leachates. In 1996, PCD carried out a national survey to identify the extent of contamination resulting from hazardous landfill leachates. Based on the results of evaluating 86 municipal landfills using the US EPA Hazardous Ranking System, it was reported that landfills are generally substandard in design, construction and operation. Twenty-one landfills require further detailed investigation to establish whether urgent improvements are necessary (Yoosook, 1999).

Due to earlier industrial activities in Thailand, there are few historically-contaminated sites. However, with the introduction of new industrial processes and underdeveloped waste management strategies, contaminated sites are an emerging issue that needs to be addressed. Some effort has started to document incidences of site contamination. However, most site reports are in the initial phase of study and information has not yet been released during 1996–1998. Even without documentation, it is already apparent from visual observation that industrial development has caused environmental contamination in Thailand and an increasing number of contamination issues will emerge in the coming years. Some of these problems will need the implementation of remediation technologies for site cleanup. However remediation research is a relatively new research field in Thailand.

WASTE MANAGEMENT POLICY AND PLANNING

The Thai government has recognized the seriousness of this environmental threat and is introducing measures to solve and prevent problems of waste pollution, waste disposal and contaminated land. The Enhancement and Conservation of National Environmental Quality Act in B.E. 2535 (1992) introduced several new concepts in environmental pollution control. Legislation is one strategy which is being enforced progressively in Thailand at present. The 20-year Policy and Plan for Enhancement and Conservation of National Environmental Quality (1997–2016) has been formulated with the aim of managing the country's natural resources and the environment, while facilitating and not obstructing the economic and social development.

In addition, the Environmental Quality Management Plan 1999–2006 prepared by OEPP (Office of Environmental Policy and Planning) set out the following goals for the nation's waste management (Office of Environmental Policy and Planning, 1998):

- reduce or control solid waste generation to a rate of not more than 1.0 kg per capita per day;
- increase the recycling rate to be not lower than 10% and 15% by the years 2001 and 2006, respectively;
- decrease the quantity of uncollected solid waste in municipal areas to lower than 10% and 5% in the years 2001 and 2006, respectively;
- every province in Thailand is to possess a master plan and action plan for the preparation of an acceptable solid waste disposal site by the year 2001. At least 50% of provinces to hold sanitary solid waste disposal facilities by the year 2006;
- install a hazardous waste management system for industrial and domestic areas that will cover the process of storage, collection, transportation, recycling, treatment and disposal by the year 2001;
- install foundation utilities in order to increase the efficiency of collection and disposal of hazardous waste for industry by not less than 80% and 95% by the year 2001 and 2006, and by 50% and 90% for domestic areas by the years 2001 and 2006, respectively;
- control hazardous waste so that the rate of generation increases by not more than 10% by the year 2001; and

- eighty per cent and 100% of governmental and private hospitals to have a completed infectious waste management system that covers separation, collection, transportation, treatment and disposal by the years 2001 and 2006, respectively.

The policies mentioned above have been developed from the guidelines in the 10- and 20-year national plans for waste management in Thailand. In order to achieve these goals, several projects have been planned and incorporated in the 1999–2006 plan. Within this plan, a large budget has been allocated for projects to be implemented in Bangkok and other provinces. Other major strategies considered for implementation in the waste management area are:

- reduce or control waste generation;
- apply the 'polluter pays' principle for both public and government organizations;
- encourage the private sector to provide services in waste management;
- invest in the construction of and provide suitable equipment for sanitary disposal facilities;
- promote investments and provide incentives to the private sector for environmental services;
- establish an emergency system to prevent and mitigate major hazardous waste accidents in the industrial sector;
- establish central waste disposal facilities that provide services to several communities; and
- improve and rehabilitate the existing unsanitary waste disposal areas.

Although the concept of privatization has been accepted and promoted for waste management in Thailand, so far, there has been little success in encouraging the private sector to participate. Although there are many private organizations working on this project, few of them participate in every stage of the waste management chain.

Weak Points of Solid and Hazardous Waste Management

Public opposition

In the past, many negative effects of solid waste management have occurred such as untidy collection, unsanitary disposal resulting in offensive smells, pests, visual impacts and devaluation of adjacent properties. Communities experiencing some of these effects have little trust in waste management, so they oppose waste projects planned for locations within their communities, particularly hazardous waste projects.

Public awareness

The environmental awareness of Thai people is currently low. Presently, low levels of community responsibility, support and contribution to activities related to solid waste management are common, resulting in unwillingness to pay appropriate service charges.

Limitations on operations, staff, budgets, and resource

Currently, many landfills are operated without being covered with fill material and also without proper maintenance of collection and disposal machinery. This inefficiency in operation and maintenance increases the cost of landfill disposal. In addition, the number of staff trained in solid waste management and engineering at the local level is limited. As a consequence, the most appropriate solid waste management solution to a particular problem is not always chosen.

Unfair prices and tariffs

For political reasons, most local government organizations do not charge the public for waste disposal. Although the 'polluter pays' principle has been declared appropriate by the government, price and tariff settings are far below the true cost of the waste disposal service.

Lack of segregation of community solid waste

Currently, co-disposal of general solid waste and locally-generated hazardous waste is a common practice. No measures have been introduced for the separation of hazardous items from domestic solid waste. As a result, these disposal practices may adversely affect environmental quality and public health in the future.

Low efficiency of solid waste recycling and minimization

In 1997, PCD reported that the recycling rate of solid waste throughout the country was only 11%, whereas the actual potential for solid waste recycling is as high as 30–45% (Pollution Control Department, 1997). If larger amounts of solid waste can be eliminated or recycled, more effective solid waste management can be achieved.

Limitations of technology and management

Treatment of hazardous waste in Thailand is currently not effective. One of the main reasons for its ineffectiveness is that the import-export/separation/collection/treatment/final-disposal chain is not managed adequately. Also, facilities for treatment and destruction of specialized wastes are limited and the lack of knowledge for the remediation of sites contaminated with hazardous material may not allow for waste pollution problems to be solved in the near future.

Regulation and enforcement

At present, the framework for licensing and regulating waste management in Thailand is extremely weak. Monitoring of compliance with legal obligations is ad hoc, and enforcement is virtually non-existent, largely as a result of political pressure not to impose a regulatory burden on business. As a result, waste management is seen as an unproductive cost rather than a legal and social responsibility, and shortcuts are taken.

Centralization

Under the current government structure, resources and expertise on waste management are held centrally in government departments, rather than with local authorities who are responsible for waste management functions. This may not allow to obtain the most appropriate solution towards implementation of solid waste management practices.

PROPOSED STRATEGIES FOR EFFECTIVE WASTE MANAGEMENT

There is a long-established consensus on the hierarchy of techniques for waste management. The first priority is to avoid waste production in the first place or, if this is not possible, to minimize the quantities of waste produced. If production of waste is unavoidable, it should be re-used, recycled, or recovered. Remaining wastes should be treated to reduce the quantities requiring final disposal. Although this hierarchy is understood in Thailand, until recently, relatively little attention has been paid to avoidance and minimization of waste production.

Successful solid and hazardous waste management can be achieved by introducing an Integrated Waste Management System and taking a holistic approach to waste prevention, recycling, waste transformation and effective disposal. The careless attitude towards waste products, which is reflected in the throw-away mentality of many Thai citizens and businesses, should be countered by a concerted campaign for the prevention of waste and conservation of resources by recycling and reusing material from our flows of waste. This will require effort and enforcement both at a political level and in industry. Waste prevention practices should be promoted in all levels of industry, business, public institutions and homes. Strategies for waste prevention that are particularly recommended include reducing the use of hazardous components, eliminating unnecessary items, using supplies and materials more efficiently, limiting and reusing packaging and composting yard trimmings on site, etc. It is essential that hazardous wastes are segregated from general waste and disposed of separately. This is an urgently-needed requirement to reduce the adverse effects of mixed solid waste and hazardous waste disposal on public health and the environment.

Waste management requires rethinking on a massive scale in the area of industrial production. The disposal of waste products needs to be taken into account in the planning and development stages. Cleaner production methods are a key element in the minimization of waste generated by industry, and should be encouraged and promoted.

Recycling also has important environmental benefits, including the conservation of energy, natural resources and valuable landfill capacity. Successful waste recycling needs:

- suitable collection programmes (convenience is the key to a successful recycling programme);
- markets for materials that are collected; and
- markets for products manufactured from recycled materials.

The interrelations between these factors exert a strong influence on the economics of the scheme. The success of recycling schemes also depends heavily upon public participation and education programmes. Recycling is expected to have a high impact in future waste management activities in Thailand. The success of schemes and policies designed to promote recycling depends as much on the commitment to the principle of encouraging public involvement as on the need to develop infrastructure and markets. Improvements in methods, the accuracy of data collection and standards of waste treatment and disposal should also encourage waste producers and handlers to develop the options available to reduce the quantity and pollution potential of wastes.

Waste transformation should be considered an important part of any integrated waste management strategy. The purpose of waste transformation is to obtain valuable outputs and to further reduce the quantity of waste to be sent for final disposal or landfill. Generally, outputs from waste transformation are conversion products such as compost, biogas, waste heat for steam or electricity generation, and animal feed. Such a strategy is particularly useful to manage putrescible wastes or organic waste. Food waste and organic waste is a major component of municipal solid wastes in Thailand, comprising 40% of the total. Therefore, the conversion of organic wastes into fertilizer or soil conditioners by composting, or the manufacture of animal feed are options with great potential, which should be actively promoted in municipal solid waste management.

After passing through the previous steps of the integrated waste management process, waste for ultimate disposal should retain no valuable content. Appropriate disposal methods for the waste should be selected and, if a landfill is to be used, the construction of central sanitary landfills should be encouraged. These could provide more efficient services for several communities, particularly TAOs, and reduce the risk from unsanitary landfill practices.

Waste management today does not allow individuals to see the consequences of their actions, or the positive results of changing their behaviour. Incentives are needed to encourage this change. In order to increase individual responsibility for waste management, institutional, legal and financial changes are needed. Multiple strategies are more suitable methods of responding to the technical, economic and social challenges of waste management than dogmatic single-strategy approaches. An intelligent combination of options for waste management, both technical and organizational, is the most productive route. A 'Pay As You Throw' approach is suggested as one element of the programme. Information dissemination, public awareness and legal instruments are necessary aspects of waste management and should be made more effective. Waste management strategies for the future must also consider steps to improve waste management linked to regulation and enforcement, for example:

- duty of care requirement for waste producers;
- licensing of companies who transport and dispose of waste to ensure they have introduced pollution prevention measures, have sufficient expertise and qualifications to run a waste disposal business properly and have sufficient insurance to pay for cleaning up after any accidents;
- site aftercare liability for operators of waste disposal sites; and
- freedom of information regulations so that the public can find out which hazardous substances or wastes are stored at particular sites.

Waste management education and training for planners, professionals, decision-makers, school teachers, students and the public is also needed. TAOs should be the first priority for such training to provide a working knowledge of waste management practices. Ongoing education to produce all levels of environmentalists and engineers who are able to understand and use the hardware and technology necessary for effective solid and hazardous waste management and contaminated site remediation should commence immediately in preparation for the future. Emphasis should be placed on training staff that is able to effectively operate sanitary landfills. In addition, training courses and programmes on basic and advanced features of disposal technology and operation should be arranged for staff and operators at the local level. This is urgently needed for municipalities that have constructed sanitary landfills, if the investment in new facilities is to be worthwhile.

Research and development is needed in order to create structures that promote the aim of integrated waste management and demonstrate the interrelated causes and effects of the complex systems that apply in the field of waste management. With the support of research and development, the current waste management system can be analyzed, with the aim of establishing a system of waste management that is more compatible with environmental requirements. Last, but not least, research, training and community outreach components must be deliberately designed to integrate with each other, to result into synergistic benefits.

CONCLUSIONS

Well-coordinated policies and strategies must be formulated to manage waste problems in Thailand. The country should find ways and means to minimize the adverse effects of wastes and find ways to educate the public to help develop an ethic for sustainable living and less waste production. Practical guidelines should be created to assist governments, industry and communities to reduce and control waste while new technologies should be use to minimize the production of wastes. Simplified and cost-effective remediation technologies should be developed as appropriate methods for solving land contamination problems in Thailand. Wastes must be managed 'from cradle to grave', from every source of production to the final point of disposal. This will involve everything from public awareness campaigns and law enforcement to engineering solutions. Strong action programmes are needed to promote community awareness. Sustainable measures for tackling waste management are complicated, but they should be based around the concept that all pollution sources are viewed together and not isolated from their surrounding environments. To solve the problems and achieve success, time and resources are needed, including human resources, know-how, technology, education, and public participation. Most important of all, consistent integrated implementation by everyone who deals with waste management is the key to success.

REFERENCES

Danteravanich, S. and Siriwong, C., 1998, Solid waste management in Southern Thailand, *The Journal of Solid Waste Technology and Management* 25: 21–26.

Office of Environmental Policy and Planning, Ministry of Science Technology and Environment, 1998, Environmental Quality Management Plan 1999–2006.

Pollution Control Department, 1997, Solid Waste Management in Thailand (in Thai).

Rotchpaiwong, S., 1999, *Environmental Situation of Thailand in 1997-98*. Amman Printing and Publishing Company, Bangkok, Thailand.

Sayamon, R, 1997, Solid Waste Problems in the Year 2000. Handout from Academic Seminar on Solid Waste Management in the Year 2000, organised by The Department of Environmental Engineering, Faculty of Engineering, Kasetsart University, 13–15 August, 1997, Bangkok (in Thai).

Yoosook, S., 1999, Municipal Solid Waste Management Systems and Facilities, Handout from Presentation in International Training Course on Wastewater Management, Water Supply and Sanitation, 19 July–6 August, 1999 at Prince of Songkla University, Thailand.

17

Utilization Potential of Agri-industry Wastes as Valuable Byproducts

K. Peverill[1], *B. Meehan*[2] and *J. Maheswaran*[3]

INTRODUCTION

The disposal of wastes to landfills is a significant cost to manufacturing industries, resulting in increased production costs and reduced profitability. Disposal of solid wastes to landfill can have serious ecological implications as well as loss of potentially-valuable resources. Further, disposal of liquid waste streams to waterways in both rural and metropolitan regions is unacceptable to society and costs of disposal to sewers are becoming increasingly stringent. Consequently, strategies for reuse of these resources need to be developed. In 1998, the Victorian State Government Environment Protection Authority (EPA) introduced an Industrial Waste Strategy specifically targeted at solid and liquid wastes generated by Victorian industries (EPA, 1998). One of the key strategic objectives announced in the strategy is to maximize the economic value of resources during their lifecycle through reuse, recycling and energy recovery in preference to disposal. In order to achieve this goal, it is necessary that options for reuse of waste streams be explored. Discussing such options is the purpose of this chapter.

Wastes from agricultural industries have considerable potential for reuse as sources of water, organic matter, nutrients, mulches and soil-conditioning agents (Rechcigl and MacKinnon, 1997). Australian agricultural soils are generally low in nutrients and organic matter, making them highly susceptible to nutrient mining, structural decline and erosion. Reuse of wastewater in the predominantly dry Australian climate is also

[1]K I P Consultancy Services Pty Ltd, 4 Collier Court, Wheelers Hill Victoria, 3150 Australia
[2]RMIT University, La Trobe St., Melbourne, Victoria, 3000 Australia
[3]Agriculture Victoria, State Chemistry Laboratory, Cnr South and Sneydes Rds., Werribee, Victoria, 3030 Australia

essential not only to conserve this limited resource but also to protect ground and surface water reserves from contamination. There is, therefore, a prima facie case for the investigation of suitable agri-industry waste streams to identify those potential products that can be applied to agricultural and horticultural soils. This chapter focuses on the reuse of solid waste streams, although one example of a product developed from a liquid waste stream is presented. There are many obvious advantages for utilization of agri-industry waste streams in this way (Cameron et al., 1996), the main ones are:

- conservation of water resources;
- protection of freshwater and marine environments;
- reduction of landfill inputs;
- reduction of waste incineration;
- recycling of nutrients;
- improved organic matter levels in soils; and
- improved nutrient status of soils.

Currently, there is little systematic information on agri-industry wastes produced in Victoria, though a scoping study was undertaken in a project sponsored by the Rural Research and Development Corporation (Meehan et al., 2001). Furthermore, wastes from such industries are often by default deemed as prescribed wastes under the current EPA legislation, making land disposal an expensive option. Surveying and characterizing these wastes is necessary before any assessment can be made of their reuse potential. If the waste streams are non-toxic and free from contamination, they can be effectively reused by careful selection and suitable pre-treatment. By combining different waste streams, composted materials of high nutrient value and with consistent physical and chemical characteristics could be tailored so as to suit various crop and soil requirements. This chapter includes the results of the recently-conducted survey (cited above) of post-farmgate agri-industry wastes produced in the Melbourne metropolitan region with potential for development as soil ameliorants or fertilizers. Studies undertaken by Agriculture Victoria, and more recently, by RMIT University have used agri-industry wastes, either directly or after pre-treatment, as supplements to, or replacements for, conventional fertilizers and sources of organic matter to improve the soil structure and water-holding capacity (Peverill et al., 1999). Liquid and solid wastes from the wool scouring industry have been developed as commercially-marketable products that could be applied to agricultural and horticultural land or used in potting media for nursery production. Recent studies with solid waste from the hide and skin processing industry have shown that this waste may be used to supplement or replace nitrogenous fertilizers (Confidential Report: Victorian Hide and Skins Producers Pty Ltd). Preliminary studies have also been undertaken with piggery wastes, abattoir wastes, sugar refinery wastes and cut-flower wastes. This Chapter presents the results of a number of laboratory, glasshouse and field investigations on the application of some of these materials to agricultural and horticultural soils. The results of four case studies are included later, together with the waste stream survey described above.

AGRI-INDUSTRY WASTES

Solid wastes from agricultural industries have high potential for reuse, and represent a high proportion of all prescribed wastes disposed of in landfills (EPA, 1996). Many of these wastes have economically-useful concentrations of essential nutrients for plant growth, are organic in nature, and can be effectively developed as useful value-added resources for agricultural industries. This, however, is contingent on demonstrating that agri-industrial wastes represent an attractive alternative as fertilizers or soil ameliorants.

Wastes from post-farmgate agricultural operations that have the potential to be applied directly or after composting include materials such as:

- animal manure and farm effluents
- chicken litter and manure
- crop residues
- feedlot wastes
- food-processing wastes
- wool-processing effluents and sludges
- animal hair
- meat-processing effluents
- fruit and vegetable processing
- cut-flower wastes

In the western region of the Melbourne metropolitan area, green organics alone generate about 120,000 tonnes per annum (Personal Communication with Western Region Waste Management Group, 1996). Extrapolating from this, the amount of wastes from agri-industries in Victoria could at least be four times this quantity, with correspondingly high disposal costs. In the vicinity of the metropolitan area, there are several high-value agricultural enterprises, such as vineyards, vegetable production, turf grass industry, cut-flower industry, and ornamental nursery industry. Potentially, such industries can utilize these wastes in their direct form or after pre-treatment for their nutrient or soil ameliorant value. Studies undertaken by the Institute of Horticultural Development, Knoxfield (Agriculture Victoria) have shown that green waste is an ideal raw material for the production of mulches, composts and growing media.

In the vegetable production area of Werribee in the Western Metropolitan area, one of the most prolific areas for the production of export-quality crucifers, continuous cropping for over 30 years has depleted the soil of organic matter, threatening the phasing out of this industry in the next ten years. At a rate of application that could sustain production (about 30 tonnes per hectare), the potential market for suitable compost as a soil conditioner in Werribee alone is estimated 50,000–150,000 tonnes per annum. If parts of the northern and eastern sections of the metropolitan area are included, the market potential for value-added wasteproducts could easily exceed 500,000 tonnes per annum.

Research currently underway at RMIT University and Agriculture Victoria is directed towards the reuse potential of agri-industry wastes being produced in Victoria. The research project, funded by the Rural Industries Research and Development Corporation and supported by EcoRecycle and the EPA, aims to quantify and qualify these wastes to determine whether they can be developed as value-added resources.

As part of the project, an analysis of prescribed waste transport data has been undertaken. The data supplied by EPA (EPA, Personal Communication 1997) cover all biodegradable organic waste transported in the Melbourne metropolitan area from January to December 1997 (Meehan et al., 1998). The data were sorted into waste-types and collated, providing information on the amounts and types of biodegradable waste generated annually. Table 17.1 summarizes the estimated total putrescible waste transported over the 12-month period. The data in the table are recorded in volume; some data were provided by EPA as mass (kilograms), and these have been converted into volumes based on estimated waste densities.

Samples of these wastes were collected from more than 20 sources around Melbourne and chemically analyzed to identify possible nutrient

Table 17.1: Estimated total putrescible waste transported (January–December 1997)

Waste type	*Volume* (m^3)	*Percentage of total (%)*
Wool scour	8100	9.3
Poultry litter	1450	1.7
Poultry waste	2400	2.7
Seafood	3450	3.9
Tannery	4700	5.4
Meat	8150	9.3
Dairy	19 550	22.3
Potato crisps	5250	6.0
Fruit and vegetable	1050	1.2
Grease and oil	33 450	38.2
Total	**87 550**	**100.0**

sources and contaminants. The results of the study are listed in Table 17.2 and each analyte has been calculated on a wet weight basis.

On the basis of this survey several wastes were selected for further investigation of suitable methods of pretreatment and utilization as soil ameliorants. Most of the wastes were found to be free of heavy-metal contamination. Tannery sludges were particularly high in chromium, and, consequently, could not be used directly in agricultural applications. Tannery hair was found to be high in nitrogen, although the chromium levels detected in the sample obtained suggest that this material would need to be pre-treated to remove this contaminant before converting it into a nutrient source for land application. However, tannery waste from a single hide-and-skin processing plant, which uses a process free of chromium, was selected to look at the feasibility of utilizing waste hair as a nitrogenous source for vegetable production. Food production wastes such as potato skins could be streamed into existing green waste composting operations as a means of diverting this material from landfill.

CASE STUDIES

Four case studies have been carried out to evaluate the use of several agri-industry wastes as soil ameliorants or nutrient sources, or both. The results of these studies are presented below.

Case Study 1—Wool Processing Effluent

It has been estimated that a single wool scour line can produce over 0.4 ML per day. The wool scouring process removes high quantities of salt and grease from the wool. Since disposal charges are based on the Biological Oxygen Demand (BOD) and salt content of the effluent, disposal is a major cost for the wool-scouring industry. A study, supported by Business Victoria and conducted in conjunction with CSIRO Division of Wool Technology (Geelong) was undertaken in 1994 to investigate the alternative uses for the effluent from wool scours (Maheswaran et al., 1994).

The effluent generated from the wool scouring process is called suint. The effluent from the first wash of the wool with hot water is free of detergent and other chemical additives. CSIRO Division of Wool Technology had developed a technique to reduce the volume of this wash by evaporation and reduction of suspended soil particles, grease and wool by centrifugation. This concentrate had a high potassium content (about 11% w/w). Other nutritive elements were comparatively low in concentration and offered little agronomic value. Levels of heavy metals and organic chemical residues were inconsequential in relation to environmental pollution and toxicity.

Table 17.2: Analysis of selected waste streams in the Melbourne metropolitan area (May 1999)

Industry waste	*Selected nutrients (%)*			*Selected heavy metals (mg/l)*							
	N	*P*	*K*	*Cd*	*Cr*	*Cu*	*Co*	*Pb*	*Zn*	*As*	*Hg*
Wool/Sludge (Av. of 3 sites)	0.84	<0.06	0.43	<1.0	24	<0.001	<0.001	<10.0	0.007	2.3	0.025
Wool/Fibres (Av. of 3 sites)	2.6	0.15	1.38	<1.0	17	0.002	<0.001	<10.0	0.008	2.1	0.025
Tannery/Sludge (Av. of 2 sites)	1.2	0.33	0.07	<1.0	3500	0.002	<0.001	<10.0	0.014	1.5	0.02
Tannery/Sludge (Cr) (Av. of 2 sites)	0.79	<0.06	0.11	<1.0	64000	0.002	0.003	<10.0	0.088	1	0.08
Tannery/Sludge (fatty) (Av. of 2 sites)	0.67	<0.06	<0.04	<1.0	65	<0.001	<0.001	<10.0	0.002	ND	<10.0
Tannery/Hair (Av. of 2 sites)	4.95	<0.06	<0.04	<1.0	81	<0.001	<0.001	<10.0	0.009	0.1	0.07
Food waste/Beans (Av. of 2 sites)	0.93	0.06	<0.04	<1.0	<10	<1.0	<0.001	<10.0	0.009	<0.1	0.03
Potato waste (1 company)	0.12	<0.06	0.18	<1.0	<10	<0.001	<0.001	<10.0	<0.001	<0.1	0.04

ND = Not determined

Replicated field trials were conducted with potatoes and pastures where potassium nutrition is important. The concentrate was diluted to several concentrations and applied to pasture and potatoes at comparable rates to regular potassic fertilizers such as potash. The results from the pasture trials have shown that the suint with appropriate dilution can be used as an alternative potassium source and pasture yields and potassium uptake obtained are comparable to those obtained using conventional potassic fertilizers (Table 17.3).

Table 17.3: Pasture yield and potassium uptake under fertilizer and suint treatments as sources of potassium

Trial	*P fertilizer only (control)*	*P + Potash as K source*	*P + suint as K source*
Pasture yield (Tonnes/ha.) at various sites			
Portarlington	1.9[a]	2.5[b]	2.4[b]
Simpson	8.1[a]	9.1[b]	9.2[b]
Larpent	5.3	5.6	5.7
Potassium uptake (K/ha.) at various sites			
Portarlington	26.5[a]	43.6[b]	42.7[b]
Simpson	128.9[a]	237.0[b]	226.5[b]
Larpent	111.3[a]	162.9[b]	166.5[b]

Values with different superscripts are significantly different

Trials with potatoes showed that yields of potatoes and potassium concentration in potatoes, treated with different sources of potassium, were comparable, but not significantly different from the nil-K treatment (Table 17.4). A possible reason for the lack of difference was the leaching of potassium at this sandy site.

Table 17.4: Yield and potassium uptake in potatoes treated with different forms of potassium

	Control (nil-K)	*Potash as K source*	*Suint as K source*
Yield (tonnes/ha.)	41.7	45.6	41.9
K concentration (%)	1.9	2.0	1.9

These studies show that:

- suint could be used in agriculture as an alternative K source without adversely affecting the yield of the crop;
- the uptake of K and yield of crops in some circumstances (for example, pasture) can be comparable to conventional potassic fertilizers such as potash; and
- the cost of production of conventional fertilizers and suint should be compared to determine whether suint is a financially-viable alternative.

Case Study 2—Wool-scour Sludge Composts

Disposal of wool-scour waste can cost up to AU$0.8 m (US$0.4 m) per scour-line per annum. The pollution load of each scour-line has a population equivalent of over 30,000 people. Generally, during wool scouring, 35% of the total initial weight ends up as waste in either the liquid or solid streams. Disposal of this type of waste has both financial and environmental consequences for scouring operations. The wool-scouring industry has taken positive steps to explore the possibilities to find alternative uses for its wastes. The Agriculture Victoria—State Chemistry Laboratory has assisted the Geelong Wool Combing Pty Ltd. (GWC), one of the major wool processors in Australia, to work towards containing all of its wastes generated within its premises and find alternative uses for them. A zero-waste output policy initiated by GWC has helped in systematically isolating each of its waste sources and then treating, characterizing, and modifying them to suit alternative uses.

During the scouring process, much of the suspended solids is removed in the first wash of the wool. Subsequent washes are centrifuged to remove the remaining suspended solids and the wool grease. Reduction of the BOD and the suspended solids in this way renders the effluent clean enough for disposal to the sewer without incurring additional charges. The remaining waste streams and the removed solids, are treated biologically to partially break down the materials and to further separate the suspended solids and the wool grease. While the effluent could be discharged safely, a composting procedure was developed to treat the highly-modified solid wastes. The ingredients used in the composting procedure are themselves sourced from other wastes generated from industries in the vicinity. To compost the wool-scour wastes, GWC incorporates woodchip wastes as the bulking agent (carbon source) and waste hair from the hide and skin processing industry (nitrogen source). Effluent, high in potassium and from its own plant, was used both for watering during composting and as a potassium source.

The resulting compost has been used in trials in the vegetable-growing area of Werribee with promising results. The soils from this area have been continuously cropped for over 30 years, and have poor organic matter content and structure. Application of soil organic ameliorants has been recommended to increase the sustainability of these soils. Trials on broccoli with composted wool-scour wastes (Table 17.5) have shown that application of compost can:

- assist in the conservation of soil moisture resulting in reduced water application by about 12% (equivalent to about AU$22.50 per ha per annum);
- decrease soil bulk density, resulting in a 12% increase in aeration

Table 17.5: Soil properties and broccoli harvest results after land application of composted wool—scour waste

Soil properties (measured after week 4/5)	*Compost applied (t/ha.)*		
	0	*20*	*80*
Soil moisture (%)	15.4[a]	17.9[b]	20.8[c]
Loss on ignition of soil (%)	5.5[a]	7.1[b]	10.4[c]
Soil bulk density (g/cm)	1.31[a]	1.26[b]	1.08[c]
Aeration porosity (% v/v) (measured after 12 weeks)	30.1[a]	28.9[a]	35.3[b]
No. of broccoli heads harvested	22,100[a]	30,100	32,800[b]

Values with different superscripts are significantly different

porosity that contributes to a more conducive environment for root penetration and growth;

- significantly contribute to the nutrient input, thus reducing the need for other fertilizers; and
- significantly reduce the residence time of the crop, contributing to early harvest.

Case Study 3—Animal Hair Wastes

During the processing of hide and skins for leather manufacturing, the hair removed is disposed as a non-prescribed waste, at a cost of about $30–40 per tonne. The Victorian Hide and Skin Producers Pty Ltd (VHSP) currently process upto 90 tonnes per week and the cost of disposal of waste hair is a significant proportion of total operating cost. The Agriculture Victoria—State Chemistry Laboratory, supported by the Commonwealth Department of Industry, Science and Tourism, enabled the VHSP to characterize its waste, identify suitable markets, and conduct trials to study the suitability of the wastes for alternative uses (Maheswaran and Tee, 1997).

Analysis identified that the hair contained about 12% nitrogen and negligible levels of heavy metals, organochlorines or organophosphates. The nitrogen in hair was organically bound and was not available for immediate release. Initial glasshouse trials indicated that about 25% of nitrogen from hair was immediately available for uptake by the first crop. Later, field trials with lettuce were conducted in the Werribee market garden area to evaluate the suitability of hair as an alternative source of nitrogen fertilizer. Milled waste hair was surface-applied to the soil at rates equivalent to some four times the amount nitrogen normally applied as conventional fertilizers. The application of nitrogen as hair proved to be useful in two ways: first as an organic soil ameliorant, preventing moisture loss from the soil; and second as an alternative source of nitrogen fertilizer.

Moisture retention under hair treatment was greater than under the control with no nitrogen (Table 17.6). Similar differences were found for plant characteristics such as plant N concentration, plant size, yield and N uptake (Table 17.6).

Table 17.6: Soil and lettuce characteristics, as affected by application of waste hair as an alternative source of nitrogen

Properties measured (after week 3)	*Nitrogen applied as hair or fertilizer*			
	0	*100% (as fertilizer)*	*100% (as hair)*	*400% (as hair)*
Soil moisture (%)	15.6[a]	15.6[a]	15.8[a]	16.4[b]
Plant N content (%)	4.7[a]	4.9[a]	4.7[a]	5.2[b]
Plant size—width (cm) (at harvest)	29.8[a]	29.8[a]	30.2[a]	31.4[b]
Yield (t/ha.)	76.3[a]	83.3[ab]	90.8[bc]	94.7[c]
N uptake (t/ha.)	2.74[a]	3.23[b]	3.24[b]	3.39[b]

Values with different superscripts are significantly different

The study has shown that waste hair from the hide and skin processing industry can also be used as an alternative source of nitrogen in vegetable production.

Case Study 4—Cut-flower Wastes

Another waste identified as going to landfill, yet having potential for reuse, was off-cuts and waste from the flower-growing industry. Cut-flower production is a major industry in Melbourne, with two main regions located in the Dandenong Mountains and in the outer southeastern suburbs. This industry produces 3000–4000 m^3 of waste per year. Growers are concerned about pathogen and pesticide transfer, so wastes are generally not re-worked back into the soil, but dumped on site or sent to landfill. A bioremediation trial using windrow composting was undertaken to assess the breakdown of pathogens and pesticides, and thereby assess the possibility of reusing cut-flower wastes as a soil-conditioning agent either by the growers themselves or through the nursery retail outlets.

A combined waste sample from several growers was analyzed for pesticide residues, with several common biocides being identified at residual levels (Meehan et al., 2001). A total of 30 m^3 of waste was then collected from three growers over a period of two weeks and transported to a municipal organic processing facility. The flower waste was incorporated into the composting process and windrows were monitored over the following weeks for temperature and relative humidity. At the conclusion of the composting trials, samples were taken and analyzed for pesticide residues. The results showed that even the most persistent pesticides used in the flower-growing industry were successfully degraded during the composting process (Table 17.7). Hence, the composting process has the potential to be used to remediate similarly contaminated green waste from other horticultural or agricultural industries, and result in composted materials that could be used as soil ameliorants or mulches.

Table 17.7: Effect of composting on pesticide residue levels in cut-flower waste

Pesticide	*Mean pesticide concentration (mg/kg)*	
	In fresh cut-flower waste	*Post-composting, with 50% cut-flower waste*
Captan	1.14	<0.01
Dichlran	<0.01	<0.01
Propargite	0.90	<0.01
Methamidophos	0.15	<0.05
Omethoate	<0.05	<0.05
Fluvalinate	<0.05	<0.05
Permethrin	0.26	<0.05

CONCLUSIONS

Studies by various institutes of Agriculture Victoria, often in conjunction with RMIT University, have shown that there is a high potential for the development of strategies for disposal of a wide range of agri-industry wastes onto agricultural land. By developing a detailed characterization of the various waste streams, it has been shown that with sound practices, it is usually possible to gain benefits from water, organic matter and nutrients in wastes through land disposal. However, care must be taken to guard against any adverse effects from waste application such as induced salinity and sodicity, nutrient imbalances, high BOD or contamination from organic residues and heavy metals. The studies have also shown that there are direct cost savings to industries through reduction in disposal costs to landfills or sewers. An indirect industry benefit is the portrayal of an environmentally-friendly image. There is also potential for industries to fully explore the potential for the development of value-added, environmentally-friendly products such as organically-based fertilizers.

ACKNOWLEDGEMENTS

The authors gratefully acknowledge the financial assistance provided by the Rural Industries Research and Development Corporation, Business Victoria and the Commonwealth Department of Industry, Science and Tourism, to undertake the various case studies. Without industry financial support and collaboration with Geelong Wool Combing Pty Ltd, the

Victorian Hide and Skin Producers Pty Ltd, F&I Baguley Flower and Plant Growers, and MulchMaster the research would not have been possible. Victorian Environment Protection Authority and EcoRecycle Victoria provided valuable information, which permitted an analysis of the prescribed waste transport data within the Melbourne Metropolitan area.

REFERENCES

Cameron, K.C., Di, H.J. and McLaren, R.G., 1996, Is Soil an Appropriate Dumping Ground for Our Wastes? ASSSI and NZSSS National Soils Conference, July 1996.

EPA Bulletin [Victorian Environment Protection Authority], 1996, (May). Government of Victoria.

EPA [Victorian Environment Protection Authority], 1998, Industrial Waste Strategy, Zeroing in on Waste. EPA, Melbourne.

Maheswaran, J., Peverill, K.I. and Brown, A.J., 1994, Agricultural Trials with Suint: An Alternative Source of Potassium. Confidential Report.

Maheswaran, J. and Tee, E., 1997, Investigation of the Potential for the Utilisation of Hair as a Fertilizer and Soil Ameliorant. Confidential Report.

Meehan, B.J., Maheswaran, J. and Phung, K., 2001, 'Reuse Potential of Agri-Industry Wastes', Rural Industries Research and Development Corporation Publication No. 01/144 (ISBN 0 642 58364 1), November.

Meehan, B.J., Baxter, F. and Maheswaran, J., 1998, Reuse Potential of Agri-industry Wastes in the Melbourne Metropolitan Region. Rural Industries Research and Development Corporation (RIRDC) Project No. RMI-10A, Progress Report.

Peverill, K I., Meehan, B J., Maheswaran, J., Baxter, F., Phung, K.P., Cody, J. and Dziedzic, A., 1999, 'Conversion Opportunities for Agri-industry Wastes', *Proceedings of the Contaminated Wastes Conference,* p. 187, Melbourne, November.

Rechcigl, J.E. and MacKinnon, H.C., 1997, *Agricultural Uses of By-products and Wastes.* American Chemical Society, Washington, DC.

18

Development of a Centralized Soil Bioremediation Facility in South Australia

N. McClure[1], *R. Bentham*[2] and *A. Hall*[3]

INTRODUCTION

In Australia, estimates of the number of contaminated sites are as high as 80 000 (Natusch, 1997). Legislation and information on the site numbers and levels of contamination vary considerably between the Australian states. Methods for assessing these sites have been standardized across Australia with the implementation of the National Environmental Protection Measure for Assessment of Contaminated Sites. In the US and Europe, there is a growing emphasis on the use of site-specific risk assessment in order to determine the need for active remediation programmes. There is also an increasing reliance on natural attenuation (intrinsic remediation) if it can be shown that natural removal rates for contaminants will ensure that there is no long-term risk to human health or the environment. Where intervention is necessary, remediation can be conducted on-site, or at the facilities established for treatment of contaminated materials, usually contaminated soils. In the past, excavation and landfill disposal or on-site containment have dominated the treatment methods for contaminated soils (Natusch, 1997; Moss, 1997). There are many alternative treatment technologies now available for the remediation of contaminated soils, including bioremediation, soil washing, and thermal desorption (Martin and Bardos, 1996).

In most cases in Australia where active soil remediation programmes have been instigated, these have been conducted on-site. Overseas, the use of centralized contaminated soil treatment facilities is more common.

[1]Flinders Bioremediation Pty Ltd, GPO Box 2100, Adelaide SA 5001, Australia
[2]Flinders University of South Australia, GPO Box 2100, Adelaide SA 5001, Australia
[3]BC Tonkin & Associates, 5 Cooke Terrace, Wayville SA 5034, Australia

WMX Technologies Inc. have developed and operate 25 fixed-location soil bioremediation facilities across the US and Europe and these have treated over three million tonnes of petroleum-contaminated soils (Hamblin and Hater, 1997). In Germany, Umweltschutz Nord have set up about 20 soil treatment facilities, located in most German States, and these facilities have treated over one million tonnes of contaminated soils using a range of treatment technologies including bioremediation and thermal desorption.

In southern Australia, options for the treatment and disposal of contaminated soils are limited. There is no landfill licensed to dispose of high level contaminated soils based on disposal criteria established by the South Australian Environment Protection Authority (SA EPA, 1997). Currently, there is only one landfill which is licensed to dispose of low-level contaminated soils, the Southern Waste Depot at Maslins. With very few mobile treatment units for soil remediation using thermal desorption available in Australia, there are limited available options for the owners of sites contaminated with high levels of pollutants. The increasing emphasis on treatment rather than landfill disposal or on-site containment has led to growing interest in the establishment of soil-treatment facilities, particularly for the remediation of soils from small inner city sites, where on-site treatment may be difficult or impossible to conduct due to site constraints.

ON-SITE TREATMENT VS CENTRALIZED SOIL TREATMENT FACILITIES

There are significant costs and regulatory hurdles associated with the transport of large volumes of contaminated soil from a site to a treatment facility; so, where possible, on-site treatment is normally the preferred option. However, for some sites, on-site treatment is either not possible or not a preferred option due to factors such as:

- lack of available space;
- time constraints (for example, a site may require immediate development, whilst a bioremediation process may require a treatment period in excess of 12 months: In addition extensive trials may have to be undertaken to optimise remediation processes prior to full scale application);
- proximity of sensitive receptors, normally local residents with concerns over health issues associated with contaminated site management;
- high cost of setting up an on-site treatment for relatively low volumes of soil; and
- constraints on continuing site use or development by the treatment process itself.

The use of centralized facilities also allows the establishment of a permanent infrastructure to allow on-line monitoring and control of bioremediation processes.

THE SOUTHERN WASTE DEPOT BIOREMEDIATION FACILITY

The bioremediation facility at the Southern Waste Depot, Maslins, combines both research and commercial soil treatment capabilities at a single site which was established by Lucas Earthmovers Pty Ltd and the Flinders University of South Australia in 1997. The facility, licensed to dispose of low-level contaminated soils in specified lined cells, is the only facility in South Australia with such a permit. The bioremediation facility is set up on a series of former sand mining tailings dams with one pad being established as a pilot scale research and feasibility testing facility. Being a former sedimentation basin, the site has the advantage of being underlain by a very low-permeability silty clay, so that it was effectively lined. The pilot-scale pad consists of ten individually-lined biopile/ composting test cells with a capacity of approximately 15 cubic metres of soil or compost and integral aeration and leachate collection systems. Test cells are covered with impermeable covers during operation. There are also 30 cubic metre lined pads larger-scale testing. The cells are aerated using either positive air blowers or vacuum pumps with treatment of exhaust gases using biofilters.

The commercial bioremediation area consists of pads of approximately 400 square metres, bituminized and graded to a sump for leachate and stormwater collection directed to the landfill leachate evaporation pond. The bioremediation facility utilizes a forced aeration system (vacuum) for both biopile and composting remediation processes. The aeration system currently used has been set up with slotted PVC piping in a gravel diffuser bed connected to a high capacity vacuum pump via a moisture knock-out drum, with exhaust gases directed through a compost biofilter. Process performance is monitored by temperature probes and dataloggers, and aeration efficiency and metabolic activity are measured using multiple gas monitors.

Environmental control measures which form a part of the ongoing operations of the site include leachate collection, treatment and monitoring, groundwater monitoring and surface water monitoring. Potential adverse air quality effects are minimal in practice, due to the distance between the site and the closest residential properties (500 metres, minimum). Management measures are routinely undertaken to control noise, odour and dust.

BIOREMEDIATION PROCESSES USED AT THE FACILITY

Bioremediation of simple petroleum hydrocarbons is carried at the bioremediation facility using either standard landfarming or biopile processes. For more difficult pollutants such as organochlorine pesticides, PCBs and PAHs, more vigorous remediation methods or combinations of technologies may be needed; results from pilot and full-scale applications have varied in their success. For example, large-scale remediation of PAH-contaminated soil at the East Perth gasworks site using a conventional bioremediation process (nutrient-amended soil piles with intermittent turning) has been conducted successfully (Bird, 1997). Processes which have shown promise for recalcitrant pollutants include the use of white rot fungi, joint chemical and physical treatment combined with biological treatment (for example, chemical oxidation followed by microbial degradation), and composting. Of these methods, composting has been applied at both pilot and full scale in Australia for contaminated soil remediation, and is being utilized at the Southern Waste Depot facility.

Recently, composting has been utilized as a bioremediation process for contaminated soils at the Southern Waste Depot facility. Contaminated soils are usually mixed with suitable bulking agents, such as woodchips and chicken manure, to improve the texture, aeration, and nutrient availability (Skladany and Metting, 1992; Riggle, 1995). The use of composting has been investigated as a means of remediation for less-volatile pollutants, mixtures of pollutants, and those more resistant to microbial degradation. This is because the composting process is known to involve the degradation of complex mixtures of organic compounds, such as lignin and polyphenols (Skladany and Metting, 1992; Tomati et al., 1996). Composting presents a strategy in which a variety of different physical and chemical environments and conditions are made available for microbial interaction with the pollutant. Combined with the metabolic diversity and high rates of degradation observed in composting processes, these factors can result in rapid and extensive removal of pollutants from the initial substrate.

Composting has been used at bench scale to investigate the potential for hazardous waste treatment. Many pollutants resistant to conventional bioremediation have been successfully treated. TNT (Tri-nitro-toluene) and RDX (Royal Demolition Explosives) compounds, petroleum hydrocarbons, PAHs and chlorophenols have all been effectively remediated using bench-scale systems (Kastner and Mahro, 1996; Laine and Jørgensen, 1997). Chlorophenol wastes, petroleum, jet fuel and explosives have also been remediated in full-scale processes (Rhodes, 1997). Composting may have several advantages over other bioremediation strategies including:

- reduction in treatment time; pollutants may be reduced to the desired target concentrations or completely remediated within 3–6 months;
- average costs for composting are between US$20 and $80 per cubic metre; and
- unlike landfarms and biopiles, composting processes can be controlled and manipulated so that remediation is not limited or prevented by external environmental conditions.

BIOREMEDIATION CASE STUDY—COMPOSTING OF PENTACHLOROPHENOL-CONTAMINATED SOIL

In Port Adelaide, a major residential development was proposed at the site of a disused timber treatment plant as part of a State Government initiative. Soil at the site was contaminated with pentachlorophenol (PCP) and turpentine, with PCP concentrations ranging from 50 to >600 ppm and turpentine concentrations of 100–1000 ppm present in the soil (McClure et al., 1997). As there was a preference for rapid development, an off-site treatment strategy was proposed. Approximately, 4000 tons of contaminated soil was transported to the Southern Waste Depot facility.

The soil was stored on a levelled and sealed (bituminized) area, with drainage to the landfill leachate collection system on the site. Representative samples of the soil were tested to investigate the feasibility of using composting as a bioremediation strategy at both laboratory and pilot scale prior to instigating the full-scale treatment process.
Composting was selected as a remediation strategy because:

- PCP is not appreciably volatile under alkaline conditions, so losses by volatilization could be expected to be minimal during the composting process;
- previous laboratory and field studies have shown successful use of composting for remediation of PCP-contaminated soil (Laine and Jørgensen, 1997; Laine et al., 1997);
- the composting process involves degradation of complex organic compounds, including phenols and substituted aromatic rings; complex and heavily-substituted xenobiotic compounds, such as PCP, appear well suited to this system; and
- the PCP had persisted at the contaminated site for many years after its use had ceased (this may have been due to nutrient limitation or to inherent resistance to degradation—previous studies have indicated that landfarm/biopile type systems may be ineffective unless augmented with PCP-degrading organisms).

LABORATORY-SCALE TRIALS

Laboratory scale, insulated, forced-aeration compost microcosms of approximately 25 litres were used for laboratory scale tests (McClure et al., 1997). Eight vessels were constructed and airflow through the vessels was adjusted to 1 litre per minute with the exhaust air passing through a compost biofilter. Eight composting microcosms containing a ratio of contaminated soil:green waste of 1:1 were set up and run for 43 days to assess the removal of PCP and determine at which stage of the composting process PCP was being removed. The level of PCP in the compost/soil mix, determined by dichloromethane (DCM) extraction and gas chromatography (GC), was reduced from an initial concentration of 68 ± 23 ppm (100%) to 11 ± 4 ppm over a period of seven weeks (Figure 18.1). The maximum PCP removal rate was observed after the thermophilic phase (day 7 onwards).

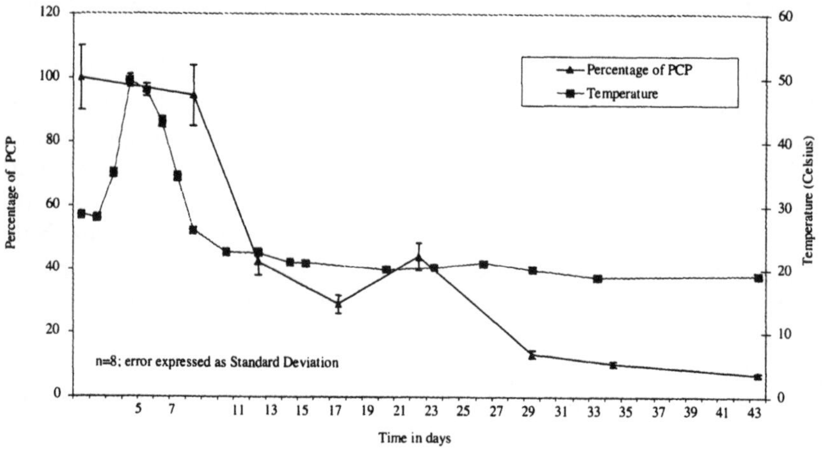

Fig. 18.1: Mean degradation of PCP in laboratory-scale compost microcosms

PILOT-SCALE COMPOSTING TRIALS

Two pilot-scale composting trials were conducted with an initial ratio of 1:1 green waste:contaminated soil in 10–12-tonne batches conducted in duplicate. An additional one tonne of chicken litter as a supplementary nutrient source was added in the first pilot trial and aeration via the integral forced-aeration system was only carried out intermittently using a generator (approximately one hour of aeration at 2–3-day intervals). In the second set of trials, aeration was carried out for one hour every six hours to accelerate the composting process. In the first trial, the thermophilic composting phase (>40°C) continued for approximately 4–6 weeks with maximum temperatures of 60°C reached after one week. PCP concentration within the pile, measured using MeOH/KOH extractions, reduced from 35–45 ppm to approximately 10 ppm over the 16-week period (Figure 18.2). After that, there was a small but less rapid decline to a final residual of 5–7 ppm. No PCP was detectable after 16 weeks using the DCM extraction procedure. Analysis of the MeOH/KOH extracts showed that there was no significant reduction of PCP during the thermophilic phase and that concentrations were reduced during the cooling off phase, which is consistent with laboratory results.

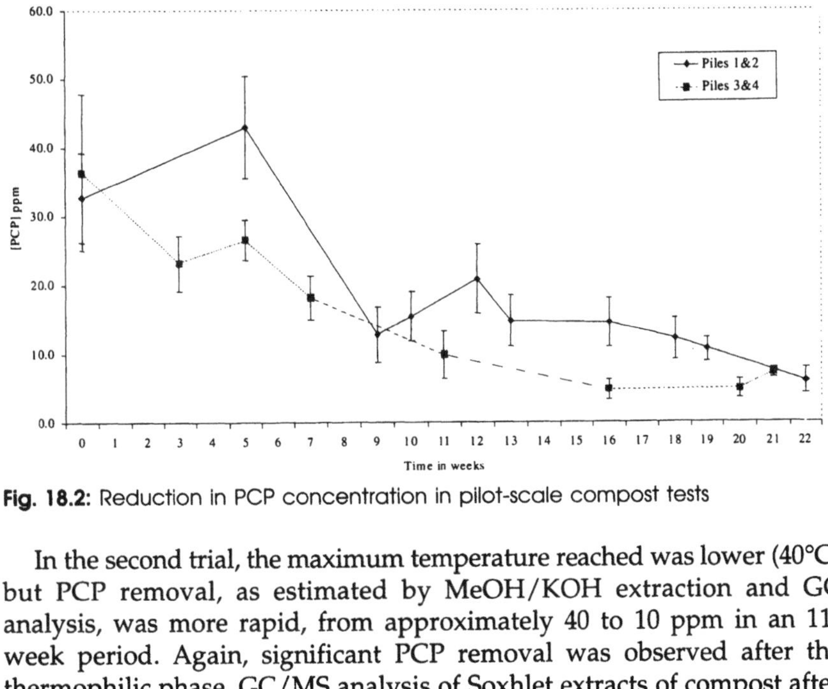

Fig. 18.2: Reduction in PCP concentration in pilot-scale compost tests

In the second trial, the maximum temperature reached was lower (40°C) but PCP removal, as estimated by MeOH/KOH extraction and GC analysis, was more rapid, from approximately 40 to 10 ppm in an 11-week period. Again, significant PCP removal was observed after the thermophilic phase. GC/MS analysis of Soxhlet extracts of compost after 11 weeks showed the presence of low concentrations (<10 ppb) of pentachloroanisole, a metabolite of PCP which has been linked to fungal degradation. GC/MS analysis using MeOH/KOH extraction confirmed a residual PCP concentration of 7–12 ppm.

Successful remediation of the PCP to target levels of 10 ppm was thus demonstrated, both at laboratory and pilot scale. Laboratory compost microcosms demonstrated the removal of upto 80% of PCP within eight weeks in a simple co-composting process with green waste. Pilot-scale tests confirmed these results with target values of 10 ppm PCP being achieved in approximately 12 weeks. Chemical analysis of residual PCP concentrations was confirmed with ecotoxicity tests that showed extensive reduction in the toxicity of the remediated soils after pilot-scale composting. As a result, full-scale bioremediation of the PCP-contaminated soil was initiated in June 1998 with approximately 2,000 tonnes of contaminated soil. PCP reductions from approximately 33 ppm (initial concentration after mixing 1:1 with green organics) to below 10 ppm were achieved over a period of five months.

CONCLUSIONS

Bioremediation is now an accepted technology for the remediation of organic pollutants in soils and is being used increasingly in Australia as also elsewhere. Whilst the bioremediation of simple hydrocarbons such as petrol and diesel has been carried out in many projects utilizing landfarming or biopile approaches, the bioremediation of resistant organic pollutants such as highly-chlorinated aromatic compounds and PAHs is still considered to be unreliable as compared to alternative chemical and physical processes or on-site or off-site containment. However, bioremediation can be conducted in situ, ex situ on site, or off-site at central facilities. Remediation off site allows rapid redevelopment of the contaminated sites and extended remediation periods, allowing residual levels of pollutants to be minimized.

The successful use of bioremediation processes such as composting to remediate soil contaminated with resistant pollutants such as PCP, PAHs and potentially PCBs, should extend the application of bioremediation and allow more sites to be cleaned up in a safe, cost-effective manner. There are still many questions to be answered to fully understand these processes, especially in the case of composting, which involves very many metabolic activities and interactions between biotic and abiotic factors. Once the parameters which affect the efficacy of these processes are better understood, opportunities will arise to develop quicker, more consistent bioremediation, systems, which can be applied to solve a wider range of environmental problems worldwide.

ACKNOWLEDGEMENTS

In preparing this chapter, the assistance of Lucas Earthmovers Pty Ltd, the Flinders University of South Australia, and BC Tonkin & Associates is gratefully acknowledged.

REFERENCES

Bird, S., 1997, Clean up of the East Perth Gasworks—A case example. WasteTECH '97, *The 4th National Hazardous and Solid Waste Convention Proceedings*: 55–67.

Hamblin, G.M. and Hater, G., 1997, Results from treating a million tons of soil. In: Alleman, B.C. and Leeson, A. (eds.). *In situ and on Site Bioremediation, Volume* 2: Papers from the Fourth International in situ and On-site Bioremediation Symposium. Batelle Press, Columbus, Ohio: 481–486.

Kastner, M. and Mahro, B., 1996, Microbial degradation of polycyclic aromatic hydrocarbons in soils affected by the organic matrix of compost. *Applied Microbiology and Biotechnology* 44: 668–675.

Laine, M.M., Haario, H. and Jørgensen, K.S., 1997, Microbial functional activity during composting of chlorophenol-contaminated sawmill soil. *Journal of Microbiological Methods* 30: 21–32.

Laine, M.M. and Jørgensen, K.S., 1997, Effective and safe composting of chlorophenol-contaminated soil in pilot scale. *Environmental Science and Technology* 31: 371–378.

Martin, I. and Bardos, P., 1996, *A Review of Full Scale Treatment Technologies for the Remediation of Contaminated Soil*. EPP Publications, Surrey, UK.

McClure, N.C., Dandie, C., Bentham, R.H., Franco, C. and Singleton, I., 1997, Establishment of a bioremediation facility in South Australia—Research and commercial potential. *Australian Biotechnology* 7(6): 345–349.

Moss, D., 1997, Current clean up technologies and practices—Directions being taken in Australia. WasteTECH '97, *The 4th National Hazardous and Solid Waste Convention Proceedings*: 14–21.

Natusch, J., 1997, Application and development of contaminated site remediation technologies in Australia. ANZAC Fellowship Report.

Rhodes, S.H., 1997, Bioremediation of Contaminated Soil—Current Practice in Australia. WasteTECH '97, *The 4th National Hazardous and Solid Waste Convention Proceedings*: 39–46.

Riggle, D., 1995, Successful bioremediation with compost. *Biocycle* 36(2): 57–59.

SA EPA [Environment Protection Authority], 1997, Disposal criteria for contaminated soil. Technical Bulletin No. 5. EPA, Adelaide, Australia.

Skladany, G.J. and Metting, F.B. Jr., 1992, Bioremediation of contaminated soil. In: Metting F.B. Jr. (Ed), *Soil Microbial Ecology: Applications for Agricultural and Environmental Management*. Marcel Dekker, New York: 483–513.

Tomati, U., Galli, U., Pasetti, L. and Volterra, E., 1996, Olive-mill waste-water bioremediation: evolution of a composting process and agronomic value of the end product. In: de Bertoldi, M., Sequi, P, Lemmes, B. and Papi, T. (Eds), *The Science of Composting*. Blackie Academic & Professional, Glasgow: 637–647.

Index

Milton Keynes UK
Ingram Content Group UK Ltd.
UKHW020322111024
449327UK00041B/2824

9 780367 446604